STATISTICAL MECHANICS

From First Principles to Macroscopic Phenomena

Based on the author's graduate course taught over many years in several physics departments, this book takes a "reductionist" view of statistical mechanics, while describing the main ideas and methods underlying its applications. It implicitly assumes that the physics of complex systems as observed is connected to fundamental physical laws represented at the molecular level by Newtonian mechanics or quantum mechanics. Organized into three parts, the first section describes the fundamental principles of equilibrium statistical mechanics. The next section describes applications to phases of increasing density and order: gases, liquids and solids; it also treats phase transitions. The final section deals with dynamics, including a careful account of hydrodynamic theories and linear response theory.

This original approach to statistical mechanics is suitable for a 1-year graduate course for physicists, chemists, and chemical engineers. Problems are included following each chapter, with solutions to selected problems provided.

J. Woods Halley is Professor of Physics at the School of Physics and Astronomy, University of Minnesota, Minneapolis.

# STATISTICAL MECHANICS

## From First Principles to Macroscopic Phenomena

J. WOODS HALLEY
*University of Minnesota*

CAMBRIDGE
UNIVERSITY PRESS

CAMBRIDGE UNIVERSITY PRESS
Cambridge, New York, Melbourne, Madrid, Cape Town,
Singapore, São Paulo, Delhi, Mexico City

Cambridge University Press
The Edinburgh Building, Cambridge CB2 8RU, UK

Published in the United States of America by Cambridge University Press, New York

www.cambridge.org
Information on this title: www.cambridge.org/9780521825757



First published 2007

*A catalogue record for this publication is available from the British Library*

ISBN 978-0-521-82575-7 Hardback

# Contents

# Preface

This book is based on a course which I have taught over many years to graduate students in several physics departments. Students have been mainly candidates for physics degrees but have included a scattering of people from other departments including chemical engineering, materials science and chemistry. I take a "reductionist" view, that implicitly assumes that the basic program of physics of complex systems is to connect observed phenomena to fundamental physical laws as represented at the molecular level by Newtonian mechanics or quantum mechanics. While this program has historically motivated workers in statistical physics for more than a century, it is no longer universally regarded as central by all distinguished users of statistical mechanics[1,2] some of whom emphasize the phenomenological role of statistical methods in organizing data at macroscopic length and time scales with only qualitative, and often only passing, reference to the underlying microscopic physics. While some very useful methods and insights have resulted from such approaches, they generally tend to have little quantitative predictive power. Further, the recent advances in first principles quantum mechanical methods have put the program of predictive quantitative methods based on first principles within reach for a broader range of systems. Thus a text which emphasizes connections to these first principles can be useful.

The level here is similar to that of popular books such as those by Landau and Lifshitz,[3] Huang[4] and Reichl.[5] The aim is to provide a basic understanding of the fundamentals and some pivotal applications in the brief space of a year. With regard to fundamentals, I have sought to present a clear, coherent point of view which is correct without oversimplifying or avoiding mention of aspects which are incompletely understood. This differs from many other books, which often either give the fundamentals extremely short shrift, on the one hand, or, on the other, expend more mathematical and scholarly attention on them than is appropriate in a one year graduate course. The chapters on fundamentals begin with a description of equilibrium for classical systems followed by a similar description for quantum mechanical systems. The derivation of the equilibrium aspects of thermodynamics is then presented followed by a discussion of the semiclassical limit.

In the second part, I progress through equilibrium applications to successively more dense states of matter: ideal classical gases, ideal quantum gases, imperfect classical gases (cluster expansions), classical liquids (including molecular dynamics) and some aspects of solids. A detailed discussion of solids is avoided because, at many institutions, solid state physics is a separate graduate course. However, because magnetic models have played such a central role in statistical mechanics, they are not neglected here. Finally, in this second part, having touched on the main states of matter, I devote a chapter to phase transitions: thermodynamics, classification and the renormalization group.

The third part is devoted to dynamics. This consists first of a long chapter on the derivation of the equations of hydrodynamics. In this chapter, the fluctuation–dissipation theorem then appears in the form of relations of transport coefficients to dynamic correlation functions. The second chapter of the last part treats stochastic models of dynamics and dynamical aspects of critical phenomena.

There are problems in each chapter. Solutions are provided for many of them in an appendix. Many of the problems require some numerical work. Sample codes are provided in some of the solutions (in Fortran) but, in most cases, it is advisable for students to work out their own solutions which means writing their own codes. Unfortunately, the students I have encountered recently are still often surprised to be asked to do this but there is really no substitute for it if one wants a thorough mastery of simulation aspects of the subject.

I have interacted with a great many people and sources during the evolution of this work. For this reason acknowledging them all is difficult and I apologise in advance if I overlook someone. My tutelage in statistical mechanics began with a course by Allan Kaufman in Berkeley in the 1960s. With regard to statistical mechanics I have profited especially from interactions with Michael Gillan, Gregory Wannier (some personally but mainly from his book), Mike Thorpe, Aneesur Rahman, Bert Halperin, Gene Mazenko, Hisao Nakanishi, Nigel Goldenfeld and David Chandler. Obviously none of these people are responsible for any mistakes you may find, but they may be given some credit for some of the good stuff. I am also grateful to the many classes that were subjected to these materials, in rather unpolished form in the early days, and who taught me a lot. Finally I thank all my Ph.D. students and postdocs (more than 30 in all) through the years for being good company and colleagues and for stimulating me in many ways.

J. Woods Halley
Minneapolis
July 2005

## References

1. For example, P. Anderson, Seminar 8 in http://www.princeton.edu/complex/site/
2. P. Anderson, *A Career in Theoretical Physics*, London: World Scientific, 1994.
3. L. D. Landau and E. M. Lifshitz, *Statistical Physics*, 3rd edition, *Part 1, Course of Theoretical Physics*, Volume 5, Oxford: Pergamon Press, 1980.
4. K. Huang, *Statistical Mechanics*, New York: John Wiley, 1987.
5. L. E. Reichl, *A Modern Course in Statistical Physics*, 2nd edition, New York: John Wiley, 1998.

# Introduction

The problems of statistical mechanics are those which involve systems with a larger number of degrees of freedom than we can conveniently follow explicitly in experiment, theory or simulation. The number of degrees of freedom which can be followed explicitly in simulations has been changing very rapidly as computers and algorithms improve. However, it is important to note that, even if computers continue to improve at their present rate, characterized by Moore's "law," scientists will not be able to use them for a very long time to predict many properties of nature by direct simulation of the fundamental microscopic laws of physics. This point is important enough to emphasize.

Suppose that, $T$ years from the present, a calculation requiring computation time $t_0$ at present will require computation time $t(T) = t_0 2^{-T/2}$ (Moore's "law,"[1] see Figure 1). Currently, state of the art numerical solutions of the Schrödinger equation for a few hundred atoms can be carried out fast enough so that the motion of these atoms can be followed long enough to obtain thermodynamic properties. This is adequate if one wishes to predict properties of simple homogeneous gases, liquids or solids from first principles (as we will be discussing later). However, for many problems of current interest, one is interested in entities in which many more atoms need to be studied in order to obtain predictions of properties at the macroscopic level of a centimeter or more. These include polymers, biomolecules and nanocrystalline materials for example. In such problems, one easily finds situations in which a first principles prediction requires following $10^6$ atoms dynamically. The first principles methods for calculating the properties increase in computational cost as the number of atoms to a power between 2 and 3. Suppose they scale as the second power so the computational time must be reduced by a factor $10^8$ in order to handle $10^6$ atoms. Using Moore's law we then predict that the calculation will be possible $T$ years from the present where $T = 16/\log_{10}2 = 53$ years. In fact, this may be optimistic because Moore's "law" may not continue to be valid for that long and also because $10^6$ atoms will not be enough in many cases. What this means is that,

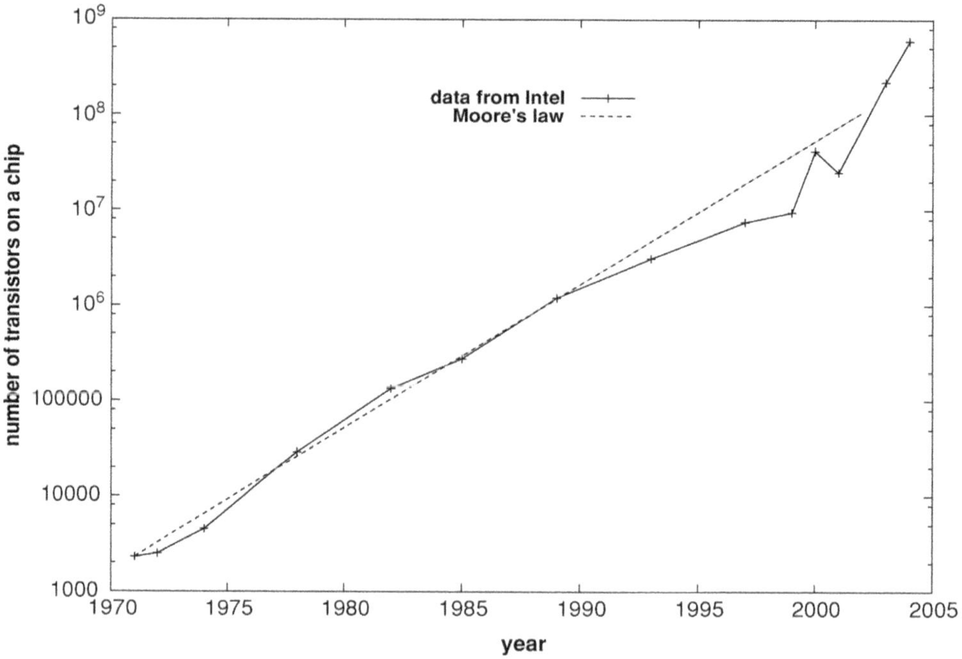


Figure 1 One version of Moore's "law."

for a long time, we will need means beyond brute force computation for relating the properties of macroscopic matter to the fundamental microscopic laws of physics.

Statistical mechanics provides the essential organizing principles needed for connecting the description of matter at large scales to the fundamental underlying physical laws (Figure 2). Whether we are dealing with an experimental system with intractably huge numbers of degrees of freedom or with a mass of data from a simulation, the essential goal is to describe the behavior of the many degrees of freedom in terms of a few "macroscopic" degrees of freedom. This turns out to be possible in a number of cases, though not always. Here, we will first describe how this connection is made in the case of equilibrium systems, whose average properties do not change in time. Having established (Part I) some principles of equilibrium statistical mechanics, we then provide (Part II) a discussion of how they are applied in the three most common phases of matter (gases, liquids and solids) and the treatment of phase transitions. Part III concerns dynamical and nonequilibrium methods.

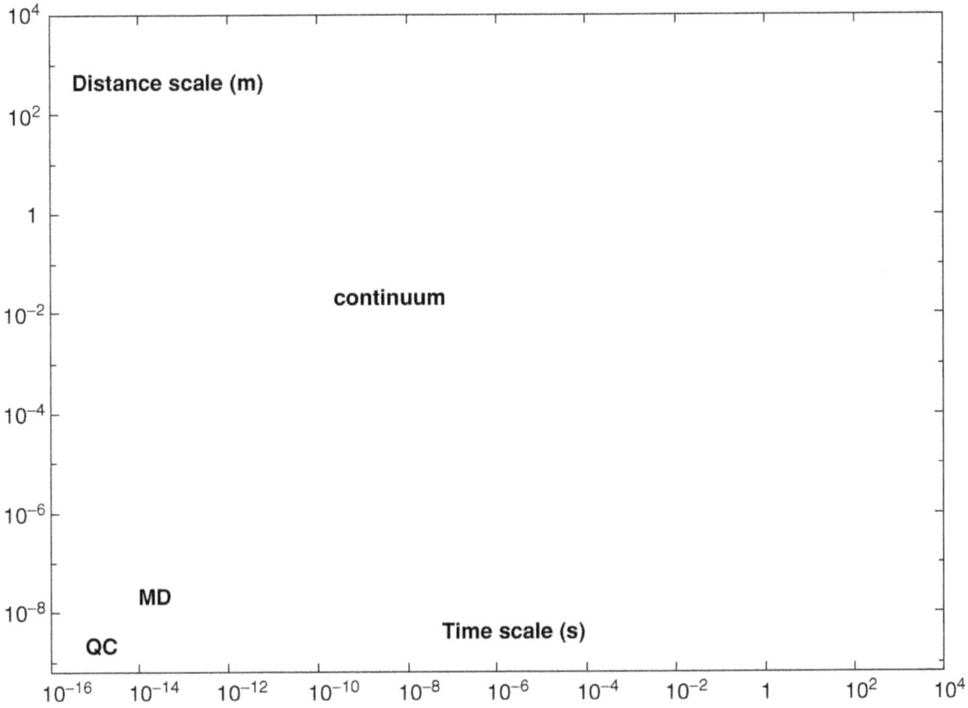


Figure 2 Computational length and time scales. QC stands for quantum chemistry methods in which the Schrödinger equation is solved. MD stands for molecular dynamics in which classical equations of motion for atomic motion are solved. Continuum includes thermodynamics, hydrodynamics, continuum mechanics, micromagnetism in which macroscopic variables describe the system. Statistical mechanics supplies the principles by which computations at these different scales are connected.

## Reference

1. C. E. Moore, *Electronics*, April 19 (1965).

# Part I

## Foundations of equilibrium statistical mechanics

# 1
# The classical distribution function

Historically, the first and most successful case in which statistical mechanics has made the connection between microscopic and macroscopic description is that in which the system can be said to be in equilibrium. We define this carefully later but, to proceed, may think of the equilibrium state as the one in which the values of the macroscopic variables do not drift in time. The macroscopic variables may have an obvious relation to the underlying microscopic description (as for example in the case of the volume of the system) or a more subtle relationship (as for temperature and entropy). The macroscopic variables of a system in equilibrium are found experimentally (and in simulations) to obey historically empirical laws of thermodynamics and equations of state which relate them to one another. For systems at or near equilibrium, statistical mechanics provides the means of relating these relationships to the underlying microscopic physical description.

We begin by discussing the details of this relation between the microscopic and macroscopic physical description in the case in which the system may be described classically. Later we run over the same ground in the quantum mechanical case. Finally we discuss how thermodynamics emerges from the description and how the classical description emerges from the quantum mechanical one in the appropriate limit.

## Foundations of equilibrium statistical mechanics

Here we will suppose that the systems with which we deal are nonrelativistic and can be described fundamentally by $3N$ time dependent coordinates labelled $q_i(t)$ and their time derivatives $\dot{q}_i(t)$ $(i = 1, \ldots, 3N)$. A model for the dynamics of the system is specified through a Lagrangian $L(\{q_i\}, \{\dot{q}_i\})$ (not explicitly time dependent) from which the dynamical behavior of the system is given by the principle of least action

$$\delta \int L \, \mathrm{d}t = 0 \tag{1.1}$$

or equivalently by the Lagrangian equations of motion

$$\frac{\partial L}{\partial q_i} - \frac{\mathrm{d}}{\mathrm{d}t}\left(\frac{\partial L}{\partial \dot{q}_i}\right) = 0 \tag{1.2}$$

Alternatively one may define momenta

$$p_i = \frac{\partial L}{\partial \dot{q}_i} \tag{1.3}$$

and a Hamiltonian

$$H = \sum_{i=1}^{N} p_i \dot{q}_i - L \tag{1.4}$$

Expressing $H$ as a function of the momenta $p_i$ and the coordinates $q_i$ one then has the equations of motion in the form

$$\frac{\partial H}{\partial p_i} = \dot{q}_i \tag{1.5}$$

$$-\frac{\partial H}{\partial q_i} = \dot{p}_i \tag{1.6}$$

In examples, we will often be concerned with a system of identical particles with conservative pair interactions. Then it is convenient to use the various components of the positions of the particles $\vec{r}_1, \vec{r}_2, \ldots$ as the quantities $q_i$, and the Hamiltonian takes the form

$$H = \sum_k \vec{p}_k^{\,2}/2m + (1/2) \sum_{k \neq l} V(\vec{r}_k, \vec{r}_l) \tag{1.7}$$

where the sums run over particle labels and $\vec{p}_k = \nabla_{\dot{\vec{r}}_k} H$. Then the Hamiltonian equations reduce to simple forms of Newton's equation of motion. It turns out, however, that the more general formulation is quite useful at the fundamental level, particularly in understanding Liouville's theorem, which we will discuss later.

In keeping with the discussion in the Introduction, we wish to relate this microscopic description to quantities which are measured in experiment or which are conveniently used to analyze the results of simulations in a very similar way. Generically we denote these observable quantities as $\phi(q_i(t), p_i(t))$. It is also possible to consider properties which depend on the microscopic coordinates at more than one time. We will defer discussion of these until Part III. Generally, these quantities, for example the pressure on the wall of a vessel containing the system, are not constant in time and what is measured is a time average:

$$\bar{\phi}_t = \frac{1}{\tau}\int_{t-\tau/2}^{t+\tau/2} \phi(q_i(t'), p_i(t'))\,\mathrm{d}t' \tag{1.8}$$

$\tau$ is an averaging time determined by the apparatus and the measurement made (or chosen for analysis by the simulator). Experience has shown that for many systems, an experimental situation can be achieved in which measurements of $\bar{\phi}_t$ are independent of $\tau$ for all $\tau > \tau_0$ for some finite $\tau_0$. It is easy to show that, in such a case, $\bar{\phi}_t$ is also independent of $t$. If this is observed to be the case for the macroscopic observables of interest, then the system is said to be in equilibrium. A similar operational definition of equilibrium is applied to simulations. In practice it is never possible to test this equilibrium condition for arbitrarily long times, in either experiment or simulation. Thus except in the rare cases in which mathematical proofs exist for relatively simple models, the existence and nature of equilibrium states are hypothesized on the basis of partial empirical evidence. Furthermore, in experimental situations, we do not expect any system to satisfy the equilibrium condition for arbitrarily long times, because interactions with the surroundings will inevitably change the values of macroscopic variables eventually. Making the system considered ever larger and the time scales longer and longer does not help here, because there is no empirical evidence that the universe itself is in equilibrium in this sense. Nevertheless, the concept of equilibrium turns out to be an extremely useful idealization because of the strong evidence that many systems do satisfy the relevant conditions over a very wide range of averaging times $\tau$ and that, under sufficiently isolated conditions, many systems spontaneously evolve rapidly toward an approximately equilibrium state whose characteristics are not sensitive to the details of the initial microscopic conditions. These empirical statements lack mathematical proofs for most systems of experimental or engineering interest, though mathematicians have made progress in proving them for simple models.

For systems in equilibrium defined in this way we are concerned with the calculation of averages of the type

$$\bar{\phi}_t = \lim_{\tau\to\infty} \frac{1}{\tau}\int_0^{\tau} \phi(\{q_i(t')\}, \{p_i(t')\})\,\mathrm{d}t' \tag{1.9}$$

We will show that it is always possible in principle to write this average in the form

$$\bar{\phi}_t = \int \rho(\{q_i\}, \{p_i\})\phi(\{q_i\}, \{p_i\})\,\mathrm{d}^{3N}q\,\mathrm{d}^{3N}p \tag{1.10}$$

in which $\rho(\{q_i\}, \{p_i\})$ is called the classical distribution function. The demonstration provides useful insight into the meaning of $\rho(\{q_i\}, \{p_i\})$. We consider the $6N$ dimensional space of the variables $\{q_i\}, \{p_i\}$, called phase space. In this space the time evolution of the system is described by the motion of a point. Take a small region of this space whose volume is denoted $\Delta^{3N}p\Delta^{3N}q$ centered at the point $(p, q)$. (Henceforth we denote $(p, q) \equiv (\{q_i\}, \{p_i\})$ and similarly $(\Delta p, \Delta q) \equiv (\{\Delta q_i\}, \{\Delta p_i\})$.) Consider the interval of time $\Delta t$ defined as

$$\Delta t(q_0, p_0, t_0; q, p, t; \Delta p, \Delta q) \tag{1.11}$$

equivalent to the time which the point describing the system spends in the region $\Delta^{3N}p\Delta^{3N}q$ around $(q, p)$ between $t_0$ and $t$ if it started at the point $(q_0, p_0)$ at time $t_0$.

Now consider the fraction of time that the system point spends in $\Delta^{3N}p\Delta^{3N}q$, denoted $\Delta w$:

$$\Delta w(q_0, p_0; q, p; \Delta p, \Delta q) = \lim_{t\to\infty}\left(\frac{\Delta t}{t - t_0}\right) \tag{1.12}$$

which is the fraction of the total time between $t_0$ and $t \to \infty$ which the system spends in the region $\Delta^{3N}p\Delta^{3N}q$ around $(q, p)$.

Now we express the time average $\bar{\phi}_t$ of equation (1.9) in terms of $\Delta w$ by dividing the entire phase space into small regions labelled by an index $k$ and each of volume $\Delta^{3N}p\Delta^{3N}q$:

$$\bar{\phi}_t = \sum_k \phi(q_0, p_0; q_k, p_k)\Delta w(q_0, p_0; q_k, p_k; \Delta p, \Delta q) \tag{1.13}$$

We then suppose that $\Delta w(q_0, p_0; q, p; \Delta p, \Delta q)$ is a well behaved function of the arguments $(\Delta p, \Delta q)$ and write

$$\Delta w = \left[\frac{\partial^{6N}\Delta w}{\partial^{3N}\Delta q\partial^{3N}\Delta p}\right]_{\Delta p=\Delta q=0} \Delta^{3N}q\Delta^{3N}p + \cdots \tag{1.14}$$

Defining

$$\rho(q_0, p_0; q, p) = \left[\frac{\partial^{6N}\Delta w}{\partial^{3N}\Delta q\partial^{3N}\Delta p}\right]_{\Delta p=\Delta q=0} \tag{1.15}$$

we then have in the limit $\Delta p\Delta q \to 0$ that

$$\bar{\phi}_t = \int \rho(q_0, p_0; q, p)\phi(q, p)\,\mathrm{d}^{3N}q\,\mathrm{d}^{3N}p \tag{1.16}$$

which is of the form (1.10). Several of the smoothness assumptions made in this discussion are open to question as we will discuss in more detail later.

Equation (1.16) is most useful if $\bar{\phi}_t$ depends only on a few of the $6N$ initial conditions $q_0, p_0$. Experimentally (and in simulations) it is found that the time averages of many macroscopic quantities measured in equilibrium systems are very insensitive to the way the system is prepared. We will demonstrate that under certain conditions, the only way in which these averages can depend on the initial conditions is through the values of the energy, linear momentum and angular momentum of the entire system. The general study of the dependence of averages of the form (1.16) on the initial conditions is part of ergodic theory. An ergodic system is (loosely from a mathematical point of view) defined as an energetically isolated system for which the phase point eventually passes through every point on the surface in phase space consistent with its energy. It is not hard to prove that the averages $\bar{\phi}_t$ in such an ergodic system depend only on the energy of the system. It is worth pointing out that the existence of ergodic systems in phase space of more than two dimensions is quite surprising. The trajectory of the system in phase space is a topologically one dimensional object (a path, parametrized by one variable, the time) yet we want this trajectory to *fill* the $6N - 1$ dimensional surface defined by the energy. The possibility of space filling curves is known mathematically (for a semipopular account see reference 1). However, for a large system, the requirement is extreme: the trajectory must fill an enormously open space of the order of $10^{23}$ dimensions! By contrast the path of a random walk has dimension 2 (in any embedding dimension)! (Very briefly, the (fractal or Hausdorff–Besicovitch) dimension of a random walk can be understood to be 2 as follows. The dimension of an object in this sense is determined as $D_{\mathrm{H}}$ defined so that when one covers the object in question with spheres of radius $\eta$ a minimum of $N(\eta)$ spheres is required and

$$L_{\mathrm{H}} = \lim_{\eta \to 0} N(\eta)\eta^{D_{\mathrm{H}}}$$

is finite and nonzero. For a random walk of mean square radius $\langle R^2 \rangle$, $N(\eta) = \langle R^2 \rangle / \eta^2$ and $D_{\mathrm{H}} = 2$. See reference 1 for details.) Nevertheless something like ergodicity is required for statistical mechanics to work, and so the paths in phase space of large systems must in fact achieve this enormous convolution in order to account for the known facts from experiment and simulation. It is not true that every system consisting of small numbers of particles is ergodic. Some of the problems at the end of this section illustrate this point. For example, a one dimensional harmonic oscillator is ergodic, but a billiard ball on a two dimensional table is not (Figure 1.1). On the other hand, in the latter case, the set of initial conditions for which it is not ergodic is in some sense "small." Another instructive example is a two dimensional harmonic oscillator (Problem 1.1).

There are several rationally equivalent ways of talking about equation (1.10). These occur in textbooks and other discussions and reflect the history of the subject as well as useful approaches to its extension to nonequilibrium systems. What we have discussed so far may be termed the Boltzmann interpretation of $\rho$ (in which $\rho$ is related to the time which the system phase point spends in each region of phase space). This is closely related to the *probability* interpretation of $\rho$ because the

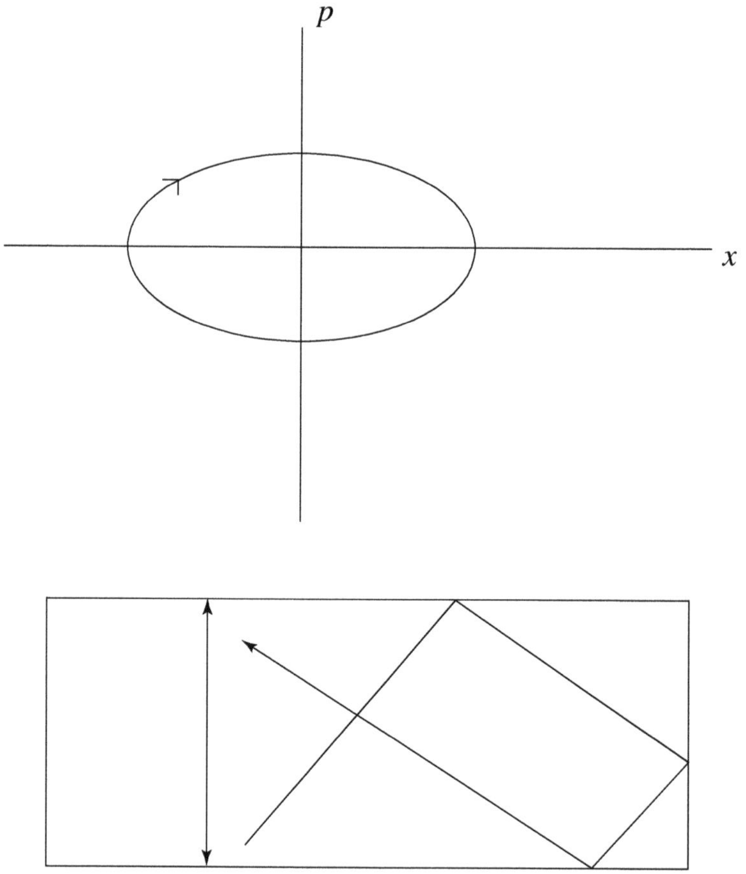


Figure 1.1 Phase space trajectory of a one dimensional oscillator fills the energy surface. For some initial conditions, a ball on a billiard table with elastic specularly reflecting walls is not ergodic.

probability that the system is found in $d^{3N}q d^{3N}p$ is just $\rho d^{3N}q d^{3N}p$ according to the standard observation frequency definition of probability. In such an interpretation, one takes no interest in the question of how the system got into each phase space region and could as well imagine that it hopped discontinuously from one to another for some purposes. Indeed such discontinuous hops (which we do not believe occur in real experimental systems obeying classical mechanics to a good approximation) do occur in certain numerical methods of computing the integrals (1.10) once the form of $\rho$ is known. Regarding $\rho d^{3N}q d^{3N}p$ as a probability opens the way to the use of information theoretic methods for approximating its form under all sorts of conditions in which various constraints are applied. For mechanical systems in equilibrium this approach leads to the same forms which we will obtain and use here. The reader is referred to the book by Katz[2] and to many papers by Jaynes for accounts of the information theoretic approach.[3,4] A third interpretation regards the integral (1.10) as describing an average over a large number (an *ensemble*) of different systems, all specified macroscopically in the same way. Specifically we may suppose that there are $\mathcal{N}$ systems with $\mathcal{N}\rho d^{3N}q d^{3N}p$ in each small region. Then the right hand side of (1.10) may be regarded as averaging $\phi$ over all $\mathcal{N}$ systems and the equality in (1.10) as stating the equality of time averages and ensemble averages. This was the approach taken by Gibbs in the first development of the foundations of the subject.[5] Gibbs regarded the equivalence of temporal and ensemble averages as a postulate and did not attempt a proof. The ensemble interpretation is of mainly historical interest but we will find its language useful in discussing Liouville's theorem below and the language of statistical mechanics contains many vestiges of it.

In statistical physics, we are mainly interested in large systems and will usually make assumptions appropriate for them. The path we will follow in order to obtain the standard forms (microcanonical, canonical and grand canonical) for the distribution function $\rho$ which successfully describe experimental and simulated equilibrium systems is as follows. (These materials come from a variety of sources but follow mainly the lines in Landau and Lifshitz' book.[6])

(1) We prove (in a physicist's manner, but following lines which can be made rigorous) the Liouville theorem, which shows that $\rho$ must be invariant in time, that is it is a constant of the motion.
(2) For large enough systems with finite range interactions, we then establish that $\rho$ can depend only on *additive* constants of the motion.
(3) Accepting that the additive constants are energy, linear and angular momentum (only) we obtain the canonical distribution. This leads to an apparent contradiction for an isolated system.
(4) We resolve this by demonstrating that the fluctuations in the energy in the canonical distribution become arbitrarily small in large enough systems.

Before proceeding let me explain why I think it worthwhile to spend time on these aspects of fundamentals. Most books of this sort simply write down the canonical distribution function and start calculating. Firstly, simulation usually uses an approach related to the microcanonical distribution, not the canonical one, whereas analytical theories almost always work with the canonical or grand canonical distribution function. Thus a firm grasp of why and when these are equivalent is of daily use in theoretical work which combines theory and simulation. Second, the proofs (inasfar as they exist) of the legitimacy of the standard distribution functions depend at several points on the largeness of the system involved, whereas simulations are necessarily constricted to quite finite systems and experiments too are increasingly interested in small systems for technical reasons. Finally, research on

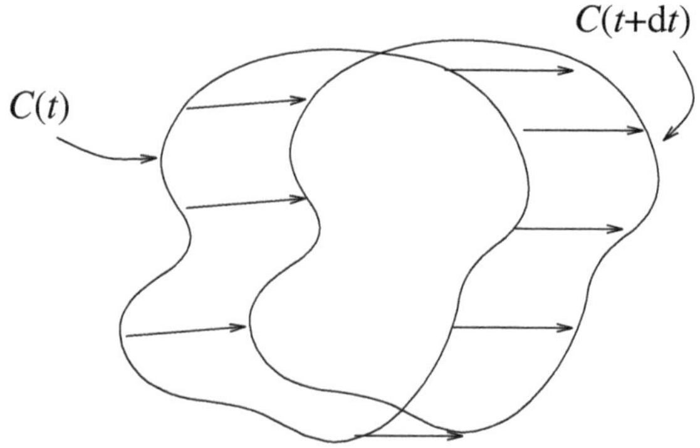


Figure 1.2 Schematic sketch of the evolution of the boundary $C(t)$ in phase space.

nonequilibrium systems will be informed by an understanding of the conditions under which an equilibrium description is expected to work.

## Liouville's theorem

The theorem states that the function $\rho(q_0, p_0; q, p)$ does not change if the phase point $q, p$ evolves in time as it does when the coordinates and momenta obey the Hamiltonian equations of motion in time. (When we actually use (1.10) to calculate an average, we do not regard the arguments $q, p$ as functions of time, but just integrate over them.) To demonstrate this, we use the ensemble interpretation. Consider a cloud of $\mathcal{N}$ phase points distributed over the phase space with density $\rho$. Consider a small but finite region in the phase space surrounded by a $6N - 1$ dimensional surface $C(t)$ around the point $p(t), q(t)$. The volume of the small region is

$$\Delta p \Delta q(t) = \int_{\text{inside } C(t)} \mathrm{d}^{3N} q(t)\, \mathrm{d}^{3N} q(t) \tag{1.17}$$

The surface $C(t)$ may be regarded as defined by the system points on it, which we regard as moving along trajectories according to Hamilton's equations as well. Thus the surface will move in time and so will the points inside it. At time $t$, the number of system points inside $C(t)$ is

$$\Delta\mathcal{N}(t) = \mathcal{N}\rho(q(t), p(t))\Delta q(t)\Delta p(t) \tag{1.18}$$

if the region is small.

Now let time evolve to $t + \mathrm{d}t$ (Figure 1.2).

The points in the boundary $C(t)$ move to form a new boundary $C(t + \mathrm{d}t)$ . The points inside $C(t)$ also move along their trajectories. But, because the solutions to the Hamiltonian equations are unique, *no trajectories cross*. Therefore the same points that lay inside $C(t)$ now lie inside $C(t+\mathrm{d}t)$ and the number of points $\Delta\mathcal{N}(t+\mathrm{d}t)$ lying inside $C(t+\mathrm{d}t)$ is the same as the number $\Delta\mathcal{N}(t)$. But by the same argument used at time $t$,

$$\Delta\mathcal{N}(t+\mathrm{d}t) = \mathcal{N}\rho(q(t+\mathrm{d}t), p(t+\mathrm{d}t))\Delta q(t+\mathrm{d}t)\Delta p(t+\mathrm{d}t) \tag{1.19}$$

where

$$\Delta q(t+\mathrm{d}t)\Delta p(t+\mathrm{d}t) = \int_{\text{inside } C(t+\mathrm{d}t)} \mathrm{d}^{3N}q(t+\mathrm{d}t)\,\mathrm{d}^{3N}q(t+\mathrm{d}t) \tag{1.20}$$

Combining (1.17), (1.18), (1.19), and (1.20) with the condition $\Delta\mathcal{N}(t+\mathrm{d}t) = \Delta\mathcal{N}(t)$ gives

$$\begin{aligned} &\rho(q(t), p(t)) \int_{\text{inside } C(t)} \mathrm{d}^{3N}q(t)\,\mathrm{d}^{3N}q(t) \\ &\quad = \rho(q(t+\mathrm{d}t), p(t+\mathrm{d}t)) \int_{\text{inside } C(t+\mathrm{d}t)} \mathrm{d}^{3N}q(t+\mathrm{d}t)\,\mathrm{d}^{3N}q(t+\mathrm{d}t) \end{aligned} \tag{1.21}$$

Thus to show that $\rho$ is constant we need to show that the integrals on the two sides of (1.21) are equal. To do that we transform the variables of integration on the right hand side to those on the left by use of the Jacobian:

$$\begin{aligned} \frac{\partial(q(t+\mathrm{d}t), p(t+\mathrm{d}t))}{\partial(q(t), p(t))} &= \begin{vmatrix} 1+\frac{\partial \dot{q}_1(t)}{\partial q_1(t)}\mathrm{d}t & \frac{\partial \dot{q}_1(t)}{\partial q_2(t)}\mathrm{d}t & \cdots & \cdots \\ \frac{\partial \dot{q}_1(t)}{\partial q_2(t)}\mathrm{d}t & 1+\frac{\partial \dot{q}_2(t)}{\partial q_2(t)}\mathrm{d}t & \cdots & \cdots \\ \cdot & \cdot & \cdot & \cdots \\ \cdot & \cdot & \cdot & \cdots \\ \cdot & \cdot & \cdot & \cdots \\ \cdots & \cdots & \cdots & \text{same for } p \end{vmatrix} \\ &= \prod_{i=q}^{3} N \left(1+\frac{\partial \dot{q}_i(t)}{\partial q_i(t)}\mathrm{d}t\right)\left(1+\frac{\partial \dot{p}_i(t)}{\partial p_i(t)}\mathrm{d}t\right) + \mathcal{O}((\mathrm{d}t)^2) \\ &= 1+\sum_i^{3N}\left(\frac{\partial \dot{q}_i(t)}{\partial q_i(t)} + \frac{\partial \dot{p}_i(t)}{\partial p_i(t)}\right)\mathrm{d}t + \mathcal{O}((\mathrm{d}t)^2) \end{aligned} \tag{1.22}$$

From the Hamiltonian equations of motion

$$\frac{\partial \dot{q}_i(t)}{\partial q_i(t)} = \frac{\partial^2 H}{\partial q_i \partial p_i} \tag{1.23}$$

$$\frac{\partial \dot{p}_i(t)}{\partial p_i(t)} = -\frac{\partial^2 H}{\partial p_i \partial q_i} \tag{1.24}$$

Thus if the Hamiltonian is analytic

$$\frac{\partial(q(t+\mathrm{d}t), p(t+\mathrm{d}t))}{\partial(q(t), p(t))} = 1 + \mathcal{O}((\mathrm{d}t)^2) \tag{1.25}$$

Thus from (1.21)

$$\begin{aligned}\frac{\mathrm{d}\rho(q(t), p(t))}{\mathrm{d}t} &= \lim_{\mathrm{d}t\to 0} \frac{\rho(q(t+\mathrm{d}t), p(t+\mathrm{d}t)) - \rho(q(t), p(t))}{\mathrm{d}t} \\ &= \lim_{\mathrm{d}t\to 0} \mathcal{O}(\mathrm{d}t^2/\mathrm{d}t) = 0\end{aligned} \tag{1.26}$$

With suitable mathematical tightening of the various steps, this line of reasoning rigorously proves the Liouville theorem (see for example Kurth[7]). The proof depends essentially on the choice of the variables $q_i$ and $p_i$ as the coordinates of phase space. For example, if one were to work in the space $\{q_i\}, \{\dot{q}_i\}$, the corresponding density would not be constant for every Lagrangian system.

### The distribution function depends only on additive constants of the motion

The preceding section sketches the proof that the density distribution function $\rho$ is a constant of the motion defined by Hamilton's equations of motion. That theorem is quite robust and in particular does not require that the system be large for its validity. To go further we need to suppose that the system has a large number of degrees of freedom. Furthermore we will assume that the interactions between the entities, usually atoms or molecules, in the system are short range in the following sense. We imagine dividing the system when it is in equilibrium into two parts both containing a large number of entities, say by designating a smooth two dimensional surface which divides the region of accessible values of each of the $(q_i, p_i)$ in two and assigning all the variables on one side of the surface at some time to one subsystem and all those on the other side to the other. If the interactions are of short range then the effects of the partition are only felt over a finite distance from the partition (which is actually somewhat larger than the range of the interaction, but which can be made much smaller than the dimension of each part). Let this distance be $d$ and the size of each partition be of order $L$. Then the magnitude of the effects of inserting the partition to the magnitude of effects from the bulk of each subsystem is roughly $L^2d/L^3 \to 0$ as $L \to \infty$. Thus, effectively, we can calculate average properties as well from the partitioned system as from the original system, as long as the properties $\phi$ which we are averaging treat every allowed region of phase space with equal weight. (The last condition means, for example, that $\phi$ could be the total energy or the average density, but not the density near the partition.)

Let the distribution function for the entities on one side of the partition be $\rho_1$ and let it depend on coordinates and momenta $q_1, p_1$ and similarly for the other side of the partition let the corresponding quantities be $\rho_2$ and $q_2, p_2$. This physical argument just given means that, in the limit of large systems

$$\rho(q, p) = \rho_1(q_1, p_1)\rho_2(q_2, p_2) \tag{1.27}$$

where $\rho(q, p)$ is the distribution function for the original unpartitioned system. Now each of these distribution functions obeys the Liouville theorem and is therefore a constant of the motion. Taking the ln of (1.27),

$$\ln \rho(q, p) = \ln \rho_1(q_1, p_1) + \ln \rho_2(q_2, p_2) \tag{1.28}$$

This means that $\ln \rho$ must be an *additive* constant of the motion for the system. In general, in a system with a phase space of $6N$ dimensions, there are $6N$ constants of the motion (of which one is a trivial choice of the origin of time). It is clear that the subset of these which is additive in the sense of (1.28) is much smaller. It is easy to show (see Landau and Lifshitz[8]) that these include the three components of the total linear momentum and the three components of the angular momentum of a system of particles. If, in the sense just discussed, the interactions between the elements of the partition can be ignored, then they include the energy as well. Though it is stated in some textbooks that these are the *only* additive constants of the motion, I do not know a proof. A large collection of evidence, not least the wide applicablity of the resulting forms for the distribution function to simulation and experiment, strongly suggests that it is often, if not always, true and we will suppose it to be so here. Some insight into this assumption is provided by the fact that conservation of energy, momentum and angular momentum can be shown to arise as a consequence of the invariance of the Hamiltonian to translations in time and space and to rotations respectively. Conversely, in cases in which the system is presumed to be constrained so that the Hamiltonian is not invariant to one of these operations, the corresponding conservation law does not hold. The most common case of this sort is that in which the system is confined to a container with fixed walls, so that the system is not invariant to spatial translation and rotation. Then the only additive conservation law of which we will take account is that of energy. If we suppose that energy, linear momentum and angular momentum are the only additive constants of the motion and use the fact that (1.28) plus Liouville's theorem shows that $\ln \rho(q, p)$ is an additive constant of the motion, then it follows that $\ln \rho(q, p)$ can only be a linear combination of these seven constants of the motion:

$$\ln \rho(q, p) = \alpha - \beta H(q, p) + \vec{\gamma} \cdot \vec{P} + \vec{\delta} \cdot \vec{L} \tag{1.29}$$

where $\alpha$, $\beta$, $\vec{\gamma}$ and $\vec{\delta}$ are constants independent of the $q, p$. The sign of the second term is chosen to conform to convention. We have sketched an argument for this form using, to reiterate, the following elements: (i) Liouville's theorem, true for systems of any size, (ii) (1.27) true only for large systems with short range interactions, and

(iii) the assumption, very likely to be correct for most systems but unproved to my knowledge, that energy, total linear momentum and total angular momentum are the only additive constants of the motion. In most of our studies we will confine attention to the case mentioned above in which the system is confined to a container so that $\vec{P}$ and $\vec{L}$ are not conserved and (1.29) becomes simply

$$\ln \rho(q, p) = \alpha - \beta H(q, p) \tag{1.30}$$

The constant $\alpha$ is determined in terms of $\beta$ by requiring the average value of a constant give, using (1.10), the constant itself. Thus

$$\rho(q, p) = \frac{\mathrm{e}^{-\beta H(q,p)}}{\int \mathrm{e}^{-\beta H(q,p)}\, \mathrm{d}^{3N}q\, \mathrm{d}^{3N}p} \tag{1.31}$$

This is the canonical distribution function. It is the distribution postulated by Gibbs and most analytical work in equilibrium statistical mechanics postulates it as a starting point.

The canonical distribution function also arises from the following "information theoretic" point of view. Consider a series of $\mathcal{N}$ measurements of the phase point of the system. Dividing the phase space into $M$ regions labelled $\alpha$ we consider the probabilities $P_\alpha$ of finding the system in each of them. ($P_\alpha = \rho(q_\alpha, p_\alpha)\Delta q \Delta p$ where $\Delta q \Delta p$ is the phase space volume of each region.) Consider a set of $\mathcal{N}$ observations of the system in which, in $M_\alpha$ of those observations, we find the system in region $\alpha$. There are

$$N_{\text{ways}} = \mathcal{N}!/\Pi_\alpha (M_\alpha)!$$

ways to get this result. If we know nothing else, then the most probable set of observations is the one for which $N_{\text{ways}}$ is maximum subject to the constraint $\sum_\alpha M_\alpha = \mathcal{N}$. It is almost obvious that $N_{\text{ways}}$ is maximized for all the $M_\alpha$ equal, in this case giving $P_\alpha = 1/M$ corresponding to constant $\rho(p, q)$. (It is instructive to work this out by taking the ln of $N_{\text{ways}}$, using Stirling's approximation, introducing a Lagrange multiplier to fix the constraint and minimizing with respect to $M_\alpha$.) If we know the value of the energy then we should make a guess consistent with that information. However, we only obtain the canonical distribution if we guess that the quantity to maximize subject to constant energy is not $N_{\text{ways}}$ but $\ln N_{\text{ways}}$. The choice of the ln function in this guess is justified by an argument similar to the one used earlier, but phrased in a more general way. Consider two systems (1) and (2) which may be regarded as subsystems of the original one, as in our earlier discussion. Then the number of possible ways to get a result is $N^{(1)}_{\text{ways}} N^{(2)}_{\text{ways}}$. But the "missing information" function $I(N_{\text{ways}})$ that we maximize is supposed to need the property that $I(N^{(1)}_{\text{ways}} N^{(2)}_{\text{ways}}) = I(N^{(1)}_{\text{ways}}) + I(N^{(2)}_{\text{ways}})$. This requirement, together with the requirement that $I(N_{\text{ways}})$ be monotonic in $N_{\text{ways}}$, is sufficient to show that

$I(N_{\text{ways}})$ must be $\ln N_{\text{ways}}$. It then follows quite easily by use of Lagrange multipliers that the maximum of $I(N_{\text{ways}}) - \beta \sum_\alpha \mathcal{N} P_\alpha H_\alpha$ gives the canonical distribution (using the Stirling approximation and assuming $\mathcal{N}$ to be large). Though this "information theoretic" derivation of the canonical distribution appears rather different from the one given earlier, it is in fact quite similar. The additivity requirement on the information function is seen to be almost the same as the requirement (1.27). However, the choice of the energy as the fixed quantity (or more generally the energy, momentum and angular momentum) is, if anything, even less well motivated in this argument than it is in the earlier one, where we only needed the assumption that the energy was the only additive constant of the motion. A very strong merit of this information theoretic point of view, however, is that the same general approach can be extended to very different problems involving, for example, the inference of the most likely conclusions from incomplete experimental data in a wide variety of circumstances in which we are not dealing with a simply characterized Hamiltonian system of particles. We refer to the cited book by Katz[2] and the numerous articles by Jaynes for further discussion and elaboration of this point of view.

The canonical distribution appears to run into a contradiction, however, if we consider that (1.31) appears to allow the energy to vary, whereas in fact the trajectory through phase space is on a fixed energy surface. For a large enough system which can be partitioned arbitrarily many times, we can show that this is not a problem. Suppose that, instead of partitioning the system into two parts, we partition it into $N'$ parts where $N' \gg 1$ but also $N/N' \gg 1$ so that each region of the partition has many particles in it. Notice that for a system of $10^{23}$ particles this would be easy to do. Then the surface to volume ratio of each region in the partition would be small in the sense discussed earlier, so we can still regard the interaction between regions as negligible as long as the interactions are short range. Thus we can write to adequate approximation that the Hamiltonian $H = \sum_{\alpha=1}^{N'} H_r(q_\alpha, p_\alpha)$ where $H_r$ is the Hamiltonian of a region. If we use the canonical form for the distribution function, we then obtain

$$\rho(p, q) = \prod_\alpha \frac{e^{-\beta H_r(q_\alpha, p_\alpha)}}{\int e^{-\beta H_r(q_\alpha, p_\alpha)}\, dq_\alpha\, dp_\alpha} \tag{1.32}$$

Now one can compute the expectation value for the total energy and its fluctuations (which really should not occur) from this distribution function. It is easy to show that

$$\frac{\langle (H - \langle H \rangle)^2 \rangle}{\langle H \rangle^2} = \frac{1}{N'} \frac{\langle H_r^2 \rangle - \langle H_r \rangle^2}{\langle H_r \rangle^2} \tag{1.33}$$

Thus, as the system gets bigger while the size of each region remains fixed, the calculated fluctuations in the energy get relatively smaller and smaller and the canonical distribution function becomes in this respect a better and better representation of the energy conserving behavior of the system. Since we used approximations requiring a large system in deriving the canonical distribution it is not surprising that it only gives a consistent description when the system is large.

## Microcanonical distribution

For an isolated system, the energy is absolutely conserved at some value $E$. If the system is large, then the arguments of the last section show that the distribution function depends *only* on $E$ in the case that a box contains the system and prevents conservation of $\vec{P}$ and $\vec{L}$. Thus the only possible distribution function for a large isolated system is

$$\rho(q, p) = \text{constant} \times \delta(E - H(q, p)) \tag{1.34}$$

This is known as the microcanonical distribution function. Note that the argument we have given for it requires that the system be large and have short range interactions in the sense discussed in the last section. (However, the argument contains a contradiction because the ln of (1.34) for a system $H = H_1 + H_2$ is not exactly the sum of the ln of the microcanonical distributions for the subsystems. We will discuss how to consider a subsystem of a system described by the microcanonical distribution below.) Of course the *exact* $\rho(q, p)$ is also proportional to $\delta(E - H(q, p))$ for any size system. However (1.34) will not be true in general for any size system. To see why, recall that in complete generality the orbits of any Hamiltonian system involve $6N - 1$ constants of the motion. Let these be denoted $Q_\lambda(q, p)$. The last coordinate is just the time itself. Thus in this representation, an orbit with energy $E$ is fully described by

$$\rho_{\text{exact}} = (1/\tau)\delta(E - H(q, p)) \prod_{\lambda=3}^{6N} \delta\big(Q_\lambda^{(0)} - Q_\lambda(q, p)\big) \frac{\partial(\{Q_\lambda\}, H, t)}{\partial(q, p)} \tag{1.35}$$

where $\tau$ is the (extremely long Poincaré) period of the orbit. The energy has been displayed explicitly and the other $6N - 2$ constants of the motion are denoted $\{Q_\lambda^{(0)}\}$. The nontrivial statement requiring invocation of properties of large systems is that the factors after the energy delta function in (1.35) do not matter for the calculation of macroscopic averages. Though it is sometimes said that simulations work in the microcanonical ensemble, it can be seen now that this is not really exactly right. Simulations are numerically approximating $\rho_{\text{exact}}$. We can also see here how the anomaly concerning the dimension of the space filled by the orbit comes in. The orbit described by (1.35) is topologically one dimensional, but the space described by the microcanonical $\rho$ has dimension $6N - 1$. Thus the exact orbit must be space filling to an astonishing degree. One other point worth noting is that if one chooses the coordinates and momenta to be the set $\{Q_\lambda\}$, $H, t$ then the orbit is not convoluted at all in the space of these coordinates and momenta but is a straight line. This reveals the fact that ergodicity (or something equivalent) cannot hold for absolutely all choices of generalized coordinates and momenta. A full theory of how the microcanonical and canonical distributions arise from the microscopic orbit must take account of this and can only hope to show their validity for some overwhelmingly large set of choices of coordinates and momenta. Experience indicates that this set will include essentially all the choices which one would naturally make for large systems.

We can use the microcanonical form to obtain the canonical distribution for a subsystem in another way. In some respects this is unnecessary since we obtained the canonical distribution in the last section without reference to the microcanonical one. However, we gain a physically useful expression for the constant $\beta$ from this approach. Let the Hamiltonian be $H_{\rm b} + H_{\rm s}$ where "b" and "s" refer to the "bath" and the "subsystem" respectively. Then the distribution function for the whole system is

$$\rho(q_{\rm b}, p_{\rm b}; q_{\rm s}, p_{\rm s}) = \text{constant} \times \delta(E - H_{\rm b} - H_{\rm s}) \tag{1.36}$$

We get the distribution for the subsystem alone by integrating out the coordinates associated with the bath:

$$\rho_{\rm s}(q_{\rm s}, p_{\rm s}) = \text{constant} \times \int {\rm d}q_{\rm b}\, {\rm d}p_{\rm b}\, \delta(E - H_{\rm b} - H_{\rm s}) \tag{1.37}$$

Now express the integration in terms of a new set of coordinates $(H_{\rm b}(q_{\rm b}, p_{\rm b}), Q_{\rm b}, P_{\rm b})$ where the first coordinate is the Hamiltonian of the bath and the $(Q_{\rm b}, P_{\rm b})$ are any set of coordinates spanning the surfaces of constant energy in the bath phase space. The integral on bath coordinates is then

$$\begin{aligned}\int {\rm d}q_{\rm b}\, {\rm d}p_{\rm b}\, \delta(E - H_{\rm b} - H_{\rm s}) &= \int {\rm d}H_{\rm b}\, \delta(E - H_{\rm b} - H_{\rm s}) \int {\rm d}P_{\rm b}\, {\rm d}Q_{\rm b} \frac{\partial(q_{\rm b}, p_{\rm b})}{\partial(H_{\rm b}, Q_{\rm b}, P_{\rm b})} \\ &\equiv \int {\rm d}H_{\rm b}\, \delta(E - H_{\rm b} - H_{\rm s}) \Omega_{\rm b}(H_{\rm b}) \\ &= \Omega_{\rm b}(E - H_{\rm s})\end{aligned} \tag{1.38}$$

in which

$$\Omega_{\rm b}(E - H_{\rm s}) \equiv \int_{H_{\rm b}=E-H_{\rm s}} {\rm d}P_{\rm b}\, {\rm d}Q_{\rm b} \frac{\partial(q_{\rm b}, p_{\rm b})}{\partial(H_{\rm b}, Q_{\rm b}, P_{\rm b})} \tag{1.39}$$

is the "area" (or volume) of the $6N_b - 1$ dimensional region in the phase space of the bath which is associated with bath energy $E - H_{\rm s}$. Requiring that $\rho_{\rm s}(q_{\rm s}, p_{\rm s})$ be normalized then gives

$$\rho_s(q_s, p_s) = \frac{\Omega_b(E - H_s)/\Omega_b(E)}{\int dq_s\, dp_s\, (\Omega_b(E - H_s)/\Omega_b(E))} \tag{1.40}$$

We rewrite

$$\Omega_b(E - H_s)/\Omega_b(E) = e^{\ln(\Omega_b(E-H_s)/\Omega_b(E))} = e^{S_b(E-H_s)/k_B} \tag{1.41}$$

where, in anticipation of common usage we define

$$S_b(E - H_s) = k_B \ln \frac{\Omega_b(E - H_s)}{\Omega_b(E)} \tag{1.42}$$

which will be called the entropy of the bath when the bath has energy $E - H_s$. $k_B$ is Boltzmann's constant and gives the entropy its usual units. As usual in classical physics, an additive constant in the definition of the entropy is arbitrary here. Now supposing that $E \gg H_s$, it is reasonable to take the first term in an expansion of $S_b(E - H_s)$ about $H_s = 0$ giving

$$S_b(E - H_s) = S_b(E) + \left(\frac{\partial S_b(E - H_s)}{\partial H_s}\right)_{H_s=0} H_s = + \left(\frac{\partial S_b(E - H_s)}{\partial H_s}\right)_{H_s=0} H_s \tag{1.43}$$

where the second equality follows only with our choice of additive constant in $S_b$. Then (1.40) becomes

$$\rho_s(q_s, p_s) = \frac{e^{-\beta H_s(q_s, p_s)}}{\int dq_s\, dp_s\, e^{-\beta H_s(q_s, p_s)}} \tag{1.44}$$

if we identify

$$\beta = \left(\frac{\partial S_b(E + x)}{\partial x}\right)_{x=0} \tag{1.45}$$

Thus we obtain the canonical distribution function for the subsystem again with the added benefit that an expression for $\beta$ is obtained which is recognizable as related to thermodynamics.

On the other hand this argument contains a swindle. The swindle occurs at equation (1.43). Why should we expand $S_b$ and not $\Omega_b$? Or to put it another way, when we get to thermodynamics, we will want $\beta$ to be independent of system size, so we will need to have $S_b$ proportional to system size. But the argument provides no guarantee that this will be so. One way to answer is to go back to the previous section: (1.45) is consistent with the results of this section and the latter depended on the divisibility of the system into subsystems whose properties could be added to get the properties of the whole system. Therefore the requirement that we expand $S_b$ and not $\Omega_b$ must also be related to additivity.

One gains some insight into how this happens by consideration of the case of a bath which is a perfect gas. The Hamiltonian is

$$H_b = \sum_{i=1}^{3N_b} p_i^2/2m \tag{1.46}$$

and the integral in (1.39) can be done by first transforming to spherical coordinates in momentum space. Let $P_1 = \sqrt{\sum_{i=1}^{3N_b} p_i^2}$. Then

$$\Omega_b(E) = V_b^{N_b} \frac{1}{\left|\frac{\partial H_b}{\partial P_1}\right|_{H_b=E}} \int_{P_1=\sqrt{2mE_b}} d^{3N_b-1}S \tag{1.47}$$

Here the integral is over the surface of a $3N_b$ dimensional sphere in momentum space with radius $\sqrt{2mE_b}$. Thus one obtains

$$\Omega_b(E) = \sqrt{\frac{m}{2E_b}} \frac{2\pi^{3N_b/2}}{(3N_b/2-1)!} (2mE_b)^{(3N_b-1)/2} V_b^{N_b} \tag{1.48}$$

We study this in the *thermodynamic limit* in which $N_b \to \infty$, $V_b \to \infty$ while $N_b/V_b$ and $E_b/N_b$ remain fixed. In this limit, $(3N_b/2-1)!$ may be approximated as

$$(3N_b/2-1)! \approx e^{-3N_b/2}(3N_b/2)^{3N_b/2} \tag{1.49}$$

and we have

$$\Omega_b(E) \to ((2E_b/3N_b)2\pi me)^{3N_b/2}\, V_b^{N_b} \tag{1.50}$$

Thus $\Omega_b$ is extremely rapidly diverging with $N_b$ but the entropy

$$S_b(E-H_s) = (3N_b/2)\ln((E-H_s)/E) \approx (3N_b/2)(-H_s/E) \tag{1.51}$$

so that the entropy is additive and

$$\beta = 3N_b/2E \tag{1.52}$$

in accordance with the equipartition theorem.

On the other hand, it is interesting to notice in this example that the limit $V_b \to \infty$ while $N_b/V_b$ is fixed is not well behaved in $\ln \Omega_b(E)$. In fact the term in $\ln \Omega_b(E)$ depending on the bath volume $V_b$ is $N_b \ln V_b$ which is proportional to $N_b \ln N_b$ and not to $N_b$ in the thermodynamic limit. This is a general feature of the classical distribution function as we have discussed it so far. In order to get a correct thermodynamic limit from the classical distribution function as one varies the number of particles, one must divide the definition of the classical distribution function by the factorial of the number of particles ($N_b!$ here). Gibbs first noted this, and argued, without knowledge of quantum mechanics, that the needed factor of $1/N!$ should be inserted because of the indistinguishability of particles. Although the factor $1/N!$ should be inserted, it arises from the indistinguishability of particles in quantum mechanics and is not actually consistent with classical mechanics. The factor $1/N!$ is a residue of quantum mechanics at the classical level. This will be discussed somewhat further in Chapter 4.

## References

1. B. Mandelbrot, *The Fractal Geometry of Nature*, San Francisco, CA: W. H. Freeman, 1983, p. 62.
2. A. Katz, *Principles of Statistical Mechanics, The Information Theoretic Approach*, San Francisco, CA: W. H. Freeman, 1967.
3. E. T. Jaynes, Information theory and statistical physics, *Physical Review* **106** (1957) 620.
4. E. T. Jaynes, in *Probability Theory, The Logic of Science*, ed. G. Larry Bretthorst, Cambridge: Cambridge University Press, 2004.
5. J. W. Gibbs, *Elementary Principles in Statistical Mechanics, Developed with Special Reference to the Foundations of Thermodynamics*, London: Charles Scribner, 1902, reprinted by Ox Bow Press, 1981.
6. L. D. Landau and E. M. Lifshitz, *Statistical Physics*, 3rd edition, *Part 1, Course of Theoretical Physics*, Volume 5, Oxford: Pergamon Press, 1980.
7. R. Kurth, *Axiomatics of Classical Statistical Mechanics*, New York: Pergamon Press, 1960.
8. L. D. Landau and E. M. Lifshitz, *Mechanics*, translated from the Russian by J. B. Sykes and J. S. Bell, Oxford: Pergamon Press, 1969.

## Problems

1.1 Consider a two dimensional harmonic oscillator obeying the equations of motion:

$$m\frac{\mathrm{d}^2x}{\mathrm{d}t^2} = -K_x x$$

$$m\frac{\mathrm{d}^2y}{\mathrm{d}t^2} = -K_y y$$

In fact, according to the definitions of this chapter, this system is not ergodic for *any* set of initial conditions or values of the force constants $K_x$ and $K_y$. However, for certain conditions on $K_x$ and $K_y$, the system satisfies a modified definition of ergodicity, in which the system point fills a portion of the constant energy surface. Prove these statements and illustrate by making some simulations of the motion numerically, showing computed trajectories for various cases in the $xy$ and $p_x p_y$ planes.

1.2 Show that a sufficient condition for (1.44) is that

$$\Omega_\mathrm{b}(E_\mathrm{b}) = \text{constant} \times (E_\mathrm{b}/N'_\mathrm{b})^{\eta N'_\mathrm{b}}$$

where $\eta$ is a real positive number and $N'_b$ is a real positive number going to infinity in the thermodynamics limit while $E_b/N'_b$ approaches a finite constant. Find $\beta$ in this case.

1.3 Find the classical distribution functions for the following one dimensional systems as a function of initial conditions.

A particle confined to a box of length $a$ by elastic walls.
A one dimensional harmonic oscillator, spring constant $K$.
A ball in the Earth's gravitational field bouncing elastically from a floor.
A pendulum with arbitrary amplitude.

# 2

# Quantum mechanical density matrix

For systems which obey quantum mechanics, the formulation of the problem of treating large numbers of particles is, of course, somewhat different than it is for classical systems. The microscopic description of the system is provided by a wave function which (in the absence of spin) is a function of the classical coordinates $\{q_i\}$. The mathematical model is provided by a Hamiltonian operator $H$ which is often obtained from the corresponding classical Hamiltonian by the replacement $p_i \to (\hbar/\mathrm{i})(\partial/\partial q_i)$. In other cases the form of the Hamiltonian operator is simply postulated. The microscopic dynamics are provided by the Schrödinger equation $\mathrm{i}\hbar(\partial\Psi/\partial t) = H\Psi$ which requires as initial condition the knowledge of the wave function $\Psi(\{q_i\}, t)$ at some initial time $t_0$. (Boundary conditions on $\Psi(\{q_i\}, t)$ must be specified as part of the description of the model as well.) The results of experiments in quantum mechanics are characterized by operators, usually obtained, like the Hamiltonian, from their classical forms and termed observables. Operators associated with observables must be Hermitian. In general, the various operators corresponding to observables do not commute with one another. It is possible to find sets of commuting operators whose mutual eigenstates span the Hilbert space in which the wave function is confined by the Schrödinger equation and the boundary conditions. A set of such (time independent) eigenstates, termed $\psi_\nu(q)$, is a basis for the Hilbert space. The relation between operators $\phi_{\mathrm{op}}$ and experiments is provided by the assumed relation

$$\bar{\phi}(t) = \int \Psi^*(\{q_i\}, t)\phi_{\mathrm{op}}\Psi(\{q_i\}, t)\,\mathrm{d}^{3N}q \tag{2.1}$$

where $\bar{\phi}(t)$, the quantum mechanical average, is the average value of the experimental observable associated with the operator $\phi_{\mathrm{op}}$ which is observed on repeated experimental trials on a system with the same wave function at the same time $t$. Unlike the classical case, even before we go to time averages or to large systems, only averages of the observed values of experimental variables are predicted by the theory. Consideration of time averaging will introduce a second level of averaging into the theory. We are working here in the Schrödinger representation of operators. Any time dependence which they have is explicit. In the study of equilibrium systems we will assume that the operators of interest are explicitly time independent which is to say that they are time independent in the Schrödinger representation (but not of course in the Heisenberg representation).

We suppose as in the classical case that in studying the macroscopic systems usually of interest in statistical physics, we are interested in the time averages of experimental observables, which we denote as

$$\bar{\bar{\phi}} = \frac{1}{\tau} \int_{t-\tau/2}^{t+\tau/2} \bar{\phi}(t')\,\mathrm{d}t' \tag{2.2}$$

The double bar emphasizes that, unlike the classical case, two kinds of averaging are taking place. As in the classical case, we define equilibrium to be a situation in which these averages are independent of $\tau$ for any $\tau > \tau_0$. By the same arguments as in the classical case, the averages are then independent of $t$ as well. It is convenient, as in the classical case, to move the origin of time to $t = -\tau/2$ before taking the limit $\tau \to \infty$ so that we are interested in averages of the form

$$\bar{\bar{\phi}} = \lim_{\tau\to\infty} \frac{1}{\tau} \int_0^{\tau} \bar{\phi}(t')\,\mathrm{d}t' \tag{2.3}$$

In the classical case, the analogous average was related to an integral over the classical variables $q, p$. In the quantum case the corresponding average is over a set of basis states of the Hilbert space defined briefly above. In particular, let $\psi_\nu(q)$ be a set of eigenstates of some complete set of commuting operators and expand the wavefunction in terms of it

$$\Psi(q,t) = \sum_\nu a_\nu(t)\psi_\nu(q) \tag{2.4}$$

It is possible in general to choose the $\psi_\nu(q)$ to be orthonormal and we will do so. Then the Schrödinger equation, expressed in terms of the coefficients $a_\nu$, becomes $\mathrm{i}\hbar(\partial a_\nu/\partial t) = \sum_{\nu'} H_{\nu\nu'} a_{\nu'}$ where $H_{\nu\nu'} = \int \psi_\nu^* H \psi_{\nu'} \mathrm{d}^{3N} q$. Inserting (2.4) into (2.3) and assuming that $\phi_{\mathrm{op}}$ is explicitly time independent gives

$$\bar{\bar{\phi}} = \lim_{\tau\to\infty} \frac{1}{\tau} \int_0^{\tau} \mathrm{d}t' \bar{\phi}(t') = \sum_{\nu,\nu'} \left\{ \lim_{\tau\to\infty} \frac{1}{\tau} \int_0^{\tau} a_\nu^*(t) a_{\nu'}(t)\,\mathrm{d}t \right\} \phi_{\nu\nu'} \tag{2.5}$$

in which

$$\phi_{\nu\nu'} = \int \psi_\nu^* \phi_{\mathrm{op}} \psi_{\nu'}\,\mathrm{d}^{3N} q \tag{2.6}$$

The interchange of the order of summation and integration should be proved to be legitimate in a rigorous treatment. If we define

$$\rho_{\nu'\nu} = \lim_{\tau\to\infty} \frac{1}{\tau} \int_0^{\tau} a_\nu^*(t) a_{\nu'}(t)\,\mathrm{d}t \tag{2.7}$$

then (2.6) becomes

$$\bar{\bar{\phi}} = \sum_{\nu,\nu'} \rho_{\nu'\nu}\phi_{\nu\nu'} \tag{2.8}$$

or in matrix notation

$$\bar{\bar{\phi}} = \mathrm{Tr}\rho\phi \tag{2.9}$$

where $\rho$ and $\phi$ are the (generally infinite dimensional) square matrices $\rho_{\nu'\nu}$ and $\phi_{\nu\nu'}$. (The inversion of the order of indices in (2.7) is made in order to permit (2.8) to be written as a matrix multiplication.) Tr means trace. Equation (2.7) is the quantum version of (1.15) while (2.9) is the quantum form of (1.16). The matrix $\rho$ is called the density matrix. Just as the classical distribution function can be described in terms of various sets of canonical coordinates and momenta related by contact transformations, the density matrix can be expressed in terms of various complete sets of orthonormal functions which span the space of wave functions and these are related by unitary transformations. Just as the classical $\rho(q, p)$ depended in principle on all of the initial classical conditions $p_0$, $q_0$, the density matrix depends, again in principle, on the initial wave function $\psi(q, t_0)$ or, equivalently, on all of the coefficients $a_\nu(t = 0)$. Here we come, however, to a significant practical difference: whereas in the classical case, the amount of data associated with specifying the initial conditions, while large, is in principle countable and finite ($6N$ numbers), the initial condition specifying the wave function requires at the numerical level an uncountably large amount of data, even for a system with a modest number of particles. This apparently academic distinction has the very practical effect that simulations of classical systems of thousands of particles are feasible while for quantum systems they are extremely difficult and not very reliable, even for simple systems.

As in the classical case, the density matrix may be interpreted in various ways. For example, one obtains the probability interpretation if one supposes only that the probability that the system has wave function $\Psi^{(j)}(q) = \sum_\nu a_\nu^{(j)}\psi_\nu(q)$ is $P_j$, thus avoiding any assumptions about the dynamics (but consistent with the known dynamics of quantum mechanics). Then the value of the double average $\bar{\bar{\phi}}$ would be

$$\bar{\bar{\phi}} = \sum_j P_j \int \Psi^{(j)*}(q)\phi_{\mathrm{op}}\Psi^{(j)}(q)\,\mathrm{d}^{3N}q = \sum_{\nu,\nu'} \rho_{\nu'\nu}\phi_{\nu\nu'} \tag{2.10}$$

where

$$\rho_{\nu'\nu} = \sum_{\text{all } j} P_j a_{\nu}^{(j)*} a_{\nu'}^{(j)} \tag{2.11}$$

These equations are equivalent to (2.9) and (2.7). This point of view was emphasized in the book by Tolman.[1]

Before proceeding to the analysis which leads to the quantum mechanical analogue to the canonical distribution function, we note some facts which follow from the definition of $\rho_{\nu'\nu}$. First, $\rho_{\nu'\nu}$ is Hermitian since

$$\rho_{\nu'\nu}^{*} = \lim_{\tau\to\infty} \frac{1}{\tau} \int_0^{\tau} a_{\nu}(t) a_{\nu'}(t)^{*}\, \mathrm{d}t = \lim_{\tau\to\infty} \frac{1}{\tau} \int_0^{\tau} a_{\nu'}^{*}(t) a_{\nu}(t)\, \mathrm{d}t = \rho_{\nu\nu'} \tag{2.12}$$

This means that, for some purposes, $\rho$ can be regarded as an operator which is an observable. Secondly, consider $\mathrm{Tr}\rho$:

$$\mathrm{Tr}\rho = \sum_{\nu} \rho_{\nu\nu} = \sum_{\nu} \lim_{\tau\to\infty} \frac{1}{\tau} \int_0^{\tau} a_{\nu}^{*}(t) a_{\nu}(t)\, \mathrm{d}t = \lim_{\tau\to\infty} \frac{1}{\tau} \int_0^{\tau} \sum_{\nu} |a_{\nu}(t)|^2\, \mathrm{d}t \tag{2.13}$$

But

$$\int \mathrm{d}^{3N} q |\Psi(q,t)|^2 = \sum_{\nu,\nu'} a_{\nu}^{*}(t) a_{\nu'}(t) \int \mathrm{d}^{3N} q\, \psi_{\nu}^{*}(q) \psi_{\nu'}(q) = \sum_{\nu} |a_{\nu}(t)|^2 = 1 \tag{2.14}$$

so

$$\mathrm{Tr}\rho = \lim_{\tau\to\infty} \frac{1}{\tau} \int_0^{\tau} \mathrm{d}t = 1 \tag{2.15}$$

The requirement that $\mathrm{Tr}\rho = 1$ (independent of basis) looks a lot like the requirement that a probability distribution be normalized and, in fact, in any given representation $\nu$ one can see from the definition that $\rho_{\nu\nu}$ is the probability of finding the system in the state $\psi_{\nu}$ when a measurement of the observables associated with the quantum numbers $\nu$ is made. But in an arbitrary basis, $\rho_{\nu\nu'}$ can have off diagonal elements which are complex and have no trivial probability interpretation.

To understand the fundamental equilibrium forms of the density matrix, we proceed much as in the classical case to prove a quantum version of the Liouville theorem and then argue on the basis of expected additive properties of large systems for a canonical form for the density matrix. The quantum mechanical version of the Liouville theorem is quite simple to obtain. First, one must decide how to define the time dependence of the density matrix. In the present case, we choose to define the time derivative of $\rho_{\nu\nu'}$ as

$$\frac{\mathrm{d}\rho_{\nu\nu'}}{\mathrm{d}t} = \lim_{\tau\to\infty} \frac{1}{\tau} \int_0^{\tau} \frac{\mathrm{d}}{\mathrm{d}t}(a_{\nu'}^{*}(t) a_{\nu}(t))\, \mathrm{d}t \tag{2.16}$$

It is immediately evident that if the limit exists it is zero:

$$\frac{\mathrm{d}\rho_{\nu\nu'}}{\mathrm{d}t} = \lim_{\tau\to\infty} \frac{(a^*_{\nu'}(\tau)a_\nu(\tau) - (a^*_{\nu'}(0))a_\nu(0))}{\tau} = 0 \tag{2.17}$$

since the coefficients $a_\nu$ must be finite if the wave functions $\Psi(q, t)$ are to be normalizable. Thus by this definition the density matrix is a constant. One can see from this that the density matrix corresponds to an operator representing a conserved quantity in the usual sense in quantum mechanics. We write the Schrödinger equation in the representation of the states $\nu$ as

$$\mathrm{i}\hbar\frac{\mathrm{d}a_\nu}{\mathrm{d}t} = \sum_{\nu'} H_{\nu\nu'}a_{\nu'} \tag{2.18}$$

which gives

$$\frac{\mathrm{d}}{\mathrm{d}t}(a^*_{\nu'}(t)a_\nu(t)) = \frac{\mathrm{i}}{\hbar}\sum_{\nu''}\left[H_{\nu''\nu'}a^*_{\nu''}(t)a_\nu(t) - a^*_{\nu'}(t)a_{\nu''}(t)H_{\nu\nu''}\right] \tag{2.19}$$

Then taking the time average $\lim_{\tau\to\infty}\frac{1}{\tau}\int_0^\tau(\ldots)\mathrm{d}t$ of both sides and *assuming that $H$ is not time dependent* gives, with the same definition (2.16) of $\mathrm{d}\rho_{\nu\nu'}/\mathrm{d}t$,

$$\frac{\mathrm{d}\rho_{\nu\nu'}}{\mathrm{d}t} = \frac{\mathrm{i}}{\hbar}\sum_{\nu''}[\rho_{\nu\nu''}H_{\nu''\nu'} - H_{\nu\nu''}\rho_{\nu''\nu'}] \tag{2.20}$$

Thus with (2.17) we have

$$\rho\mathbf{H} - \mathbf{H}\rho = 0 \tag{2.21}$$

in matrix notation so that $\rho$ can be regarded as an operator corresponding to a constant of the motion in the quantum mechanical sense. This formulation will also prove quite useful in describing time dependent phenomena in Part III. Now consider a special basis in which the density matrix has a particularly simple form which allows an unambiguous interpretation. Consider the complete set of commuting operators which includes the Hamiltonian. The operators represent all the $3N$ quantum mechanical constants of the motion of the system. In the basis $\psi_\nu(q)$ which are simultaneously eigenvalues of all these operators, $\rho$, which because it is itself a constant of the motion must be a function of these $3N$ operators, must also be diagonal. Because $\rho$ is Hermitian, its diagonal matrix elements in this basis must be real and, again from the definition, positive. Thus the quantities $\rho_{\nu\nu}$ can be interpreted as the probabilities of finding the system with values of the $3N$ constants of the motion designated by the $3N$ quantum numbers $\nu$ and there are no off diagonal elements of $\rho$ in this basis. Unfortunately, in a large interacting system, the $3N$ operators associated with all the constants of the motion are never known.

For large systems, partitionable in the same sense discussed for classical systems, we can now construct arguments for a canonical density matrix very similar to those in the last chapter. In particular, we suppose that for a partition into two large systems, the density matrix is a product in the following sense. We first work formally in the special basis discussed in the last paragraph, in which the density matrix is rigorously diagonal and in which the quantum numbers $\{\nu\}$ denote eigenvectors of $3N$ linearly independent operators, including $H$ which commute with the Hamiltonian $H$. Because $\rho$ itself commutes with the Hamiltonian it must itself be diagonal in this representation. If the Hamiltonian of the partitioned system can, to a good approximation (and using arguments completely analogous to those used in the classical case), be written as $H_1 + H_2$, ignoring interaction terms in the thermodynamic limit, then in this representation the diagonal elements of the density matrix (which are the only nonzero ones) can be written

$$\rho_{\nu,\nu} = \rho^{(1)}_{\nu_1,\nu_1}\rho^{(2)}_{\nu_2,\nu_2} \tag{2.22}$$

where $\nu_1$ and $\nu_2$ designate bases for the two partitions which also simultaneously diagonalize all the constants of the motion of those two partitions individually. Now we may take the natural logarithm of (2.22) much as in the classical case:

$$\ln \rho_{\nu,\nu} = \ln \rho^{(1)}_{\nu_1,\nu_1} + \ln \rho^{(2)}_{\nu_2,\nu_2} \tag{2.23}$$

which shows that $\ln \rho_{\nu,\nu}$ can be a function only of the quantum numbers in $\nu$ corresponding to additive constants of the motion. If, as in the classical case, we suppose these to be energy, linear and angular momentum then we have

$$\rho_{\nu\nu'} = \delta_{\nu,\nu'}\,\mathrm{e}^{\alpha}\,\mathrm{e}^{-\beta E_\nu + \vec{\delta}\cdot\vec{P}_\nu + \vec{\gamma}\cdot\vec{L}_\nu} \tag{2.24}$$

in this representation. Here $E_\nu$, $\vec{P}_\nu$, $\vec{L}_\nu$ are the eigenvalues of energy, linear and angular momentum. We may determine $\alpha$ from the requirement that $\mathrm{Tr}\rho = 1$:

$$\rho_{\nu\nu'} = \delta_{\nu,\nu'}\frac{\mathrm{e}^{-\beta E_\nu + \vec{\delta}\cdot\vec{P}_\nu + \vec{\gamma}\cdot\vec{L}_\nu}}{\sum_\nu \left(\mathrm{e}^{-\beta E_\nu + \vec{\delta}\cdot\vec{P}_\nu + \vec{\gamma}\cdot\vec{L}_\nu}\right)} \tag{2.25}$$

Finally one can remove the restriction to a particular basis, noting that (2.25) can be written as the operator

$$\rho = \frac{\mathrm{e}^{-\beta H + \vec{\delta}\cdot\vec{P} + \vec{\gamma}\cdot\vec{L}}}{\mathrm{Tr}\left(\mathrm{e}^{-\beta H + \vec{\delta}\cdot\vec{P} + \vec{\gamma}\cdot\vec{L}}\right)} \tag{2.26}$$

and if we restrict attention to the case of systems in a stationary "box"

$$\rho = \frac{\mathrm{e}^{-\beta H}}{\mathrm{Tr}(\mathrm{e}^{-\beta H})} \tag{2.27}$$

which is called the canonical density matrix.

A similar treatment is possible in the case that we allow the number of particles in the partitions of a large system to vary. This is more convenient in the quantum mechanical case than in the classical one, because the formalism of second quantization makes it straightforward to describe the system in terms of a Hamiltonian operator with a variable number of particles. Assuming the Hamiltonian to be written in this way, we consider the case in which the number of particles is conserved, $[N, H] = 0$. Then the argument, given a partition of a large system, proceeds as before, except that the constants of the motion now include $N$ as well as energy and momenta. Thus, denoting the constant analogous to $\beta$ by $-\beta\mu$ we have

$$\rho = \frac{\mathrm{e}^{-\beta H+\beta\mu N}}{\mathrm{Tr}(\mathrm{e}^{-\beta H+\beta\mu N})} \tag{2.28}$$

This is often called the grand canonical partition function (or ensemble).

## Microcanonical density matrix

We can also study the implications of the previous arguments for the total system following the lines of the classical case. In particular note that in a representation in which the Hamiltonian is diagonal, the coefficients $a_\nu(t)$ have the time dependence

$$a_\nu(t) = a_\nu(0)\,\mathrm{e}^{-\mathrm{i}E_{\nu_E}t} \tag{2.29}$$

so that

$$\rho_{\nu\nu'} = \lim_{\tau\to\infty}\frac{1}{\tau}\int_0^\tau a^*_{\nu'}(0)a_\nu(0)\,\mathrm{e}^{\mathrm{i}(E_{\nu'_E}-E_{\nu_E})t}\,\mathrm{d}t = \begin{cases} a^*_{\nu'}(0)a_\nu(0) & E_{\nu'_E} = E_{\nu_E} \\ 0 & \text{otherwise} \end{cases} \tag{2.30}$$

This is analogous to the condition that the energy is exactly conserved in an isolated system in the classical case. Here, however, we have a difference, because the initial wave function may not be an energy eigenstate. If it is not, then the density matrix, though diagonal in the energy quantum number, may not be infinitely sharply peaked at a particular value of the energy, even for an isolated system. Analogous to the classical case, the factors $a^*_{\nu'}(0)a_\nu(0)$ in (2.30) can contribute a dependence of $\rho_{\nu\nu'}$ on quantum numbers other than those associated with the energy (or the linear and angular momenta). These dependences are only expected to be absent in the case that we have a large system with the additive properties already extensively discussed. Then the energy diagonal elements of $\rho_{\nu\nu'}$ can depend *only* on the energy and we have the closest quantum analogue to the microcanonical ensemble in classical statistical mechanics:

$$\rho_{\nu\nu'} = \begin{cases} \text{constant} & E - \delta E/2 \le E_{\nu'_E} = E_{\nu_E} \le E + \delta E/2 \\ 0 & \text{otherwise} \end{cases} \tag{2.31}$$

This is significantly more arbitrary than its analogue in the classical case as we have discussed and it is not of much practical use. We can show, exactly as before, that any discrepancy between the canonical and microcanonical expectation values of the energy can be made arbitrarily small by taking an infinitely large system and letting $\delta E \to 0$.

We note a property of (2.30) closely analogous to the one discussed for the exact classical distribution function in the last chapter. It is quite easy to show that the eigenvalues of (2.30) within the energy subspace characterized by $\nu_E$ are $\sum_{\tilde{\nu} \text{ for } \nu_E} |a_\nu(0)|^2, 0, \ldots, 0$ where the number of zeroes is 1 less than the degeneracy of the level characterized by $\nu_E$ . The eigenvectors are $a_\nu(0)/\sqrt{\sum_{\tilde{\nu} \text{ for } \nu_E} |a_\nu(0)|^2}$ for the first eigenvalue with the other eigenvectors being orthogonal to it in the degenerate subspace. This diagonalization is quite closely analogous to the contact transformation taking the classical system to the set of constant coordinates and momenta and, analogously to that classical case, it leads to a form of the density matrix which is clearly in conflict with the microcanonical and canonical forms. Analogously to the classical case, we conclude that the canonical and microcanonical forms cannot be good approximations for evaluation of averages in absolutely all quantum mechanical bases but only, in some sense, in "almost all" of them for large systems. The bases which we implicitly select in making measurements are presumably overwhelmingly likely to be among the bases for which the standard ensembles are a good description. Of course, the actual diagonalization of the density matrix in the subspaces in a real large system will in general be completely impractical because the initial quantum mechanical state is not known.

## Reference

1. R. C. Tolman, *The Principles of Statistical Mechanics*, London: Oxford University Press, 1967.

## Problems

2.1 Use a representation $\nu$ in which $H$ is diagonal to show that $\rho_{\nu\nu'}$ defined by (2.12) is always diagonal in the energy.

2.2 Write $a_\nu(t=0) = r_\nu e^{i\phi_\nu}$ in a representation in which the Hamiltonian is diagonal, Here $r_\nu$ and $\phi_\nu$ are real. Show that the assumptions of Chapter 2 leading to (2.24) mean that the density matrix is independent of the phases $\phi_\nu$. In some treatments of the foundations of quantum statistical mechanics this is elevated to a postulate, termed the hypothesis of random a priori phases.

2.3 Work out the density matrix in the case that the wave function is an energy eigenstate for the case of a particle in a box. Use it to derive general expressions for the time averages of arbitrary functions of the momentum and of the coordinate, expressing the result as a sum of the term arising from the classical distribution function (derived in Problem 1.3) and a correction term associated with quantum mechanics. Under what circumstances is the extra term small? Generalize to the case of an arbitrary initial wave function.

# 3
# Thermodynamics

With the form of the density matrix established it now becomes possible to extract the fundamental features of thermodynamics from the theory, thus establishing a relation between thermodynamics and mechanics. The main remaining concept required for this is a general definition of entropy, to which we turn first below. From this we can easily extract the familiar general relations of equilibrium thermodynamics, which we then review.

## Definition of entropy

We carry through the discussion for the canonical, quantum mechanical case. We start with the idea that the equilibrium density matrix, when expressed in terms of the quantum constants of the motion, is a function only of the energy in the case of greatest interest. We denote such a representation $\nu_E$, $\nu'$ where $\nu_E$ designates the quantum number specifying the energy and $\nu'$ is an abbreviation for all the other $3N - 1$ constants of the quantum motion. The density matrix is then diagonal and its diagonal matrix elements are denoted $\rho_{\nu_E,\nu';\nu_E,\nu'}$. The entropy is related to the number of states associated with the system when it is in equilibrium. To make sense of this we first sum $\rho_{\nu_E,\nu';\nu_E,\nu'}$ on all of its quantum numbers except $\nu_E$. Because $\rho_{\nu_E,\nu';\nu_E,\nu'}$ depends only on $\nu_E$ this gives

$$\sum_{\nu'} \rho_{\nu_E,\nu';\nu_E,\nu'} = \rho(E_{\nu_E}) \sum_{\nu' \text{ with energy } E_{\nu_E}} 1 \tag{3.1}$$

where for example in the case of the canonical density matrix

$$\rho(E_{\nu_E}) = \mathrm{e}^{-\beta E_{\nu_E}} / \mathrm{Tr}\, \mathrm{e}^{-\beta H} \tag{3.2}$$

The factor $\sum_{\nu' \text{ with energy } E_{\nu_E}} 1$ is nearly what we want because it measures the number of states consistent with the system having energy $E_{\nu_E}$. However, in a system described by the canonical density matrix, the energy is not fixed, so it is not immediately transparent what energy we should take. To resolve this question, we denote

$$\Omega(E_{\nu_E}) = \sum_{\nu' \text{ with energy } E_{\nu_E}} 1 \tag{3.3}$$

and use the fact that, from the normalization of the density matrix,

$$\sum_{\nu_E} \rho(E_{\nu_E})\Omega(E_{\nu_E}) = 1 \tag{3.4}$$

Consider the nature of the summand: $\rho(E_{\nu_E})$ is an exponentially decreasing function of $E_{\nu_E}$ while $\Omega(E_{\nu_E})$ is a rapidly increasing function so that this summand will have a sharp peak at the average energy $\bar{E}$. Thus the sum should only depend on $\rho(E_{\nu_E})$ evaluated at the energy $\bar{E}$ and it is reasonable to write

$$\rho(\bar{E})\Delta\Gamma = \sum_{\nu_E} \rho(E_{\nu_E})\Omega(E_{\nu_E}) \tag{3.5}$$

where $\Delta\Gamma$ is the number of states associated with the equilibrium density matrix. But using the normalization condition (3.4) this gives

$$\Delta\Gamma = 1/\rho(\bar{E}) \tag{3.6}$$

We identify the entropy as

$$S = k_{\mathrm{B}} \ln \Delta\Gamma \tag{3.7}$$

Some other perspectives on this definition will be illustrated in the problems. Using (3.6) this gives

$$S = -k_{\mathrm{B}} \ln \rho_{\mathrm{c}}(\bar{E}) \tag{3.8}$$

where the subscript c has been added to $\rho$ to specify the canonical density matrix. In the case that the number of particles can vary, a virtually identical argument gives

$$S = -k_{\mathrm{B}} \ln \rho_{\mathrm{gc}}(\bar{E}, \bar{N}) \tag{3.9}$$

## Thermodynamic potentials

We define the canonical partition function $Z_{\mathrm{c}}$ as

$$Z_{\mathrm{c}} \equiv \mathrm{Tr}\, \mathrm{e}^{-\beta H} \tag{3.10}$$

Then (3.8) becomes

$$S = -k_{\mathrm{B}} \ln(\mathrm{e}^{-\beta\bar{E}}/Z_{\mathrm{c}}) = k_{\mathrm{B}}\beta\bar{E} + k_{\mathrm{B}} \ln Z_{\mathrm{c}} \tag{3.11}$$

Using $\beta = 1/k_B T$ to define the temperature, we then obtain

$$\bar{E} - TS = -k_B T \ln Z_c \tag{3.12}$$

If the quantity $S$ is indeed the thermodynamic entropy, then $\bar{E} - TS$ is the Helmholtz free energy, denoted $F$ (or $A$ in the chemical literature). Thus

$$F = -k_B T \ln Z_c = -k_B T \ln \mathrm{Tr}\, e^{-\beta H} \tag{3.13}$$

This establishes the needed relation between a thermodynamic quantity and the microscopic, quantum mechanical model. Familiar relationships of thermodynamics follow from this if we suppose that the energy $E_\nu$ of the system depends on experimentally controlled variables $X_i$ (for example the volume, which fixes the boundary conditions on the wave functions, or a field, which fixes a term in the Hamiltonian). We now vary $F$ with respect to $T$ and to the variables $X_i$:

$$dF = \left( -k_B \ln Z_c - \frac{k_B T}{Z_c} \frac{\partial Z_c}{\partial T} \right) dT + \frac{1}{Z_c} \sum_i \sum_\nu e^{-\beta E_\nu} \frac{\partial E_\nu}{\partial X_i} dX_i \tag{3.14}$$

In the second term on the right one can evaluate

$$\frac{\partial Z_c}{\partial T} = \sum_\nu \frac{E_\nu}{k_B T^2} e^{-\beta E_\nu} \tag{3.15}$$

so that the expression in (...) on the right hand side becomes

$$-k_B \ln Z_c - \bar{E}/T \equiv -S \tag{3.16}$$

using (3.11). Thus

$$dF = -S dT + \sum_i \left\langle \frac{\partial \mathcal{H}}{\partial X_i} \right\rangle dX_i \tag{3.17}$$

in which $\mathcal{H}$ is the Hamiltonian. The most common example is that in which the only $X_i$ is the volume $V$. Then $\langle \partial \mathcal{H}/\partial X_i \rangle = \partial \bar{E}/\partial X_i$ is minus the pressure and (3.17) becomes

$$dF = -S\, dT - P\, dV \tag{3.18}$$

If $X$ is a magnetic field intensity $H$ then $\partial \bar{E}/\partial X_i$ is minus the magnetization. In another common example, $X$ is an electric field and $\partial \bar{E}/\partial X_i$ is the negative of the polarization. (Note that the $X_i$ can be either intensive (independent of the number of degrees of freedom) or extensive (proportional to the number of degrees of freedom). The key question is whether the $X_i$ can be interpreted directly in the microscopic calculation of eigenvalues and eigenvectors (as can $V$, $H$ and $E$ whereas $P$, $M$ and $\mathcal{P}$ the polarization cannot be directly used in this way).) We will deal in the rest of this section only with the case in which the only relevant variable $X$ is the volume.

The extension to other cases is not difficult. In the case that $X$ is the volume, the Gibbs free energy $G$ is defined to be

$$G = F + PV \tag{3.19}$$

and by use of (3.18)

$$\mathrm{d}G = -S\,\mathrm{d}T + V\mathrm{d}P \tag{3.20}$$

The enthalpy $W$ (sometimes this is denoted $H$) is defined by

$$W = E + PV \tag{3.21}$$

giving

$$\mathrm{d}W = T\,\mathrm{d}S + V\,\mathrm{d}P \tag{3.22}$$

Finally inserting $F = \bar{E} - TS$ into (3.18) one obtains

$$\mathrm{d}\bar{E} = T\,\mathrm{d}S - P\,\mathrm{d}V \tag{3.23}$$

Each of these various thermodynamic potentials can be seen to be constant when a different pair of external variables is held constant. It is often convenient to regard each potential as a function of those variables. Thus we regard $F$ as depending on $T$ and $V$, $G$ on $T$ and $P$, $\bar{E}$ on $S$ and $V$, and $W$ on $S$ and $P$. The relation (3.13) permits all these potentials to be calculated using the microscopic Hamiltonian in the case of the canonical density matrix.

We now go over a similar discussion for the case of the grand canonical density matrix. The entropy is

$$S = -k_{\mathrm{B}} \ln \rho(\bar{E}, \bar{N}) = -k_{\mathrm{B}}(\beta \bar{N}\mu - \beta \bar{E} - \ln Z_{\mathrm{gc}}) \tag{3.24}$$

where

$$Z_{\mathrm{gc}} \equiv \sum_{N,\nu} \mathrm{e}^{\beta\mu N - \beta E_{\nu,N}} \tag{3.25}$$

or from (3.24)

$$TS = -\bar{N}\mu + \bar{E} + k_{\mathrm{B}}T \ln Z_{\mathrm{gc}} \tag{3.26}$$

The quantity $\bar{E} - TS - \bar{N}\mu$ is called the thermodynamic potential in thermodynamics and is denoted $\boldsymbol{\Omega}$:

$$\boldsymbol{\Omega} = -k_{\mathrm{B}}T \ln Z_{\mathrm{gc}} \tag{3.27}$$

This establishes a connection between the microscopic Hamiltonian and thermodynamics in the grand canonical case, analogous to (3.13). Again, we suppose that

$E_{\nu,N}$ vary with some experimentally controlled parameters $X_i$ and obtain

$$\mathrm{d}\boldsymbol{\Omega} = \left(-k_{\mathrm{B}} \ln Z_{\mathrm{gc}} - \frac{k_{\mathrm{B}}T}{Z_{\mathrm{gc}}}\frac{\partial Z_{\mathrm{gc}}}{\partial T}\right)\mathrm{d}T - \frac{k_{\mathrm{B}}T}{Z_{\mathrm{gc}}}\frac{\partial Z_{\mathrm{gc}}}{\partial \mu}\mathrm{d}\mu + \sum_i \left\langle \frac{\partial \mathcal{H}}{\partial X_i} \right\rangle \mathrm{d}X_i \quad (3.28)$$

The term in (...) can be shown to be $-S$ by use of (3.24). The next term is

$$-\frac{k_{\mathrm{B}}T}{Z_{\mathrm{gc}}}\frac{\partial Z_{\mathrm{gc}}}{\partial \mu} = -\frac{k_{\mathrm{B}}T}{Z_{\mathrm{gc}}}\beta \sum_{N,\nu} N \mathrm{e}^{(N\mu - E_{\nu,N})\beta} = -\bar{N} \quad (3.29)$$

Thus

$$\mathrm{d}\boldsymbol{\Omega} = -S\,\mathrm{d}T - \bar{N}\,\mathrm{d}\mu + \sum_i \left\langle \frac{\partial \mathcal{H}}{\partial X_i} \right\rangle \mathrm{d}X_i \quad (3.30)$$

We specialize this as before to the case of just one $X_i$ which is the volume $V$:

$$\mathrm{d}\boldsymbol{\Omega} = -S\,\mathrm{d}T - \bar{N}\,\mathrm{d}\mu - P\mathrm{d}V \quad (3.31)$$

From (3.26) and (3.27), $\boldsymbol{\Omega} = \bar{E} - TS - \bar{N}\mu$. Then we have, since $F = \bar{E} - TS$, that

$$\mathrm{d}F = \mathrm{d}\boldsymbol{\Omega} + \mathrm{d}(\bar{N}\mu) = -P\,\mathrm{d}V - S\,\mathrm{d}T + \mu\,\mathrm{d}\bar{N} \quad (3.32)$$

This is consistent with (3.18) which was derived in the case that $\bar{N} =$ constant. Similarly the expressions for $G$, $W$ and $\bar{E}$ become

$$\mathrm{d}G = V\,\mathrm{d}P - S\,\mathrm{d}T + \mu\,\mathrm{d}\bar{N} \quad (3.33)$$

$$\mathrm{d}W = V\,\mathrm{d}P - T\,\mathrm{d}S + \mu\,\mathrm{d}\bar{N} \quad (3.34)$$

$$\mathrm{d}\bar{E} = T\,\mathrm{d}S - P\,\mathrm{d}V + \mu\,\mathrm{d}\bar{N} \quad (3.35)$$

Note that we have not exhausted the list of possible thermodynamic potentials for the case in which the number of particles varies. $\boldsymbol{\Omega}$ is the Legendre transform of $F$ with respect to $\mu$ and $N$ and is constant when $T$, $V$, $\mu$ are fixed. We may define similar Legendre transforms of $\bar{E}$ and $W$. I do not know names for these and will call them $\boldsymbol{\Omega}_{\mathbf{E}}$ and $\boldsymbol{\Omega}_{\mathbf{W}}$ which are defined as

$$\boldsymbol{\Omega}_{\mathbf{E}}(S, V, \mu) = \bar{E} - \mu\bar{N} = -PV + ST \quad (3.36)$$

$$\boldsymbol{\Omega}_{\mathbf{W}}(S, P, \mu) = W - \mu\bar{N} = \bar{E} + PV - \mu\bar{N} = ST \quad (3.37)$$

but if we try to do the same thing with $G$ we get (see (3.44) below)

$$\boldsymbol{\Omega}_{\mathbf{G}}(T, P, \mu) = G - \mu\bar{N} \equiv 0 \quad (3.38)$$

The corresponding differential relations are

$$\mathrm{d}\boldsymbol{\Omega}_{\mathbf{E}}(S, V, \mu) = T\,\mathrm{d}S - P\,\mathrm{d}V - \bar{N}\,\mathrm{d}\mu \quad (3.39)$$

$$\mathrm{d}\boldsymbol{\Omega}_{\mathbf{W}}(S, P, \mu) = T\,\mathrm{d}S + V\,\mathrm{d}P - \bar{N}\,\mathrm{d}\mu \quad (3.40)$$

The last relation can be rewritten using $\boldsymbol{\Omega}_{\mathbf{W}}(S, P, \mu) = ST$ as

$$V\,\mathrm{d}P - S\,\mathrm{d}T = \bar{N}\,\mathrm{d}\mu \tag{3.41}$$

This is very well known and is usually written in a slightly different form by dividing by $\bar{N}$, defining $s = S/\bar{N}$, $v = V/\bar{N}$ as the entropy and volume per particle:

$$\mathrm{d}\,\mu = v\,\mathrm{d}P - s\,\mathrm{d}T \tag{3.42}$$

In this form it is known as the Gibbs–Duhem relation. Using $\boldsymbol{\Omega}_{\mathbf{E}}(S, V, \mu) = -PV + ST$ one can easily show that (3.39) *also* reduces to this same Gibbs–Duhem relation. To summarize, the only new information in these last three Legendre transforms is the Gibbs–Duhem relation (3.42).

## Some thermodynamic relations and techniques

Here we review some thermodynamic relations and methods. We will follow common usage in thermodynamics arguments and drop the bar on $\bar{N}$ in this section. We will only include the bar on $\bar{N}$ later when its absence is likely to cause confusion. Note first that, generally, the differential relations just listed may be used to write expressions for the first derivatives of the thermodynamic potentials in terms of their independent variables. For example, from (3.33) we have

$$\left(\frac{\partial G}{\partial N}\right)_{P,T} = \mu \tag{3.43}$$

However, since $\mu$ must be independent of system size this equation can be integrated on $N$ to give

$$G = \mu N \tag{3.44}$$

One can use this in the definition of $\boldsymbol{\Omega}$

$$\boldsymbol{\Omega} = \bar{E} - TS - \bar{N}\mu = G - PV - \mu N = -PV \tag{3.45}$$

Using (3.27), this is an equation of state. One may use the same differential relations to express any thermodynamic potential explicitly in terms of derivatives of the eigenvalues of the underlying quantum mechanical problem. For example from (3.32) we have

$$\left(\frac{\partial F}{\partial V}\right)_{T,N} = -P \tag{3.46}$$

which is also an equation of state since the left hand side has been expressed in (3.13) in terms of the microscopic model. From this

$$G = F + PV = -k_{\rm B}T \ln \sum_{\nu} {\rm e}^{-E_\nu/k_{\rm B}T} - \frac{\sum_\nu \frac{\partial E_\nu}{\partial V} {\rm e}^{-E_\nu/k_{\rm B}T}}{\sum_\nu {\rm e}^{-E_\nu/k_{\rm B}T}} V \tag{3.47}$$

providing a prescription for calculating $G$ from first principles.

Directly measurable thermodynamic quantities are in most cases second derivatives of the thermodynamics potentials. For example the specific heat at constant volume

$$C_v \equiv T \left(\frac{\partial S}{\partial T}\right)_{V,N} = -T \left(\frac{\partial^2 F}{\partial T^2}\right)_{V,N} \tag{3.48}$$

However, $C_v$ may also be expressed as a first derivative by writing the relation for $\mathrm{d}\bar{E}$ in terms of independent variables $T$, $V$ and $N$:

$$\begin{aligned} \mathrm{d}\bar{E} &= T\,\mathrm{d}S - P\,\mathrm{d}V + \mu\,\mathrm{d}N \\ &= T\left(\left(\frac{\partial S}{\partial T}\right)_{V,N} \mathrm{d}T + \left(\frac{\partial S}{\partial V}\right)_{T,N} \mathrm{d}V + \left(\frac{\partial S}{\partial N}\right)_{T,V} \mathrm{d}N\right) - P\,\mathrm{d}V + \mu\,\mathrm{d}N \end{aligned} \tag{3.49}$$

from which

$$\left(\frac{\partial E}{\partial T}\right)_{V,N} = T\left(\frac{\partial S}{\partial T}\right)_{V,N} = \left(\frac{\partial E}{\partial S}\right)_{V,N} \left(\frac{\partial S}{\partial T}\right)_{V,N} = C_v \tag{3.50}$$

giving another expression for $C_v$. From the transformation (3.49) we also obtain the relations

$$\left(\frac{\partial \bar{E}}{\partial V}\right)_{T,N} = -P + T\left(\frac{\partial S}{\partial V}\right)_{T,N} = \left(\frac{\partial \bar{E}}{\partial V}\right)_{S,N} + \left(\frac{\partial \bar{E}}{\partial S}\right)_{V,N} \left(\frac{\partial S}{\partial V}\right)_{T,N} \tag{3.51}$$

and finally

$$\left(\frac{\partial \bar{E}}{\partial N}\right)_{T,V} = \mu + T\left(\frac{\partial S}{\partial N}\right)_{T,V} = \left(\frac{\partial \bar{E}}{\partial N}\right)_{S,V} + \left(\frac{\partial \bar{E}}{\partial S}\right)_{V,N} \left(\frac{\partial S}{\partial N}\right)_{T,V} \tag{3.52}$$

Equation (3.50) is an example of the use of the chain rule but the other two relations represent the somewhat more subtle relation which arises between two partial derivatives of the same quantity with respect to the same variable when different quantities are held fixed during the differentiation. One way of expressing this relation more generally is to consider a thermodynamic function $w(x, z)$ and transform its total differential so that it is expressed in terms of independent variables $x$, $y$

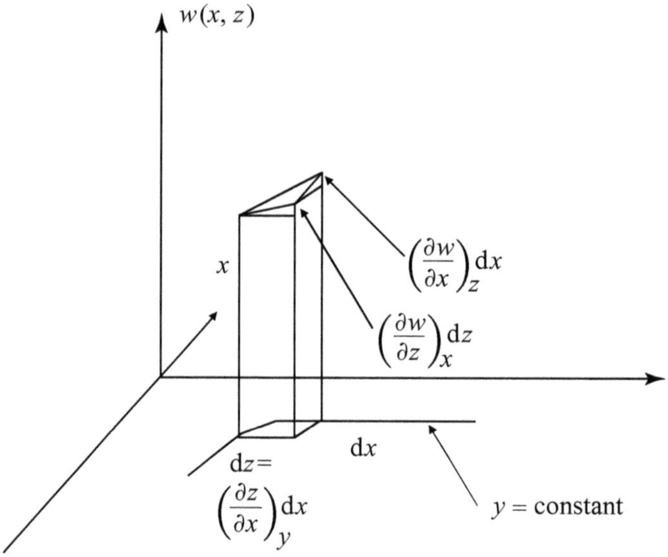


Figure 3.1 Geometrical interpretation of equation (3.54).

instead of $x, z$:

$$\begin{aligned}
\mathrm{d}w &= \left(\frac{\partial w}{\partial x}\right)_z \mathrm{d}x + \left(\frac{\partial w}{\partial z}\right)_x \mathrm{d}z = \left(\frac{\partial w}{\partial x}\right)_z \mathrm{d}x + \left(\frac{\partial w}{\partial z}\right)_x \left\{\left(\frac{\partial z}{\partial y}\right)_x \mathrm{d}y + \left(\frac{\partial z}{\partial x}\right)_y \mathrm{d}x\right\} \\
&= \left[\left(\frac{\partial w}{\partial x}\right)_z + \left(\frac{\partial w}{\partial z}\right)_x \left(\frac{\partial z}{\partial x}\right)_y\right] \mathrm{d}x + \left(\frac{\partial w}{\partial z}\right)_x \left(\frac{\partial z}{\partial y}\right)_x \mathrm{d}y \\
&= \left(\frac{\partial w}{\partial x}\right)_y \mathrm{d}x + \left(\frac{\partial w}{\partial y}\right)_x \mathrm{d}y
\end{aligned} \tag{3.53}$$

Thus in general by equating the terms proportional to $\mathrm{d}x$ on each side of the last equality

$$\left(\frac{\partial w}{\partial x}\right)_y = \left(\frac{\partial w}{\partial x}\right)_z + \left(\frac{\partial w}{\partial z}\right)_x \left(\frac{\partial z}{\partial x}\right)_y \tag{3.54}$$

This can be seen to give the relations for derivatives of $\bar{E}$ above. For example, (3.51) follows from (3.54) by taking $w = \bar{E}$, $x = V$, $z = S$ and $y = T$. Whether one chooses to remember (3.54) or to rederive it as needed is a matter of taste. The meaning of (3.54) is illustrated in Figure 3.1.

Another set of useful relations follows from our forms for the total derivatives by requiring that the second cross derivatives be well defined, as they must be. Thus, for example, by requiring that

$$\left(\frac{\partial^2 F}{\partial T\,\partial V}\right)_N = \left(\frac{\partial^2 F}{\partial V\,\partial T}\right)_N \tag{3.55}$$

one obtains the identity

$$\left(\frac{\partial S}{\partial V}\right)_{N,T} = \left(\frac{\partial P}{\partial T}\right)_{V,N} \tag{3.56}$$

These are well known as Maxwell relations.

Another useful relation may be obtained by considering just three variables $z$, $x$ and $y$ of which two are independent. Then we may express the total differential d$z$

$$\mathrm{d}z = \left(\frac{\partial z}{\partial x}\right)_y \mathrm{d}x + \left(\frac{\partial z}{\partial y}\right)_x \mathrm{d}y \tag{3.57}$$

This must be consistent with the relation that is obtained by expressing d$y$ on the right hand side in terms of d$z$ and d$x$ whence

$$\begin{aligned}\mathrm{d}z &= \left(\frac{\partial z}{\partial x}\right)_y \mathrm{d}x + \left(\frac{\partial z}{\partial y}\right)_x \left[\left(\frac{\partial y}{\partial x}\right)_z \mathrm{d}x + \left(\frac{\partial y}{\partial z}\right)_x \mathrm{d}z\right] \\ &= \left[\left(\frac{\partial z}{\partial x}\right)_y + \left(\frac{\partial z}{\partial y}\right)_x \left(\frac{\partial y}{\partial x}\right)_z\right]\mathrm{d}x + \left(\frac{\partial z}{\partial y}\right)_x \left(\frac{\partial y}{\partial z}\right)_x \mathrm{d}z\end{aligned} \tag{3.58}$$

But by the chain rule

$$\left(\frac{\partial z}{\partial y}\right)_x \left(\frac{\partial y}{\partial z}\right)_x = 1 \tag{3.59}$$

so we require

$$\left(\frac{\partial z}{\partial x}\right)_y + \left(\frac{\partial z}{\partial y}\right)_x \left(\frac{\partial y}{\partial x}\right)_z = 0 \tag{3.60}$$

or

$$\left(\frac{\partial z}{\partial y}\right)_x \left(\frac{\partial y}{\partial x}\right)_z \left(\frac{\partial x}{\partial z}\right)_y = -1 \tag{3.61}$$

Because of its usefulness, I will also describe one other way to manipulate these relations (which is equivalent to the foregoing). One defines the Jacobian determinant in the usual way as

$$\begin{vmatrix} \left(\frac{\partial x}{\partial w}\right)_z & \left(\frac{\partial x}{\partial z}\right)_w \\ \left(\frac{\partial y}{\partial w}\right)_z & \left(\frac{\partial y}{\partial z}\right)_w \end{vmatrix} = \frac{\partial(x, y)}{\partial(w, z)} \tag{3.62}$$

It is easy to show by direct substitution that

$$\frac{\partial(x, y)}{\partial(w, y)} = \left(\frac{\partial x}{\partial w}\right)_y \tag{3.63}$$

It is somewhat less obvious that

$$\frac{\partial(x, y)}{\partial(z, w)} \frac{\partial(z, w)}{\partial(t, s)} = \frac{\partial(x, y)}{\partial(t, s)} \tag{3.64}$$

The easiest way to see this is to consider reexpressing the differential element, say $\mathrm{d}x\,\mathrm{d}y$ in terms of the element $\mathrm{d}s\,\mathrm{d}t$. It cannot matter whether one does this directly or by passing through the pair of variables $z, w$ on the way:

$$\mathrm{d}x\,\mathrm{d}y = \frac{\partial(x, y)}{\partial(t, s)}\mathrm{d}t\,\mathrm{d}s = \frac{\partial(x, y)}{\partial(z, w)}\mathrm{d}z\,\mathrm{d}w = \frac{\partial(x, y)}{\partial(z, w)} \frac{\partial(z, w)}{\partial(t, s)}\mathrm{d}t\,\mathrm{d}s \tag{3.65}$$

If this seems too abstract, one can prove (3.64) by direct substitution, using the relation (3.54) and the chain rule. As an example of the use of (3.64), a compact proof of (3.61) is produced by use of this formulation:

$$\left(\frac{\partial y}{\partial x}\right)_z \left(\frac{\partial z}{\partial y}\right)_x = \frac{\partial(y, z)}{\partial(x, z)} \frac{\partial(z, x)}{\partial(y, x)} = -\frac{\partial(y, z)}{\partial(y, x)} = -\left(\frac{\partial z}{\partial x}\right)_y \tag{3.66}$$

## Constraints on thermodynamic quantities

From this formulation one can obtain some well known constraints on thermodynamic quantities. For example, the temperature, which is related to the thermodynamic potentials through

$$\left(\frac{\partial \bar{E}}{\partial S}\right)_V = T \tag{3.67}$$

is also $k_B/\beta$ where $\beta$ is the factor appearing in the density matrix. Because the quantum mechanical energy spectrum of any system must be bounded from below (i.e. there must be a lowest energy level) but not from above, the partition function will not be finite unless $\beta$, and hence $T$, is positive. Actually, if the energy spectrum has a large gap, it can sometimes appear to be effectively bounded from above and this makes metastable states possible in which the effective temperature is negative. Such conditions occur in some nuclear magnetic resonance systems, for example.

The condition that the temperature be the same throughout the system follows trivially from our formulations, in which the parameter $\beta$ is the same for every subsystem. This condition, stated as the condition that two systems in thermal contact have the same temperature, is sometimes called the zeroth law of thermodynamics.

We also have the usual formulation of the first law of thermodynamics for example from the form (3.35).

The second law of thermodynamics, stating that the entropy always increases in time, is not really a statement about equilibrium statistical mechanics but one about dynamics. Our formulation has nothing to say about it. Note that, to make sense of it, one has first to define entropy in a way that does not require a long time in order to determine it. We have not done this here, but that need cause no serious problem if there is an empirically short time for the establishment of a state which looks approximately like an equilibrium one. Even granting a useful definition, the question of the status of the second law is very subtle. There are cases in which the entropy, suitably defined, decreases for very short times, but there are no known experimental cases in which it does not increase eventually. The theoretical status of this fact is still discussed and debated. A widespread, but not universal, consensus is that the origin of the second law lies in the low entropy initial state of the universe. Those interested in pursuing these issues are encouraged to study the conference proceedings edited by Halliwell, Perez-Mercader and Zurek[1] and, particularly, for a briefer discussion, the article by Lebowitz in that volume.[2]

The statement that the specific heat of a system goes to zero as the temperature goes to zero is known as the third law of thermodynamics. It follows quite simply from the grand canonial formulation:

$$C_v = -T\left(\frac{\partial^2 F}{\partial T^2}\right)_{V,N} \tag{3.68}$$

and using

$$F = -k_{\mathrm{B}} T \ln Z_{\mathrm{c}} \tag{3.69}$$

Suppose that the ground state has degeneracy $G_0$ and energy $E_0$. Then at low enough temperatures we may write

$$Z_{\mathrm{c}} \approx \mathrm{e}^{-E_0\beta}\left(G_0 + G_1\,\mathrm{e}^{-(E_1-E_0)\beta} + \cdots\right) \tag{3.70}$$

whence

$$C_v \to= k_{\mathrm{B}}\left(\frac{G_1}{G_0}\right)\left(\frac{(E_1-E_0)^2}{(k_{\mathrm{B}}T)^2}\right)\mathrm{e}^{-\beta(E_1-E_0)} \to 0 \tag{3.71}$$

as $T \to 0$. In practice, it is difficult to achieve temperatures low enough to satisfy the conditions of this proof. In many cases, for higher temperatures, the specific heat goes toward zero as a power law $C_V \propto T^x$.

We may prove a constraint on the specific heat $C_v$ as follows:

$$S = -\left(\frac{\partial F}{\partial T}\right)_V = \left(\frac{\partial(k_{\mathrm{B}}T\ln Z_{\mathrm{c}})}{\partial T}\right)_V = k_{\mathrm{B}}\ln Z + \frac{\bar{E}}{T} \tag{3.72}$$

Taking a second derivative

$$\left(\frac{\partial S}{\partial T}\right)_V = \frac{1}{k_B T^3}(\overline{E^2} - \bar{E}^2) \tag{3.73}$$

where

$$\overline{E^2} = \frac{\sum_\nu E_\nu^2 \mathrm{e}^{-\beta E_\nu}}{\sum_\nu \mathrm{e}^{-\beta E_\nu}} \tag{3.74}$$

by direct use of the expression for the partition function. But $\overline{E^2} - \bar{E}^2 > 0$ for any distribution of energy levels, so $C_v = T\,(\partial S/\partial T)_V > 0$ for any system obeying the canonical density matrix. Because $(\partial S/\partial T)_V = -(\partial^2 F/\partial T^2)_V$ this means that $F$ has negative curvature in the $T$ direction.

We may similarly consider the curvature in the $V$ direction which is related to the compressibility. We evaluate

$$\left(\frac{\partial P}{\partial V}\right)_T = -\left\langle\frac{\partial^2 H}{\partial V^2}\right\rangle + \frac{1}{k_B T}\left(\left\langle\frac{\partial H}{\partial V}^2\right\rangle - P^2\right) \tag{3.75}$$

in which

$$\left\langle\frac{\partial H}{\partial V}^2\right\rangle = \frac{\sum_\nu \left(\frac{\partial E_\nu}{\partial V}\right)^2 \mathrm{e}^{-\beta E_\nu}}{\sum_\nu \mathrm{e}^{-\beta E_\nu}} \tag{3.76}$$

and

$$\left\langle\frac{\partial^2 H}{\partial V^2}\right\rangle = \frac{\sum_\nu \left(\frac{\partial^2 E_\nu}{\partial V^2}\right) \mathrm{e}^{-\beta E_\nu}}{\sum_\nu \mathrm{e}^{-\beta E_\nu}} \tag{3.77}$$

This relation has been the subject of some rather obscure discussion in the literature. The last two terms on the right hand side of (3.75) give the mean square fluctuation in the pressure (times $1/k_B T$). The quantity $\langle\partial H^2/\partial V\rangle - P^2 = \langle\partial H^2/\partial V\rangle - \langle\partial H/\partial V\rangle\langle\partial H/\partial V\rangle$ is positive definite. Rearranging we have

$$\left(\frac{1}{k_B T}\right)\left(\left\langle\left(\frac{\partial H}{\partial V}\right)^2\right\rangle - \left\langle\frac{\partial H}{\partial V}\right\rangle\left\langle\frac{\partial H}{\partial V}\right\rangle\right) = \left(\frac{\partial P}{\partial V}\right)_T + \left\langle\frac{\partial^2 H}{\partial V^2}\right\rangle > 0 \tag{3.78}$$

As long as the system is macroscopically homogeneous, the right hand side is easily seen to depend on the number of particles as $N^{-1}$ so the fluctuations in the pressure are of order $N^{-1/2}$ as expected. (A system containing more than one phase requires more discussion in this regard.) From the last inequality

$$\left\langle\frac{\partial^2 H}{\partial V^2}\right\rangle > -\left(\frac{\partial P}{\partial V}\right)_T \tag{3.79}$$

Further, for a mechanically stable homogeneous system we must have $(\partial P/\partial V)_T < 0$ so we require

$$\left\langle \frac{\partial^2 H}{\partial V^2} \right\rangle > 0 \tag{3.80}$$

for mechanical stability of a homogeneous system.

## References

1. J. J. Halliwell, J. Perez-Mercader and W. H. Zurek, *Physical Origins of Time Asymmetry*, Cambridge: Cambridge University Press, 1994.
2. J. Lebowitz, in *Physical Origins of Time Asymmetry*, ed. J. J. Halliwell, J. Perez-Mercader and W. H. Zurek, Cambridge: Cambridge University Press, 1994, p. 131.

## Problems

3.1 Carry through the argument for the canonical case in the grand canonical one to show that (3.9) is an appropriate expression for the entropy in that case.

3.2 Show that if one assumes for a large system that the product $\rho(E_{\nu_E})(\mathrm{d}\Omega(E_{\nu_E})/\mathrm{d}E_{\nu_E})$ is constant over a range $\Delta E$ around $\bar{E}$ and zero elsewhere then (3.5) gives $\Delta\Gamma = (\mathrm{d}\Omega/\mathrm{d}E)(E = \bar{E})\Delta E$.

3.3 Estimate the width of the peak in the summand of the right hand side of (3.5) in the case of a perfect gas (neglecting any effects of exchange).

3.4 Find explicit expressions for the thermodynamics potentials $F$, $G$, $W$ and $\bar{E}$ in terms of the energy level spectrum of the system in the grand canonical case.

3.5 Evaluate the terms in (3.78) for a classical ideal gas and illustrate thereby the various points of the general discussion. (The energy levels may be taken to be $E_{\{\vec{p}_i\}} = \sum_i^N \vec{p}_i^{\,2}/2m$. The components of the momenta can be taken to have the values $\hbar \times$ integers$/V^{1/3}$ as can be seen from the discussion of the semiclassical limit in the next chapter, so the momenta depend on volume as $\vec{p}_i = (V_0/V)^{1/3}\vec{p}_i^{\,(0)}$ where $V^{(0)}$ is a reference volume. Thus derivatives can be evaluated and then $V$ can be set back to $V_0$.)

3.6 Express $C_p - C_v$ as a function of derivatives involving $P$, $V$ and $T$. Use your expression to explain qualitatively why this quantity is zero for low temperature solids which do not experience phase transitions at low temperatures but is very large for systems near a gas–liquid critical point in which the liquid and the gas are nearly in equilibrium with each other.

# 4

# Semiclassical limit

In Chapter 1 we dealt with some foundational questions for systems described by classical mechanics and in Chapter 2 we discussed similar questions for systems obeying quantum mechanics. In Chapter 3 we connected the results of Chapter 2 to thermodynamics. A point left hanging by this discussion is the transition from the description of Chapters 2 and 3 (quantum mechanical) to that of Chapter 1 (classical). Here we address this point. The approach will be to show circumstances in which the quantum mechanical description (nearly) reduces to the classical one. In fact this chapter will not be the last time we address this issue, since a more complete treatment must await the introduction of cluster expansions in Chapter 6.

## General formulation

Observables are related to observations in the quantum mechanical formulation by

$$\bar{\bar{\phi}} = \mathrm{Tr}\rho\phi = \sum_{\nu,\nu'} \rho_{\nu,\nu'}\phi_{\nu',\nu} \tag{4.1}$$

where

$$\rho_{\nu,\nu'} = \frac{\langle \nu \mid \mathrm{e}^{-\beta H} \mid \nu' \rangle}{Z_{\mathrm{c}}} \tag{4.2}$$

using the canonical density matrix. Here we have used a general basis $\mid \nu\rangle$ which does not necessarily diagonalize the Hamiltonian. The general idea in passing to the classical limit is to evaluate $\bar{\bar{\phi}}$ in the basis of plane wave states obeying periodic boundary conditions in a volume $V$

$$\langle \vec{r} \mid \nu\rangle = \langle \vec{r}_1, \ldots, \vec{r}_N \mid \vec{k}_1, \ldots, \vec{k}_N\rangle = \frac{1}{V^{N/2}} \frac{1}{\sqrt{\eta(\vec{k}_1, \ldots, \vec{k}_N)}} \sum_{\mathcal{P}}{}' (\pm)^{\mathcal{P}} \prod_{i=1}^{N} \mathrm{e}^{\mathrm{i}\vec{k}_i\cdot\vec{r}_{\mathcal{P}(i)}} \tag{4.3}$$

in both the interacting and noninteracting cases, even though this basis only diagonalizes the Hamiltonian in the noninteracting case. In equation (4.3), the $\mathcal{P}(i)$ are permutations of those of the numbers $1, \ldots, N$ which refer to distinct $\vec{k}_i$ such that $\vec{k}_i \neq \vec{k}_j$. This last restriction is the meaning of the prime on the sum on permutations $\mathcal{P}$. The number of such permutations is

$$\eta(\vec{k}_1, \ldots, \vec{k}_N) = \frac{N!}{N_{\vec{k}_1}!N_{\vec{k}_2}!\cdots} \tag{4.4}$$

where $N_{\vec{k}_i}$ is the number of factors with $\vec{k}_i = \vec{k}'_i$ ($\eta = N!$ for fermions). $(\pm)^{\mathcal{P}}$ is the sign of the permutation. $V$ is the volume of the system. The sum on $\nu$ in (4.1) becomes a sum on $\vec{k}_i$ in this basis. In a large system the sum on the $\vec{k}_i$ can be expressed as an integral and thence as an integral on momenta, while the matrix elements in (4.1) contain integrals on the positions $\vec{r}_1, \ldots, \vec{r}_N$. Thus $\bar{\bar{\phi}}$ can be expressed in terms of an integral on the phase space which can be compared with the classical result. The general result of this program is that, under certain conditions which we will elaborate, the partition function and the density matrix in this basis take the approximate forms

$$Z_{\mathrm{c}} \to \frac{1}{h^{3N}N!}\int \mathrm{d}^{3N}q\,\mathrm{d}^{3N}p\,\mathrm{e}^{-\beta H(p,q)} \tag{4.5}$$

while

$$\rho \to \frac{1}{h^{3N}N!Z_{\mathrm{c}}}\mathrm{e}^{-\beta H(p,q)} \tag{4.6}$$

Equation (4.5) differs from the classical expression because of the factors $1/h^{3N}N!$. Though these cancel out in (4.6) they can be seen to be relevant to the thermodynamics which involves $\ln Z_{\mathrm{c}}$. We return to this below.

## The perfect gas

To carry out this program we begin with the perfect gas. Then

$$H = \frac{-\hbar^2}{2m}\sum_i \nabla_i^2 \tag{4.7}$$

where $m$ is the mass of the particles. Inserting this in (4.1) and using the basis (4.3) gives

$$\bar{\bar{\phi}} = \frac{1}{Z_c V^N} \sum_{\{N_{k_i}\} \text{ such that } \sum_i N_{k_i}=N} (1/n!) \sum_{\vec{k}_1,\dots,\vec{k}_n} \sum_{\mathcal{P},\mathcal{P}'} (\pm)^{\mathcal{P}} (\pm)^{\mathcal{P}'} \frac{1}{\eta(\vec{k}_1,\dots,\vec{k}_N)}$$
$$\times \int d\vec{r}_1,\dots,d\vec{r}_N \, e^{-\sum_i \hbar^2 \vec{k}_i^2/2mk_B T} \phi(\hbar\vec{k}_1,\dots,\hbar\vec{k}_n,\vec{r}_1,\dots,\vec{r}_N) \prod_i^N e^{i\vec{k}_i\cdot(\vec{r}_{\mathcal{P}(i)}-\vec{r}_{\mathcal{P}'(i)})} \tag{4.8}$$

here we have written

$$\sum_\nu (\dots) = \frac{1}{n!} \sum_{\vec{k}_1,\dots,\vec{k}_n} (\dots) \tag{4.9}$$

to take account of the fact that the states obtained by permuting the $\vec{k}_i$ are not different. Here $n$ is the number of $\vec{k}_i$'s which appear at least once. (For fermions, $n = N$ and the sum $\sum_{\{N_{k_i}\} \text{ such that } \sum_i N_{k_i}=N}$ has only one term, with $N$ of $N_{k_i} = 1$ and the rest zero. .) Now we transform this by defining

$$i' = \mathcal{P}(i) \qquad \mathcal{P}'' = \mathcal{P}'\mathcal{P}^{-1} \tag{4.10}$$

so that $(\pm)^{\mathcal{P}}(\pm)^{\mathcal{P}'} = (\pm)^{\mathcal{P}''}$ and writing

$$\sum_{\vec{k}_i} (\dots) = \frac{V}{(2\pi)^3} \int d\vec{k}_i (\dots) \tag{4.11}$$

If we write $\hbar\vec{k}_i = \vec{p}_i$ and use the fact that $\phi$ must be invariant under permutations of $\vec{r}_1,\dots,\vec{r}_N$ if $\phi$ is to be an observable then we get

$$\bar{\bar{\phi}} = \frac{1}{h^{3N} Z_c} \sum_{\{N_{k_i}\} \text{ such that } \sum_i N_{k_i}=N} (1/n!) \int d^{3n}p \, d^{3N}q \, e^{-\sum_i (\vec{p}_i^2/2mk_B T)}$$
$$\times \phi(\vec{p}_1,\dots,\vec{p}_N,\vec{r}_1,\dots,\vec{r}_N) \sum_{\mathcal{P}''}{}' (\pm)^{\mathcal{P}''} \prod_{i=1}^N e^{i\frac{\vec{p}_i}{\hbar}\cdot(\vec{r}_i-\vec{r}_{\mathcal{P}''(i)})} \tag{4.12}$$

where the sum on $\mathcal{P}$ has been done. The integral on the momenta is only over those momenta which are not equal in the list $\vec{p}_1, \ldots, \vec{p}_N$ in the case of bosons. Thus

$$\bar{\bar{\phi}} = \frac{1}{h^{3N}N!}\int d^{3N}p\, d^{3N}q \frac{e^{-H(p,q)/k_BT}}{Z_c}\phi(q,p)$$
$$+ \sum_{\{N_{k_i}\}\text{ such that }\sum_i N_{k_i}=N,\text{some }N_{k_i}\neq 1} (1/h^{3n}n!)\int d^{3n}p d^{3n}q \frac{e^{-H(p,q)/k_BT}}{Z_c}\phi(q,p)$$
$$+ \sum_{\{N_{k_i}\}\text{ such that }\sum_i N_{k_i}=N} (1/h^{3n}n!) \sum_{\mathcal{P}''}{}'' (\pm)^{\mathcal{P}''}$$
$$\times \int d^{3N}pd^{3n}q \frac{e^{-H(p,q)/k_BT}}{Z_c}\phi(q,p)\prod_{i=1}^{N} e^{i\frac{\vec{p}_i}{\hbar}\cdot(\vec{r}_i-\vec{r}_{\mathcal{P}''(i)})} \tag{4.13}$$

in which the sum on permutations $\mathcal{P}''$ now excludes the identity permutation. The first term on the right hand side is the semiclassical limit in which we are interested. The second term is absent in the case of fermions. We can get more explicit expressions for the correction term in the case of the partition function which is obtained by dropping the factor $\phi/Z_c$. For bosons, the second term, which does not contain permutations, can be shown to be a factor $\lambda^3(N/V)$ smaller than the first. $\lambda$ is the thermal wavelength defined below. In the third term, the lowest order contribution is also the term, for bosons, for which all the $N_{k_i} = 0$ or 1 and we find

$$Z_c = \frac{1}{h^{3N}N!}\Bigg[\int d^{3N}pd^{3N}qe^{-H(p,q)/k_BT} + \sum_{\mathcal{P}''}{}''(\pm)^{\mathcal{P}''}$$
$$\times \int d^{3N}p\, d^{3N}q\, e^{-H(p,q)/k_BT}\prod_{i=1}^{N} e^{i\frac{\vec{p}_i}{\hbar}\cdot(\vec{r}_i-\vec{r}_{\mathcal{P}''(i)})}\Bigg] \tag{4.14}$$

The integral on $\vec{p}_i$ may be done in this case in the second term giving

$$\int \frac{d\vec{p}_i}{h^3} e^{-p_i^2/2mk_BT} e^{\frac{i}{\hbar}\vec{p}_i\cdot(\vec{r}_i-\vec{r}_{\mathcal{P}(i)})} = \frac{1}{\lambda^3}e^{-\pi|\vec{r}_i-\vec{r}_{\mathcal{P}(i)}|^2/\lambda^2} \equiv \frac{1}{\lambda^3}f(|\,\vec{r}_i - \vec{r}_{\mathcal{P}(i)}\,|) \tag{4.15}$$

where $\lambda$ is the thermal wavelength, $\lambda \equiv \sqrt{2\pi\hbar^2/mk_BT}$.

Thus

$$Z_c = \left(\frac{V}{\lambda^3 N}\right)^N \left(1 + \sum_{\mathcal{P}''}{}''(\pm)^{\mathcal{P}''}\int \frac{d^N\vec{r}}{V^N}\prod_i^N f(|\,\vec{r}_i - \vec{r}_{\mathcal{P}''(i)}\,|)\right) \tag{4.16}$$

where the lowest order Stirling approximation was used to evaluate the factorial. The question of determining the condition under which the second term can be dropped is somewhat delicate and we will defer it until Chapter 6. In fact it is sufficient to require that $\lambda^3 N/V \ll 1$ whereas a careless treatment might suggest that the left hand side of this inequality would need to be multiplied by $N$.

For interacting gases, a similar set of transformations can be worked out. We will consider the Hamiltonian

$$H = \sum_i \vec{p}_i^2/2m + \sum_{i<j} v(|\vec{r}_i - \vec{r}_j|) \tag{4.17}$$

The partition function in the canonical case is

$$Z = \frac{1}{N!V^N} \sum_{\vec{k}_1,\ldots,\vec{k}_N} \sum_{\mathcal{P}} \sum_{\mathcal{P}'} \frac{1}{\eta(\vec{k}_1,\ldots,\vec{k}_N)} \int \mathrm{d}^N\vec{r} \prod_{i=1}^N \mathrm{e}^{-\mathrm{i}\vec{k}_i\cdot\vec{r}_{\mathcal{P}(i)}} \,\mathrm{e}^{-\beta(T+V)} \prod_{j=1}^N \mathrm{e}^{\mathrm{i}\vec{k}_j\cdot\vec{r}_{\mathcal{P}'(j)}} \tag{4.18}$$

where $T = \sum_i \vec{p}_i^2/2m$. The complication in this case is that

$$\mathrm{e}^{-\beta(T+V)} \neq \mathrm{e}^{-\beta T}\mathrm{e}^{-\beta V} \tag{4.19}$$

We write

$$\mathrm{e}^{-\beta(T+V)} = \mathrm{e}^{-\beta T}\mathrm{e}^{-\beta V}\mathrm{e}^{\beta\mathcal{O}_1}\mathrm{e}^{\beta^2\mathcal{O}_2}\ldots \tag{4.20}$$

and evaluate the operators by successively differentiating (4.20) with respect to $\beta$ and setting $\beta = 0$. This gives

$$\mathcal{O}_1 = 0 \tag{4.21}$$

$$\mathcal{O}_2 = -1/2[T, V] \tag{4.22}$$

We drop the remaining terms in (4.20). Using the explicit expression (4.17) we then find

$$\mathcal{O}_2 = \frac{\hbar^2}{4m}\sum_{k\neq l} \nabla_k^2 v_{k,l} - \frac{\hbar^2}{2m}\sum_k \vec{F}_k \cdot \nabla_k \tag{4.23}$$

in which

$$\vec{F}_k = -\nabla_k \sum_{l\neq k} v_{kl} \tag{4.24}$$

We will not carry out a detailed analysis, but only note that the preceding analysis for the perfect gas could be essentially carried through without change as long as the terms $-\beta\mathcal{O}_2$, which act essentially like an additional term in the Hamiltonian, can be ignored. Thus, in addition to the requirement that $\lambda \ll a$ (the interparticle spacing), which is required in order to ignore exchange effects, we have here an added requirement that $\lambda^2\nabla^2 v \ll v$ in order to apply classical statistical mechanics. We will defer a more detailed analysis until we have described the relevant cluster expansion technique.

## Problems

4.1 Show that the first order correction to the semiclassical limit for the perfect gas can be represented by a temperature dependent effective potential of form:

$$\tilde{v}_{ij} = -k_B T \ln \left[ 1 \pm e^{\frac{-2\pi}{\lambda^2} |\vec{r}_i - \vec{r}_j|^2} \right] \tag{4.25}$$

Sketch this potential as a function of $|\vec{r}_i - \vec{r}_j|$ and discuss its meaning in the cases of fermions and bosons.

4.2 Consider a system with the Hamiltonian

$$H = \sum_i \vec{p}_i^2/2m + (K/2) \sum_i \vec{r}_i^2$$

There are $N$ particles and we will suppose that the temperature is high enough to work in the semiclassical limit.

(a) Under some physical circumstances, the volume is irrelevant in such a system. State a criterion in terms of $V$, $T$, and $K$ under which this is the case.

(b) Under the circumstances in which $V$ is irrelevant, define thermodynamic functions appropriately in terms of $T$, $K$ and other variables which are appropriately introduced through Legendre transformation. Prove differential relations for these thermodynamic potentials, and establish as many relationships analogous to the ones found in Chapter 3 for a system in which $T$ and $V$ are natural variables as you can. (It turns out here that, at fixed $K$, $T$, the chemical potential is *not* independent of $N$. This problem is best understood by calculating explicit expressions for $F$ and $\Omega$ using the Hamiltonian and the semiclassical limit expressions, in lowest order, in Chapter 4.)

# Part II

## States of matter in equilibrium statistical physics

# 5

# Perfect gases

Here we begin a discussion of applications to systems of increasing density with the application of the formalism to the simplest of all models, in which the particles have kinetic energy but do not interact. Though this sounds straightforward, we note two issues. First, if we take the Hamiltonian to be

$$H = \sum_i p_i^2/2m$$

(which is what we will use in the partition functions calculated below) then it should be clear that there are $N$ trivially identified constants of the motion in such a system, namely the energies of individual particles. Such a system cannot exchange energy between particles and cannot satisfy any reasonable ergodicity requirement. As a consequence, though we can study the properties of such an ideal gas when it obeys the canonical distribution, we have no assurance at all that it will ever be found in such a state, since a system initiated experimentally away from the equilibrium distribution will stay there. The obvious resolution to this dilemma is to include interactions between the particles which are always present (at least for massive particles and in the case of an isolated system; equilibration can also occur as a result of interaction with the environment, for example, the walls of a container containing a gas). Collisions of the molecules allow energy exchange, ergodicity and the approach to equilibrium in time but they lead to two further questions. First, how fast does the approach to equilibrium occur and second, by what criteria do we decide whether the equilibrium system can, after all, be described as an ideal gas? With respect to the first question, in a dilute gas, the general notion is that one needs a significant number (around 100 in practice) of molecular collisions per particle to achieve equilibrium. Elementary kinetic theory estimates (Problem 5.1) show that at room temperature, equilibration of most gases will occur in minutes or less when their densities are as large as $10^{23}$ cm$^{-3}$. However, this becomes a more serious problem at low temperatures. The second issue is addressed in the next chapter, in which a careful treatment of imperfect interacting gases appears. Roughly speaking, the classical criterion for ignoring corrections due to interactions is $\rho \ll 1/\sigma^{3/2}$ where $\sigma$ is the collision cross-section. In the quantum case, the criterion is less trivial to state and harder to achieve experimentally. Indeed a system which could be described as an ideal Bose gas containing massive particles was only observed very recently. For reasons to be discussed later, many fermion systems behave approximately as perfect Fermi gases at low temperatures.

## Classical perfect gas

As discussed in the last chapter, the partition function in the semiclassical limit is (equation (4.16))

$$Z_c = \left(\frac{Ve}{\lambda^3 N}\right)^N \tag{5.1}$$

where the somewhat more accurate form $N! \approx N^N e^N$ has been used. The Helmholtz free energy is

$$F = -k_B T \ln Z_c = -k_B T N \left(\ln\left(\frac{a}{\lambda}\right)^3 + 1\right) \tag{5.2}$$

where $a^3 = V/N$. The entropy and specific heat are

$$S = -\left(\frac{\partial F}{\partial T}\right)_{V,N} = N k_B \left(\ln\left(\frac{a}{\lambda}\right)^3 + \frac{5}{2}\right) \tag{5.3}$$

$$C_V = T\left(\frac{\partial S}{\partial T}\right)_V = \frac{3Nk_B}{2} \tag{5.4}$$

which is familiar as a form of the equipartition theorem.

It is important to notice that, without the factor $1/N!$ in the expression for the partition function in the semiclassical limit, the free energy and the entropy would not be proportional to the number of particles $N$. Tracing this factor through the calculations of the last chapter, one sees that it arose when we performed the sum on states associated with all possible sets of plane waves for the independent particles of the perfect gas. For a given set $\vec{k}_1, \ldots, \vec{k}_N$, the states obtained by permuting the labels on the $\vec{k}_i$ are identical because the particles are identical, so when we integrated independently on $\vec{k}_1, \ldots, \vec{k}_N$ we had to divide by $N!$. Thus the factor $1/N!$ arises from the indistinguishability of the particles. It does not arise naturally in the classical formulation and indeed the entropy and free energy obtained from the classical formulation are not intensive if this factor is not added by hand. Gibbs noticed this problem and added the factor $1/N!$ by hand and apparently by trial and error based on physical reasoning but without the benefit of quantum mechanics.

The factor associated with particle indistinguishability occurs, as expected, in a somewhat different way in a perfect gas mixture, when one has, say, two types of particles, $N_1$ of type 1 and $N_2$ of type 2. Then the partition function is easily seen to be

$$Z_{\mathrm{c,mixture}} = \frac{V^{N_1+N_2}}{\lambda_1^{3N_1}\lambda_2^{3N_2} N_1! N_2!} \tag{5.5}$$

The entropy becomes

$$S_{\mathrm{mixture}} = k_{\mathrm{B}}\left(N_1\left(\ln\left(\frac{a_1}{\lambda_1}\right)^3 + \frac{5}{2}\right) + N_2\left(\ln\left(\frac{a_2}{\lambda_2}\right)^3 + \frac{5}{2}\right)\right) \tag{5.6}$$

In terms of $\rho_1 = 1/a_1^3$, $\rho_2 = 1/a_2^3$ this can be written

$$S_{\mathrm{mixture}} = k_{\mathrm{B}} V\left(-\rho_1 \ln \rho_1 - \rho_2 \ln \rho_2 - \rho_1 \ln \lambda_1^3 - \rho_2 \ln \lambda_2^3\right) + 5 k_{\mathrm{B}} N/2 \tag{5.7}$$

The first two terms are sometimes called the entropy of mixing.

It is instructive to do some of the same calculations for a perfect gas in the grand canonical case. One has

$$Z_{\mathrm{gc}} = \sum_N \mathrm{e}^{N\beta\mu} Z_N = \sum_N \mathrm{e}^{N\beta\mu} \frac{V^N}{\lambda^{3N} N!} = \exp\left(\mathrm{e}^{\beta\mu}\frac{V}{\lambda^3}\right) \tag{5.8}$$

Thus

$$\Omega = -PV = -k_{\mathrm{B}} T \mathrm{e}^{\beta\mu}\frac{V}{\lambda^3} \tag{5.9}$$

But $\mu$ is determined by

$$\begin{aligned}\bar{N} &= \frac{\sum_N N\,\mathrm{e}^{\beta N\mu} Z_N}{\sum_N \mathrm{e}^{\beta N\mu} Z_N} = \frac{1}{\beta}\frac{\partial}{\partial\mu}\ln\left(\sum_N \mathrm{e}^{N\mu\beta} Z_N\right) \\ &= \frac{1}{\beta}\frac{\partial}{\partial\mu}\ln\left(\mathrm{e}^{\mathrm{e}^{\beta\mu} V/\lambda^3}\right) = \frac{1}{\beta}\frac{\partial}{\partial\mu}\frac{\mathrm{e}^{\beta\mu} V}{\lambda^3} = \frac{\mathrm{e}^{\beta\mu} V}{\lambda^3}\end{aligned} \tag{5.10}$$

so that $PV = N k_{\mathrm{B}} T$ in agreement with the result in the canonical case. The specific heat is found from

$$S = -\left(\frac{\partial\Omega}{\partial T}\right)_{\mu,V} = \frac{V k_{\mathrm{B}} \mathrm{e}^{\beta\mu}}{\lambda^3}\left[\frac{5}{2} - \frac{\mu}{k_{\mathrm{B}} T}\right] \tag{5.11}$$

by use of (5.9) and (5.10)

$$S = \bar{N} k_{\mathrm{B}}\left(\frac{5}{2} + \ln\left(\frac{a^3}{\lambda^3}\right)\right) \tag{5.12}$$

where $a^3 = V/\bar{N}$ (which is temperature dependent here). This is consistent with (5.3) in the canonical case. Defining $C_{v,\bar{N}} = T\,(\partial S/\partial T)_{V,\bar{N}}$ gives $C_{V,\bar{N}} = \frac{3}{2}k_B\bar{N}$ consistent with the canonical case. Note that $C_{V,\mu} \equiv T\,(\partial S/\partial T)_{V,\mu}$ would be quite different (see problem 5.2).

## Molecular ideal gas

Here we suppose that the centers of mass of the molecules obey classical mechanics but that the internal dynamics of the molecules is still quantum mechanical. Roughly, one can see that the requirements for this are that

$$\lambda = \left(\frac{2\pi\hbar^2}{Mk_BT}\right)^{1/2} \ll \left(\frac{V}{N}\right)^{1/3} \tag{5.13}$$

where $M$ is the molecular mass and $N$ is the number of molecules. On the other hand we do not assume that the separation of the energy levels of the individual molecules is $\ll k_BT$. Writing the Hamiltonian for such a system needs to be done with some care. Suppose there are $M$ molecules, each requiring $3n$ particle coordinates for a description. We consider a homonuclear gas for simplicity (like $H_2$) though the extension to the heteronuclear case is not difficult. There is a potential energy function $V(\vec{r}_1, \ldots, \vec{r}_N)$ where $N = Mn$ and the Hamiltonian in general is

$$H = \sum_{i=1}^{N} \vec{p}_i^{\,2}/2m + V \tag{5.14}$$

Now at the temperatures at which we are working we will assume that the relevant eigenvectors of the Hamiltonian can be written in the form:

$$\psi_\nu = \psi_{\{\vec{k}_i\},\{n_i\}} = \frac{1}{\sqrt{\eta}}\frac{1}{V^{M/2}}\sum_{\mathcal{P}}(\pm)^{\mathcal{P}}\prod_{i=1}^{M} e^{i\vec{k}_i\cdot\vec{\mathcal{P}R}_{(i)}}\phi_{n_i}(\mathcal{P}\{q\}_i) \tag{5.15}$$

In each term here we have grouped the particle coordinates $\vec{r}_{\mathcal{P}(1)}, \ldots, \vec{r}_{\mathcal{P}(n)}, \vec{r}_{\mathcal{P}(n+1)}, \ldots, \vec{r}_{\mathcal{P}(2n)}, \ldots, \vec{r}_{\mathcal{P}(N-n+1)}, \ldots, \vec{r}_{\mathcal{P}(N)}$, corresponding to assigning these groupings to the $M$ molecules. These groups are labelled with the index $i = 1, \ldots, M$. The center of mass of each of these groups is $\vec{\mathcal{P}R}_{(i)}$ and the remaining coordinates associated with the group $i$ after the center of mass transformation are denoted $\mathcal{P}\{q\}_i$. $\phi_{n_i}(\mathcal{P}\{q\}_i)$ is to be regarded as the wave function of the $i$th molecule and is assumed to be localized around the center of mass of the $i$th coordinate grouping. $\eta$ was defined in equation (4.4). Consider the action of the Hamiltonian on the term associated with the permutation $\mathcal{P}$ in this wave function.

We assume that the centers of mass $\vec{R}_i$ are far enough apart so that interactions between particles in separate molecules (represented by the $\phi_{n_i}(\mathcal{P}\{q\}_i)$) are negligible. Then, when the Hamiltonian acts on this term in the wave function, the only terms in the Hamiltonian which contribute significantly are those parts $V_1(\mathcal{P}\{q\}_i)$ which describe the interactions between the particles in each molecule. Then the eigenvalue can be shown to be

$$E_\nu = \sum_i \left( \frac{\hbar^2 \vec{k}_i^2}{2M} + \epsilon_{n_i} \right) \tag{5.16}$$

where

$$\left( \sum_{\mathcal{P}i} p_{\mathcal{P}i}^2/2m + V_1(\mathcal{P}q_i) \right) \phi_n(\mathcal{P}q_i) = \epsilon_n \phi_n(\mathcal{P}q_i) \tag{5.17}$$

Notice that for terms in the wave function in which particles have been interchanged by permutation between molecules, different terms in the potential energy are significant. Here $\vec{k}_i$ is the momentum associated with the center of mass of the $i$th molecule and $p_{\mathcal{P}i}$, $\mathcal{P}\{q\}_i$ are the remaining degrees of freedom associated with that molecule, which must be treated quantum mechanically.

In (5.15), $\mathcal{P}$ permutes labels associated with all the indistinguishable particles, including ones on different molecules, in principle. However, here the permutations which interchange identical particles on different molecules may be neglected. To see this consider a particle labelled $\alpha$ on the $i$th molecule and an identical particle labelled $\alpha'$ on another molecule $i'$. In the term in the partition function associated with the permutation which interchanges these two particles and does nothing else, the factors depending on $\vec{k}_i$ are

$$\frac{V}{2\pi^3} \int e^{\frac{-\hbar^2 k_i^2}{2M}} e^{\vec{k}_i \frac{m_\alpha}{M} \cdot (\vec{r}_\alpha - \vec{r}_{\alpha'})} \, d\vec{k}_i \tag{5.18}$$

where $m_\alpha$ is the mass of the particles being interchanged and $M$ is the mass of the molecule as before. The factor $m_\alpha/M$ arises from the definition of the center of mass of the molecule $\vec{R}_i = \frac{1}{M} \sum_\alpha m_\alpha \vec{r}_\alpha$. The integral is done as in Chapter 4 and we have

$$\frac{1}{\lambda^3} e^{\frac{-\pi \Delta r^2 (m_\alpha/M)^2}{\lambda^2}} \tag{5.19}$$

where $\Delta r = |\vec{r}_\alpha - \vec{r}_{\alpha'}|$. This will be small if the density is low enough but the condition is slightly more stringent than (5.13) might imply because of the factor

$m_\alpha/M$. Thus it appears that we might require

$$\lambda = \left(\frac{2\pi\hbar^2}{Mk_BT}\right)^{1/2} \ll \left(\frac{V}{N}\right)^{1/3}\left(\frac{m_\alpha}{M}\right) \tag{5.20}$$

where $m_\alpha$ is the lightest particle in the molecule. This condition would be very stringent for electrons on molecules! However, when we consider the integrals on $\vec{r}_\alpha, \vec{r}_{\alpha'}$ which enter the relevant term in the partition function, we see that the term which must be neglected is proportional to

$$\int \mathrm{d}\vec{r}_\alpha \int \mathrm{d}\vec{r}_{\alpha'} \int \mathrm{d}\{\vec{r}\}'_i \int \mathrm{d}\{\vec{r}\}'_{i'}\, \mathrm{e}^{-2\pi(m_\alpha/M)|\vec{r}_\alpha - \vec{r}_{\alpha'}|^2\lambda^2} \phi_{n_i}(\vec{r}_\alpha, \{r\}'_i)\phi^*_{n'_i}(\vec{r}_\alpha, \{r\}'_{i'})\phi^*_{n_i}(\vec{r}_{\alpha'}, \{r\}'_i)\phi_{n'_i}(\vec{r}_{\alpha'}, \{r\}'_{i'}) \tag{5.21}$$

Here $\{r\}'_i$ means all the coordinates on $i$ except $\vec{r}_\alpha$ or $\vec{r}_{\alpha'}$ and $\{r\}'_{i'}$ means all the coordinates on $i'$ except $\vec{r}_\alpha$ or $\vec{r}_{\alpha'}$. The local wave functions $\phi$ will overlap very little in dilute gas and so, in particular, the terms involving exchange of electrons will be small long before the condition (5.20) is satisfied. On the other hand, it is true that both effects, associated with the momentum averaging and with spatial averaging, indicate that other things (such as the strength of the binding of the particle to the molecule) being equal, the lightest particles in the molecule will be the easiest to exchange because the wave function overlaps will shrink exponentially with $\sqrt{m_\alpha}$ (as one can see from the WKB approximation, for example).

On the basis of these arguments we neglect terms in the partition function in which one permutation of the particle labels occurs on one side of the matrix element and another permutation, in which exchange of particles between molecules has occurred relative to the permutation on the left hand side, occurs on the right hand side. In this way one is grouping the terms involving different permutations of the coordinates in (5.15) as follows. Start with a given assignment of particle numbers to molecules and add all permutations of labels within each molecule. Now add all terms in which all the labels associated with one molecule are interchanged with all the labels associated with another. Finally add all permutations resulting in the assignment of different coordinate labels to the molecules and similarly permute the coordinates in each assignment, first within the molecules, and then interchange the labels of all the coordinates of each molecule for each such assignment. Now the approximation to be made consists of two aspects. The overlaps of terms involving different particle assignments to a given molecule can be ignored as long as the range of the local wave functions $\phi$ is much less than the mean distance between molecules. This criterion involves the temperature, because the relevant molecular wave functions will have larger size for larger energies (and at high enough energies the molecule will not be bound at all). Thus this aspect of the approximation requires that the temperature be much less than the molecular binding energy. On the other hand, if we wish to treat the centers of mass of the molecules classically, then the thermal wavelength associated with the molecular mass must be much less than the distance between molecules so that the effects of permutations of entire sets of coordinates between molecules can be ignored. This second requirement puts a lower bound on the temperatures where the approximations are valid, while the first requirement puts an upper bound on the temperature.

If we have $N$ particles, combined into $M$ molecules so that there are $n = N/M$ coordinates associated with each molecule, then the three varieties of permutations discussed above are $(n!)^M$ permutations of the internal coordinates, $M!$ permutations of all the coordinates of each molecule with all the coordinates of each other molecule and $N!/(n!)^M M!$ assignments of coordinate labels to the molecules. We have been saying that we can ignore cross terms associated with the last kind of permutations as long as the temperature and density are low enough so that the range of all the molecular wave functions is much less than the intermolecular distance. We can ignore cross terms associated with the $M!$ permutations of all the coordinates of one molecule with all those of another as long as the temperature is high enough so that the molecular thermal wavelength is much less than the intermolecular distance. We cannot ignore the permutations associated with internal degrees of freedom of the molecules. (Molecules whose internal dynamics is classical are not known to exist.) Assuming that the $\phi_{n_i}$ are already appropriately symmetrized or antisymmetrized with respect to permutations of labels within a molecule, one can work with one assignment of particle labels to molecules because each of the $(n!)^M$ terms associated with different assignments will give the same result in the partition function and cross terms are ignored. Thus one can work with the wave function

$$\psi_{\{\vec{k}_i\},\{n_i\}} = \frac{1}{\sqrt{M!}V^{M/2}} \sum_{\mathcal{P}'} \prod_{i=1}^{M} \mathrm{e}^{\mathrm{i}\vec{k}_i \cdot \vec{R}_{\mathcal{P}(i)}} \phi_{n_i}(\{q\}_i) \tag{5.22}$$

in which the sum on permutations only includes those in which all the coordinates assocated with one molecule have been interchanged with all the coordinates associated with another. The factor $\eta$ in (5.15) has been replaced by $M!$ assuming that, in the semiclassical limit which we consider, no plane wave state associated with the motion of the center of mass of the molecules is macroscopically occupied. We may now calculate $Z$ by changing sums into integrals as in the discussion of the semiclassical limit, adding a factor $M!$ to take account of the fact that a new state is not produced by permuting the $\{\vec{k}_i\}$. Then we have

$$Z_{\mathrm{c}} = \frac{V^M}{M!} \int \mathrm{d}^M \vec{k} \sum_{\{n_i\}} \mathrm{e}^{-\beta\left(\sum_i \frac{\hbar^2 k_i^2}{2m} + \epsilon_{n_i}\right)} = \frac{V^M}{M!\lambda^{3M}} \prod_{i=1}^{M} \sum_{n_i} \mathrm{e}^{-\beta \epsilon_{n_i}} \tag{5.23}$$

Thus

$$
\begin{aligned}
F &= -k_B T M \left\{ \ln\left(\frac{a^3}{\lambda^3}\right) + 1 \right\} - k_B T \sum_i \ln\left(\sum_n e^{-\beta\epsilon_n}\right) \\
&= -k_B T M \left( \ln\left(\frac{a^3}{\lambda^3}\right) + 1 + \ln \sum_n e^{-\beta\epsilon_n} \right)
\end{aligned}
\tag{5.24}
$$

in which $\lambda$ and $a^3$ have their previous definitions. The last term may be calculated from the solutions to (5.17) which describes a single molecule at rest.

As an example consider the case of homonuclear diatomic molecules in harmonic approximation for the vibrational levels. The spectrum $\epsilon_n$ of molecular energy levels is

$$
\epsilon_n = \epsilon_{n,L,M} = \hbar\omega_0(n + 1/2) + \frac{\hbar^2}{2I} L(L+1) \tag{5.25}
$$

in which $\omega_0$ is the harmonic vibrational frequency of the molecule, $\omega_0 = \sqrt{2K/m}$ where $K$ is the spring constant. Thus the free energy is (here and henceforth we denote the number of molecules by $N$)

$$
F = N k_B T \left[ \ln\left(\frac{\lambda^3}{a^3} - 1\right) + \frac{\hbar\omega_0\beta}{2} - \ln(1 - e^{-\beta\hbar\omega_0}) - \ln\left(\sum_L (2L+1) e^{\left\{\frac{-\hbar^2}{2I} L(L+1)\beta\right\}}\right)\right] \tag{5.26}
$$

The sum on $L$ must be treated with caution when the two nuclei of the diatomic molecule are identical. In practice we can assume that the electronic degrees of freedom play no role because the molecule at thermal energies is separated from its first excited electronic state in the Born–Oppenheimer approximation by an energy gap which is much larger than $k_B T$. However, the spins of the nuclei in the molecule are weakly coupled by energies much less than $k_B T$ and this has interesting effects on the physics. In effect, the various nuclear spin levels are degenerate. We consider the case of $H_2$ gas. The vibrational frequency $\omega_0$ is 4400 cm$^{-1}$ and the first rotational level at $J = 1$ is about 120 cm$^{-1}$ so at temperatures much less than about $10^3$ K we can certainly consider the molecules to be in their vibrational ground state. Then the nuclear wave function is of form

$$
\phi(\vec{r}_1, \vec{r}_2; i_z^{(1)}, i_z^{(2)}) = \phi_{\text{vib}} Y_{M_L,L}(\hat{r}_{12}) \chi_I(i_z^{(1)}, i_z^{(2)}) \tag{5.27}
$$

The wave function of the nuclear spins does turn out to be relevant. Consider the case of $H_2$. The protons are fermions and the whole wave function must be antisymmetric under interchange of 1 and 2. The ground vibrational state is even under interchange. The proton nuclear spins are $1/2$ so the allowed values of $I$ are 1 and 0. The $I = 1$ state is a triplet of which each state is even under interchange.

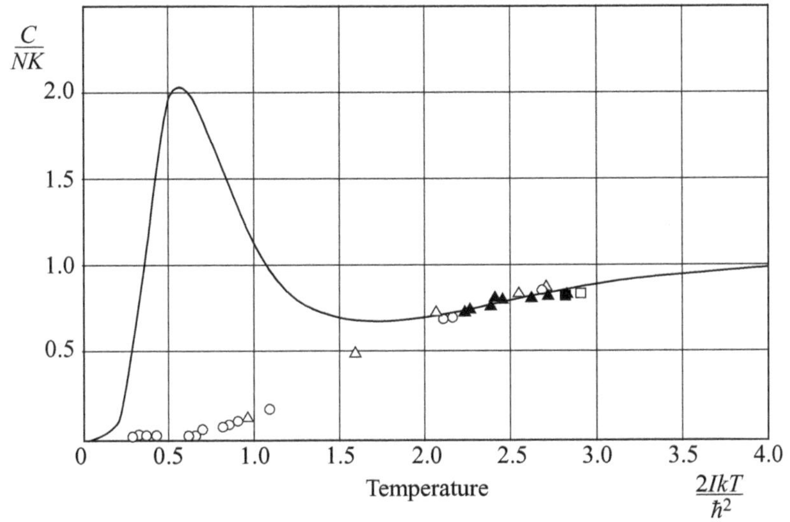


Figure 5.1 Comparison of the fully equilibrated theory for the specific heat of $H_2$ gas with experiment.

Therefore the rotational spherical harmonic must be odd under interchange. On the other hand when $I = 0$, the nuclear spin state is an odd singlet and the values of $L$ are even. Recalling that the nuclear spin levels are all degenerate we conclude that the sum on $L$ in (5.26) must be split into even and odd parts with weight 1 for even values of $L$ and weight 3 for odd values of $L$. The results of comparing this theory with experiments on specific heat are shown in Figure 5.1. Astonishingly, it does not work at all.

The problem is that this system is not ergodic. The different values of nuclear spin for the molecules are stable over extremely long times so that the needed changes in the nuclear spins which must accompany any changes in the distribution between even and odd values of $L$ cannot take place. Instead, one can regard the experimental system as a gas mixture consisting of two types of molecules which can exchange energy among themselves as the temperature changes, but not between each type. The molecules are called parahydrogen (nuclear spin 0, even $L$) and orthohydrogen (nuclear spin 1, odd $L$). The difference is, in summary, that if the system were equilibrated one would have

$$F_{\text{equil}} = Nk_{\text{B}}T\left[\ln\left(\frac{\lambda^3}{a^3}\right) - 1 + \frac{\hbar\omega_0\beta}{2} + \ln(1 - \mathrm{e}^{-\beta\hbar\omega_0})\right.$$
$$\left. - \ln\left(\sum_{L\ \text{even}}(2L+1)\mathrm{e}^{\left\{\frac{-\hbar^2}{2I}L(L+1)\beta\right\}} + 3\sum_{L\ \text{odd}}(2L+1)\mathrm{e}^{\left\{\frac{-\hbar^2}{2I}L(L+1)\beta\right\}}\right)\right] \quad (5.28)$$

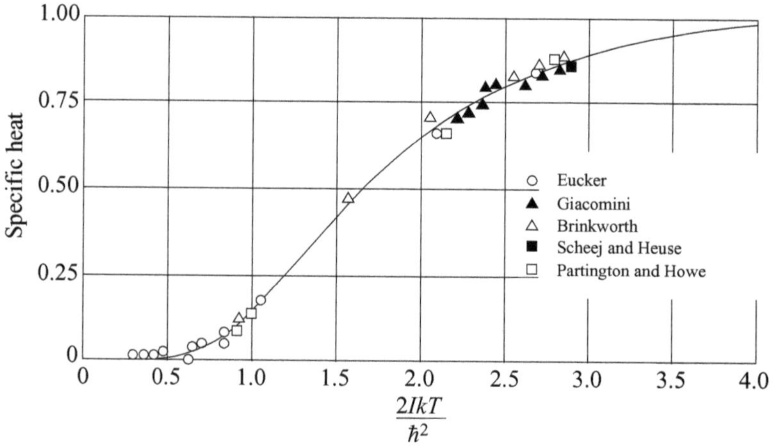


Figure 5.2 Result of the "mixture" theory of $H_2$ gas compared with experiment.

whereas

$$F_{\text{mixture}} = Nk_{\mathrm{B}}T\left[\ln\left(\frac{\lambda^3}{a^3}\right) - 1 + \frac{\hbar\omega_0\beta}{2} + \ln(1-\mathrm{e}^{-\beta\hbar\omega_0}) - (1/4) \right.$$
$$\left. \times\ln\left(\sum_{L\text{ even}}(2L+1)\mathrm{e}^{\left\{\frac{-\hbar^2}{2I}L(L+1)\beta\right\}}\right) - (3/4)\ln\left(\sum_{L\text{ odd}}(2L+1)\mathrm{e}^{\left\{\frac{-\hbar^2}{2I}L(L+1)\beta\right\}}\right)\right] \tag{5.29}$$

Figure 5.2 shows the results for the mixture theory. They agree much better with the experiments.

The situation in this ideal gas of molecules is one in which the centers of molecular mass are essentially treated classically, while the internal degrees of freedom are treated quantum mechanically. In some respects, it can serve as a "toy model" for thinking about problems of measurement and interpretation in quantum mechanics, where such mixtures of classical degrees of freedom and subsystems for which the internal degrees of freedom are inescapably quantum, occur. For example, the relationship between the quantum mechanical phases of the $M!$ terms in (5.22) is irrelevant to the calculation of the partition function and, as far as the calculation goes, those relative phases could as well be random. In the more general discussion of measurement, one considers systems which quantum mechanically "decohere" in a similar way. For many purposes, we can describe the molecular gas system by just one of the $M!$ terms in (5.22), somewhat as one is said to describe the universe in terms of one term in an enormously complex sum of terms, each associated with another "parallel universe." Inelastic collisions of molecules in such a gas have some features like measurements in the general discussion of quantum mechanical interpretation. As long as they do not involve quantum mechanical exchange of particles between molecules, inelastic collisions result in changes of the quantum mechanical state of the molecules, with attendant changes in the center of mass momenta of the collision partners. Thus changes in the quantum subsystems are associated with changes in classical variables (the centers of mass) which are playing a role here like classical "measurement apparatus." Collisions in which exchanges of particles between molecules are significant take the system from one of the terms in (5.22) to another and thus the assumption of "decoherence" breaks down in the presence of such collisions.

## Quantum perfect gases: general features

The quantum perfect gas is conceptually well defined by the same Hamiltonian studied for the classical case, suitably quantized

$$H = \sum_i \frac{-\hbar^2 \nabla_i^2}{2m} \tag{5.30}$$

and supplemented by the requirement of Bose or Fermi statistics for the wave functions. Experimental realization of this model is a much more difficult affair. This may be understood as follows. Though the reason for the success of the classical model for perfect gases is only understood in detail in terms of cluster expansions discussed in the next chapter, one can understand the physical argument as follows. In a dilute classical gas, the mean free path of the particles is much longer than the range of the interaction potentials. Thus the particles spend most of their time moving freely and very little of it in collision. Now consider the quantum case. Now the particles cannot be considered as localized. Indeed we saw in the semiclassical limit that they have an effective radius of the order of the thermal wavelength which diverges at low temperatures. Once the thermal wavelength exceeds the range of the interparticle interactions, the classical argument for the applicability of the perfect gas model to a gas of particles which are really interacting breaks down. It turns out that there is a completely different reason why the interactions can be neglected in many Fermi systems at very low temperatures. That will be discussed in Chapter 7. However, for Bose systems no such argument exists and it has been extremely difficult to find systems of atoms which act like perfect Bose gases. Finally however, there are both Fermi and Bose systems in which the interactions are essentially zero, namely neutrino and photon systems, so we do have accessible realizations of both cases. There is one more caveat, namely that these particles are massless so that the form (5.30) does not really apply. (This is not a serious problem.) But the masslessness also means that the number of particles is not conserved and, though this presents no computational difficulties, it changes the physics significantly, particularly in the case of bosons.

In studying quantum perfect gases, it is easier to work within the grand canonical density matrix. The partition function is

$$Z_{\mathrm{gc}} = \sum_{\{n_\nu\}} \mathrm{e}^{\sum_\nu n_\nu(\mu-\epsilon_\nu)\beta} \tag{5.31}$$

Here we use the fact that the eigenfunctions of (5.30) (or its generalizations to the case of massless particles) can be written as symmetrized or antisymmetrized products of $N$ one particle eigenstates $\phi_\nu$ satisfying

$$H_1\phi_\nu = \epsilon_\nu\phi_\nu \tag{5.32}$$

where we write the slightly more general form

$$H = \sum_{i=1}^{N} H_1(i) \tag{5.33}$$

for the Hamiltonian $H$. $n_\nu$ is the number of factors $\phi_\nu$ in the product and may be regarded as the number of particles "in" the state $\phi_\nu$. The number of particles $N$ is then $N = \sum_\nu n_\nu$ and the energy of a state characterized by a set $\{n_\nu\}$ is $E_{\{n_\nu\}} = \sum_\nu n_\nu\epsilon_\nu$ The equation (5.31) follows easily from $Z_{\mathrm{gc}} = \mathrm{Tr}\, \mathrm{e}^{\beta(N\mu-H)}$. In the case of symmetric wave functions (bosons) the statistics impose no constraints on the numbers $n_\nu$ but in the Fermi case they require $n_\nu = 0, 1$ only. (We include spin in the label $\nu$ here.) Thus the sums in (5.31) are easy to do

$$Z_{\mathrm{gc}} = \sum_{\{n_\nu\}} \prod_\nu \mathrm{e}^{n_\nu(\mu-\epsilon_\nu)\beta} = \prod_\nu \begin{cases} \frac{1}{1-\mathrm{e}^{(\mu-\epsilon_\nu)\beta}} & \text{bosons} \\ 1+\mathrm{e}^{(\mu-\epsilon_\nu)\beta} & \text{fermions} \end{cases} \tag{5.34}$$

In the boson case, we have summed a geometric series. The sum is convergent only if $\mathrm{e}^{(\mu-\epsilon_\nu)\beta} < 1$ for all values of $\epsilon_\nu$ (including 0). This is only possible if $\mathrm{e}^{\mu\beta} < 1$ which requires $\mu < 0$ for bosons quite generally. The thermodynamic potential is

$$\Omega = -k_{\mathrm{B}}T \ln Z_{\mathrm{gc}} = \pm k_{\mathrm{B}}T \sum_\nu \ln\left(1 \mp \mathrm{e}^{(\mu-\epsilon_\nu)\beta}\right) \tag{5.35}$$

From $\bar{N} = -(\partial\Omega/\partial\mu)_{T,V}$ we have

$$\bar{N} = \sum_{\nu} \frac{1}{\mathrm{e}^{(\epsilon_\nu-\mu)\beta} \mp 1} \tag{5.36}$$

This suggests that the summand is the average number of particles "in" the state $\nu$ (since the averaging is a linear process) and $N = \sum_\nu n_\nu$. One may also demonstrate this directly by calculating

$$\bar{n}_\nu = \frac{\sum_{\{n_{\nu'}\}} n_\nu\, \mathrm{e}^{-\sum_{\{\nu'\}} n_{\nu'}(\epsilon_{\nu'}-\mu)\beta}}{\sum_{\{n_{\nu'}\}} \mathrm{e}^{-\sum_{\{\nu'\}} n_{\nu'}(\epsilon_{\nu'}-\mu)\beta}} \tag{5.37}$$

The entropy is found from $S = -(\partial\Omega/\partial T)_{\mu,V}$

$$S = k_{\mathrm{B}} \sum_{\nu} \left( \mp \ln\left(1 \mp \mathrm{e}^{(\mu-\epsilon_\nu)\beta}\right) + \frac{\beta(\epsilon_\nu - \mu)}{\mathrm{e}^{(\epsilon_\nu-\mu)\beta} \mp 1} \right) \tag{5.38}$$

It is illuminating to rearrange this using the relation

$$\beta(\epsilon_\nu - \mu) = \ln(\bar{n}_\nu \pm 1) - \ln \bar{n}_\nu \tag{5.39}$$

which is not hard to prove, giving

$$S = k_{\mathrm{B}} \sum_{\nu} [(\bar{n}_\nu \pm 1)\ln(1 \pm \bar{n}_\nu) - \bar{n}_\nu \ln \bar{n}_\nu] \tag{5.40}$$

By use of Stirling's approximation, this form of $S$ can be shown to give the number of ways of distributing particles in the states $\nu$. To show this in detail requires a coarse graining of the energy scale in order to justify the use of Stirling's approximation (Problem 5.7).

One can now use (5.38) together with (5.35) to show that

$$\bar{E} = \Omega + ST + \mu\bar{N} = \sum_{\nu} \frac{\epsilon_\nu}{\mathrm{e}^{(\epsilon_\nu-\mu)\beta} \mp 1} \tag{5.41}$$

which is also obtained from the expression $E_{\{n_\nu\}} = \sum_\nu \epsilon_\nu n_\nu$ by use of the linear property of the average. The specific heat at fixed $\bar{N}$ (which is usually what is measured) is

$$C_{V,\bar{N}} = \left(\frac{\partial \bar{E}}{\partial T}\right)_{V,\bar{N}} = \left(\frac{\partial \bar{E}}{\partial T}\right)_{V,\mu} + \left(\frac{\partial \bar{E}}{\partial \mu}\right)_{V,T} \left(\frac{\partial \mu}{\partial T}\right)_{V,\bar{N}} \tag{5.42}$$

## Quantum perfect gases: details for special cases

To evaluate these expressions, we need to change the sum on single particle states $\nu$ to an integral and this requires some further specification of the Hamiltonian $H_1$ in

(5.33). We first consider the case of massive, nonrelativistic particles characterized by the one particle Hamiltonian (5.30). Then the eigenvalues $\epsilon_\nu$ are characterized by the wave vector $\vec{k}$ in three dimensions and we can change the sums on $\vec{k}$ to integrals as before, assuming periodic boundary conditions.

$$\sum_{\vec{k}}(\ldots) = \frac{V}{(2\pi)^3}\int \mathrm{d}\vec{k}(\ldots) \tag{5.43}$$

This step only works if there are no singularities in the summand. This is not as trivial a constraint as it might appear, as we will discuss shortly in the case of bosons. Because the various summands in the expressions for the thermodynamic quantities are functions only of the energy $\epsilon_{\vec{k}} = \hbar^2k^2/2m$, it is possible and very useful to express the integral on $\vec{k}$ as an integral on the energy $\epsilon$

$$\sum_{\vec{k}} F(\epsilon_{\vec{k}}) = \frac{V}{(2\pi)^3}\int \mathrm{d}\vec{k}\,F(\epsilon_{\vec{k}}) = \frac{V}{(2\pi)^3}\int_0^\infty 4\pi k(\epsilon)^2 F(\epsilon)\frac{\mathrm{d}k}{\mathrm{d}\epsilon}\mathrm{d}\epsilon$$
$$= \int_0^\infty \mathcal{N}(\epsilon)F(\epsilon)\,\mathrm{d}\epsilon \tag{5.44}$$

in which $\mathcal{N}(\epsilon)$ is called the density of single particle states and is given in this case by

$$\mathcal{N}(\epsilon) = \frac{Vm^{3/2}\epsilon^{1/2}}{2^{1/2}\pi^2\hbar^3} \tag{5.45}$$

Using these expressions, one can integrate the expression for $\Omega$ in (5.35) by parts in order to show that

$$\Omega = \frac{-2}{3}\bar{E} \tag{5.46}$$

so that

$$\bar{E} = \frac{3}{2}PV \tag{5.47}$$

It is probably useful to note that we can recover the semiclassical limit from these expressions. For example, if we suppose that the fugacity $z = \mathrm{e}^{\beta\mu}$ is $\ll 1$ then from (5.36), (5.44) and (5.45) one finds

$$\bar{N} = \frac{2}{\pi^{1/2}}\frac{V}{\lambda^3}z\int_0^\infty \frac{x^{1/2}\,\mathrm{d}x}{\mathrm{e}^x \mp z} \approx \frac{2}{\pi^{1/2}}\frac{V}{\lambda^3}z\int_0^\infty \frac{x^{1/2}\,\mathrm{d}x}{\mathrm{e}^x} = \frac{V}{\lambda^3}z = \frac{V}{\lambda^3}\mathrm{e}^{\beta\mu} \tag{5.48}$$

which is identical with (5.10). Using the semiclassical expression

$$\mu = -k_\mathrm{B}T\ln\left(\frac{a^3}{\lambda^3}\right) \tag{5.49}$$

shows that the condition $z \ll 1$ is identical to our previous condition $\lambda \ll a$ for the semiclassical limit.

Next we consider some useful properties of integrals which enter the theory of perfect quantum gases. We are often concerned with integrals of the form:

$$\int_0^\infty \frac{z^{x-1}}{e^z \pm 1} dz = \int_0^\infty z^{x-1} e^{-z} \sum_{n=0}^{\infty} (\mp)^n e^{-nz} dz = \sum_{n=0}^{\infty} (\mp)^n \int_0^\infty z^{x-1} e^{-z(n+1)} dz$$
$$= \sum_{n=0}^{\infty} \int_0^\infty y^{x-1} e^{-y} dy \frac{(\mp)^n}{(n+1)^x} = \sum_{n=0}^{\infty} \frac{(\mp)^n}{(n+1)^x} \Gamma(x) \qquad (5.50)$$

where the definition of the gamma function

$$\Gamma(x) = \int_0^\infty y^{x-1} e^{-y} dy \qquad (5.51)$$

has been used. By use of the definition

$$\zeta(x) = \sum_{n=1}^{\infty} \frac{1}{n^x} \qquad (5.52)$$

we have in the Bose case that

$$\int_0^\infty \frac{z^{x-1}}{e^z - 1} dz = \Gamma(x)\zeta(x) \qquad (5.53)$$

In the Fermi case one has

$$\sum_{n=1}^{\infty} (-)^{n+1} \frac{1}{n^x} = \sum_{n=1}^{\infty} \frac{1}{n^x} - 2 \sum_{n \text{ even}} \frac{1}{n^x} = \zeta(x) - \frac{2}{2^x} \sum_{n=1}^{\infty} \frac{1}{n^x} = (1 - 2^{1-x})\zeta(x) \qquad (5.54)$$

so that

$$\int_0^\infty \frac{z^{x-1}}{e^z + 1} dz = \Gamma(x)(1 - 2^{1-x})\zeta(x) \qquad (5.55)$$

These expressions are useful only if $x > 1$. The functions $\Gamma(x)$ and $\zeta(x)$ are listed in various tables and are available numerically for example in the software libraries IMSL and NAG. A few useful values are listed in the following table.

| $x$ | $\zeta(x)$ | $\Gamma(x)$ |
|---|---|---|
| 3/2 | 2.612 | $\sqrt{\pi}/2$ |
| 2 | $\pi^2/6$ | 1 |
| 5/2 | 1.314 | 3/4 |
| 3 | 1.202 | 2 |
| 5 | 1.037 | 24 |

Generally, $\Gamma(n) = (n-1)!$ for $n$ an integer $> 1$. The integrals

$$F_k(\eta) = \int_0^\infty \frac{z^k\,\mathrm{d}z}{\mathrm{e}^{(z-\eta)}+1} \tag{5.56}$$

are also tabulated.[1]

## Perfect Bose gas at low temperatures

We first consider the case of massive nonrelativistic particles for which the number is conserved. Consider the expression

$$\bar{N} = \sum_{\vec{k}} \frac{1}{\mathrm{e}^{(\epsilon_{\vec{k}}-\mu)\beta}-1} \tag{5.57}$$

where $\epsilon_{\vec{k}} = \hbar^2k^2/2m$. Making the conversion from a sum to an integral:

$$\bar{N} = \frac{Vm^{3/2}}{\sqrt{2}\pi^2\hbar^3}\int_0^\infty \frac{\epsilon^{1/2}}{\mathrm{e}^{(\epsilon-\mu)\beta}-1}\,\mathrm{d}\epsilon \tag{5.58}$$

Inspection of the integrand in the last expression shows that the integral is largest when $\mu = 0$ (recall that $\mu \le 0$ for bosons). Thus we apparently have the condition

$$\bar{N} \le \frac{Vm^{3/2}}{\sqrt{2}\pi^2\hbar^3}\int_0^\infty \frac{\epsilon^{1/2}}{(\mathrm{e}^{\epsilon\beta}-1)}\,\mathrm{d}\epsilon \tag{5.59}$$

But this is clearly unphysical, since the right hand side is finite and independent of $\bar{N}$ so that if it were correct, we could not form a Bose gas with a number density larger than the right hand side divided by $V$. Even worse, if one explores the right hand side, one sees that it decreases with decreasing temperature so that by lowering the temperature we can conclude that no Bose gas at any finite density could be formed at low enough temperature. The problem has occurred at the step taking us from (5.57) to (5.58). To see what has gone wrong consider the term in (5.57) corresponding to $\vec{k} = 0$. It is

$$n_{\vec{k}=0} = \frac{1}{e^{-\mu\beta} - 1} \tag{5.60}$$

as $\mu \to 0^-$ (which is the limit we took in getting (5.59)); this term diverges. On the other hand, in (5.59) this term has zero weight. Thus the treatment of the $\vec{k} = 0$ term by the continuum approximation must be incorrect. When, at fixed temperature, the right hand side of (5.59) is less than the number of particles, then we must treat the $\vec{k} = 0$ term in the sum (5.57) separately and take $\mu$ to have a value much less than $\hbar^2\vec{k}^2/2m$ for any finite $\vec{k}$ but such that (5.60) is big enough to make up the deficit in the number of particles left by the right hand side of (5.59). (It is useful to think about whether this can be done consistently in the case that the volume becomes large. It is not hard to show that the ratio of $\mu$ to the smallest finite value of $\hbar^2\vec{k}^2/2m$ scales as $V^{-1/3}$ so that for large enough volumes, a $\mu$ satisfying both conditions exists (Problem 5.8).) In that case, the sum on $\vec{k}$ for $\vec{k} \neq 0$ has the same value as before but there is an added term from $n_{\vec{k}=0}$ in the sum for $\bar{N}$. As before, this leads to an expression for $\mu$ in terms of $\bar{N}$ but the scaling of $\mu$ with $\bar{N}$ is unusual. All these new features must be invoked at temperatures below the temperature $T_0$ at which the right hand side of (5.59) is exactly equal to the number $\bar{N}$ of particles. $T_0$ is evaluated by making use of the integrals discussed in the last section:

$$\bar{N}/V = \frac{m^{3/2}}{\sqrt{2}\pi^2\hbar^3}\int_0^\infty \frac{\epsilon^{1/2}\,d\epsilon}{e^{\epsilon\beta_0} - 1} = \frac{m^{3/2}}{\sqrt{2}\pi^2\hbar^3}\beta_0^{-3/2}\int_0^\infty \frac{x^{1/2}\,dx}{e^x - 1} \tag{5.61}$$

Defining the number density by $\rho = \bar{N}/V$ and rearranging this one shows that

$$T_0 = \frac{2\pi}{\zeta(3/2)^{2/3}}\frac{\hbar^2}{k_B m}\rho^{2/3} \approx \frac{3.31\hbar^2}{mk_B}\rho^{2/3} \tag{5.62}$$

Apart from factors of order unity this is easily understood as the temperature at which the approximation $a^3 \gg \lambda^3$ breaks down. What is remarkable is that there is a sharp change in the properties of the model at $T = T_0$. Indeed this is our first example of a phase transition, the Bose–Einstein condensation, and remains one of the few phase transitions for which we have exact mathematical solutions. For $T < T_0$, the equation for the number of particles is

$$\bar{N} = n_{\vec{k}=0} + \sum_{\vec{k}\neq 0}\frac{1}{e^{\epsilon_{\vec{k}}\beta} - 1} = \bar{n}_0 + \frac{Vm^{3/2}}{2^{1/2}\pi^2\hbar^3}\int_0^\infty \frac{\epsilon^{1/2}\,d\epsilon}{e^{\epsilon\beta} - 1} \tag{5.63}$$

The integral is evaluated in exactly the same way as before with the result

$$\bar{n}_0 = \left(1 - \left(\frac{T}{T_0}\right)^{3/2}\right)\bar{N} \tag{5.64}$$

The specific heat for $T < T_0$ is

$$C_{V,N} = \left(\frac{\partial E}{\partial T}\right)_{V,N} = \left(\frac{\partial E}{\partial T}\right)_{V,\mu} \tag{5.65}$$

since the second term in (5.42) can be shown to be zero. Then

$$\begin{aligned} C_{V,N} &= \frac{\partial}{\partial T}\left(\frac{Vm^{3/2}}{2^{1/2}\pi^2\hbar^3}\int_0^\infty \frac{\epsilon^{3/2}\,\mathrm{d}\epsilon}{\mathrm{e}^{\epsilon\beta}-1}\right) = \frac{\partial}{\partial T}T^{5/2}\frac{Vm^{3/2}k_\mathrm{B}^{5/2}}{2^{1/2}\pi^2\hbar^3}\int_0^\infty \frac{x^{3/2}\,\mathrm{d}x}{\mathrm{e}^x-1} \\ &= \frac{5}{2}\frac{Vk_\mathrm{B}^{5/2}m^{3/2}T^{3/2}}{2^{1/2}\pi^2\hbar^3}\Gamma(5/2)\zeta(5/2) \end{aligned} \tag{5.66}$$

For $T > T_0$ the second term in (5.42) is not zero and it is necessary to calculate $\mu$ in order to obtain the specific heat at fixed particle number. This is of interest because it shows that a singularity occurs in this thermodynamic quantity, characteristic of a phase transition. One might think that it would be possible for $T$ just above $T_0$ to make an expansion in the expression for $\bar{N}$ as a function of $\mu$, which is expected to be small at those temperatures. It turns out, however, that the leading term in $\mu$ is sublinear so that this procedure does not work. Instead we add and subtract the value at $\mu = 0$:

$$\bar{N} = \frac{Vm^{3/2}}{2^{1/2}\pi^2\hbar^3}\left(\int_0^\infty \frac{\epsilon^{1/2}\,\mathrm{d}\epsilon}{\mathrm{e}^{\epsilon\beta}-1} + \int_0^\infty\left(\frac{\epsilon^{1/2}}{\mathrm{e}^{(\epsilon-\mu)\beta}-1} - \frac{\epsilon^{1/2}}{\mathrm{e}^{\epsilon\beta}-1}\right)\mathrm{d}\epsilon\right) \tag{5.67}$$

which is an identity. The second, $\mu$ dependent integral has its largest contribution for the small $\epsilon$ region of the integral. The leading term in $\mu$ is obtained by expanding the integrand for small $\mu$ and small $\epsilon$:

$$\bar{N} = \frac{Vm^{3/2}}{2^{1/2}\pi^2\hbar^3}\left(\int_0^\infty \frac{\epsilon^{1/2}\,\mathrm{d}\epsilon}{\mathrm{e}^{\epsilon\beta}-1} + k_\mathrm{B}T\mu\int_0^\infty \frac{\epsilon^{1/2}}{(\epsilon-\mu)\epsilon}\mathrm{d}\epsilon\right) \tag{5.68}$$

The second integral is transformed to a familiar form by the transformations $y^2 = \epsilon/\mid\mu\mid$ and is $\pi/\sqrt{\mid\mu\mid}$. The first term can be written in the form $\bar{N}(T/T_0)^{3/2}$. Thus

$$\sqrt{\mid\mu\mid} = \bar{N}\left(\left(\frac{T}{T_0}\right)^{3/2} - 1\right)\frac{2^{1/2}\pi\hbar^3}{Vm^{3/2}k_\mathrm{B}T} \tag{5.69}$$

Finally we use (5.61) in the form

$$\frac{\bar{N}}{V} = (k_\mathrm{B}T_0)^{3/2}\frac{m^{3/2}}{2^{1/2}\pi^2\hbar^3}\zeta(3/2)\Gamma(3/2) \tag{5.70}$$

with the result that

$$\mu = -k_{\mathrm{B}} T \left( \left( \frac{T}{T_0} \right)^{3/2} - 1 \right)^2 \left( \frac{\zeta(3/2)\Gamma(3/2)}{\pi} \right)^2 \tag{5.71}$$

It turns out that a Taylor expansion of the energy for small $\mu$ also fails. A correct result is obtained by using

$$\bar{E} = -\frac{3}{2}\Omega \tag{5.72}$$

so that

$$\left( \frac{\partial \bar{E}}{\partial \mu} \right)_{T,V} = \frac{-3}{2} \left( \frac{\partial \Omega}{\partial \mu} \right)_{T,V} = \frac{3\bar{N}}{2} \tag{5.73}$$

Then using (5.68) in the form

$$\bar{N} = \frac{V m^{3/2}}{2^{1/2}\pi^2\hbar^3} \left( \int_0^\infty \frac{\epsilon^{1/2}\,\mathrm{d}\epsilon}{\mathrm{e}^{\epsilon\beta} - 1} + \mathcal{O}(\mid \mu \mid^{1/2}) \right) \tag{5.74}$$

we obtain

$$\bar{E} = E_0(T) + 3\frac{V m^{3/2}\mu}{2^{3/2}\pi^2\hbar^3} \int_0^\infty \frac{\epsilon^{1/2}\,\mathrm{d}\epsilon}{\mathrm{e}^{\epsilon\beta} - 1} + \mathcal{O}(\mid \mu \mid^{3/2}) \tag{5.75}$$

Then using our expression for $\mu$:

$$C_V(T) = C_V^{(0)}(T) + \frac{9 V m^{3/2} k_{\mathrm{B}}^{5/2} T_0^{3/2}}{2^{3/2}\pi^4\hbar^3} \left( 1 - \left( \frac{T}{T_0} \right)^{3/2} \right) (\zeta(3/2)\Gamma(3/2))^3 \tag{5.76}$$

Note that there is no discontinuity in $C_V$ but there is a discontinuity in its slope at $T_0$.

Bose–Einstein condensation remained a theoretical curiosity, subject to some controversy in the earliest times, from the introduction of the concept in 1925 by Albert Einstein[2] until its experimental realization in a rubidium vapor in 1995.[3] As discussed briefly above, the experimental difficulty is that the stable phase of all real monatomic systems below the Bose–Einstein condensation temperature is a solid at all densities. Therefore, Bose–Einstein condensation in a dilute vapor can only be observed in a kind of metastable quasiequilibrium, in which there are sufficient interactions between the particles to achieve equilibrium of the kinetic energy between the atoms of the gas, but the collisions are rare enough and the gas is sufficiently isolated to prevent the initiation of freezing into the solid state.

In the successful experiments, isolation was achieved by use of optical and magnetic traps, and cooling was carried out in the last stages by a kind of radiation

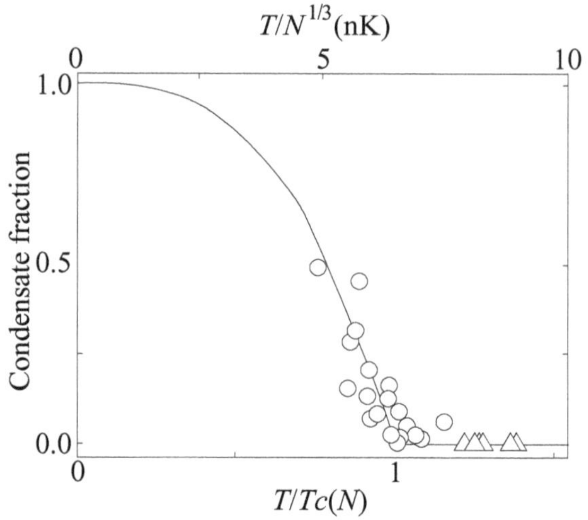


Figure 5.3 Condensate fraction $N_0/N$ measured as a function of $T/N^{1/3}$ for $5 \times 10^6$ sodium atoms trapped in a spherical harmonic well, compared with theory. The solid line is the prediction in the thermodynamic limit for atoms in a harmonic trap (see Problem 5.13). From reference 4 by permission.

induced stripping of the atoms with the highest energies from the trap. The trap imposed a harmonic oscillator potential on the gas, so the analysis just described needs to be modified to take that into account. Although the effects of interactions need to be taken carefully into account for a full analysis, it does turn out that the noninteracting theory provides a good account of the observed phase. For example, we show the measured condensate fraction compared with the noninteracting result in Figure 5.3.

## Perfect Fermi gas at low temperatures

Though there is no phase transition in the noninteracting Fermi gas model, the properties at low temperatures require careful treatment of the integrals. This is basically because the function $1/(e^{\beta(\epsilon-\mu)}+1)$ develops a singularity as $\beta \to \infty$ at $\epsilon = \mu$. As a result, low temperature expansions, though entirely possible, require some care. We consider the integral

$$I(\mu, T) = \int_0^\infty \frac{f(\epsilon)\,d\epsilon}{e^{\beta(\epsilon-\mu)}+1} \tag{5.77}$$

which is of the general type which is encountered. At zero temperature the integral is $I(\mu, T\!=\!0) = \int_0^{\mu(T=0)} f(\epsilon)\,d\epsilon$. (It is clear that $\mu(T=0) > 0$ here, because otherwise we could not satisfy requirements on the total number of particles.) The object of the analysis is to write $I(\mu, T)$ as $I(\mu, T = 0)$ + correction terms which can be written as a series in $T$. To accomplish this we change the variable to $z = \beta(\mu - \epsilon)$ and rearrange the integral as follows:

$$I(\mu, T) = -k_B T \int_{\beta\mu}^{-\infty} \frac{f(\mu - k_B T z)\,dz}{e^{-z} + 1}$$
$$= -k_B T \left[ \int_{\beta\mu}^{0} \frac{f(\mu - k_B T z)\,dz}{e^{-z} + 1} + \int_{0}^{-\infty} \frac{f(\mu - k_B T z)\,dz}{e^{-z} + 1} \right] \tag{5.78}$$

Now we rewrite

$$\frac{1}{e^{-z} + 1} = 1 - \frac{1}{e^{z} + 1} \tag{5.79}$$

in the first term giving three terms:

$$I = -k_B T \int_{\beta\mu}^{0} f(\mu - k_B T z)\,dz + k_B T \int_{\beta\mu}^{0} \frac{f(\mu - k_B T z)\,dz}{e^{z} + 1}$$
$$- k_B T \int_{0}^{-\infty} \frac{f(\mu - k_B T z)\,dz}{e^{-z} + 1} \tag{5.80}$$

Now change the variable in the first term back to $\epsilon$ and set $z \to -z$ in the third term:

$$I = \int_0^{\mu} f(\epsilon)\,d\epsilon - k_B T \int_0^{\beta\mu} \frac{f(\mu - k_B T z)\,dz}{e^{z} + 1} + k_B T \int_0^{\infty} \frac{f(\mu + k_B T z)\,dz}{e^{z} + 1} \tag{5.81}$$

The first term looks much like the $T \to 0$ limit. In the second term we may set $\beta\mu \to \infty$ to first order in $e^{-\beta\mu}$. As long as $k_B T \ll \mu(T = 0)$ this will produce errors which are much smaller than those associated with cutting off the series in $k_B T/\mu(T = 0)$ which we will find. (In particular let $x = k_B T/\mu(T = 0)$. It is not hard to show that for values of $x$ up to $x_c$, we can ignore terms of order $e^{(-1/x)}$ compared to all terms in the power series $\sum_n A_n x^n$ for which $n < -1/x_c \ln x_c$.) Next we expand the first term in $\mu(T) - \mu(0)$ giving

$$I = \int_0^{\mu(0)} f(\epsilon)\,d\epsilon + (\mu(T) - \mu(0)) f(\mu(0)) + \mathcal{O}((\mu(T) - \mu(0))^2)$$
$$+ k_B T \left[ \int_0^{\infty} \frac{f(\mu + k_B T z) - f(\mu - k_B T z)}{e^{z} + 1} dz + \mathcal{O}(e^{-\beta\mu}) \right] \tag{5.82}$$

Finally we obtain the required series by expanding the last term in powers of $T$. The integrand is odd in $z$ so only the odd terms in $z$ survive the expansion of the numerator, but the extra power of $k_B T$ outside the integral means that the expansion contains only even powers in $k_B T$:

$$I = \int_0^{\mu(0)} f(\epsilon)\, d\epsilon + (\mu(T) - \mu(0)) f(\mu(0)) + \mathcal{O}((\mu(T) - \mu(0))^2)$$
$$+ 2f'(\mu(T))(k_B T)^2 \int_0^\infty \frac{z\, dz}{e^z + 1} + \frac{2f'''(\mu(T))}{3!}(k_B T)^4 \int_0^\infty \frac{z^3\, dz}{e^z + 1} + \cdots \tag{5.83}$$

We apply this first to calculation of $\mu(T)$ at low temperatures. Then $I = N$, the number of particles. $f(\epsilon) = \mathcal{N}(\epsilon)$, the density of states which, without spin degeneracy, is $Vm^{3/2}\epsilon^{1/2}/\sqrt{2}\pi^2\hbar^3$. Then at zero temperature we have

$$N = (2/3)Vm^{3/2}\mu(0)^{3/2}/\sqrt{2}\pi^2\hbar^3 \tag{5.84}$$

which gives the standard expression for $\mu(0)$, conventionally called the Fermi energy

$$\mu(0) \equiv \epsilon_F = \left(\frac{3\pi^2 N}{V}\right)^{2/3} \left(\frac{\hbar^2}{2m}\right) 2^{2/3} \tag{5.85}$$

(The factor $2^{2/3}$ disappears in the common case that one includes a factor 2 in the density of states to account for the spin degeneracy of electrons.) To obtain the leading low temperature corrections we use the next terms in (5.83):

$$N = \int_0^{\epsilon_F} \mathcal{N}(\epsilon)\, d\epsilon + (\mu(T) - \mu(0))\mathcal{N}(\mu(0)) + 2(k_B T)^2$$
$$\times \left[\mathcal{N}'(\mu(0)) + (\mu(T) - \mu(0))\mathcal{N}''(\mu(0)) + \cdots\right] \int_0^\infty \frac{z\, dz}{e^z + 1} + \mathcal{O}((k_B T)^4) \tag{5.86}$$

The first term on the right cancels the $N$ on the left. The next two terms show that $\mu(T) - \mu(0)$ is of order $(k_B T)^2$ so that the second term in [. . .] may be dropped at low $T$. Then using $\mathcal{N}'/\mathcal{N} = 1/2\epsilon$ one has

$$\mu(T) - \mu(0) = \frac{-(k_B T)^2}{\mu(0)} \int_0^\infty \frac{z\, dz}{e^z + 1} = \frac{-(k_B T)^2}{\mu(0)} \frac{\pi^2}{12} \tag{5.87}$$

using

$$\int_0^\infty \frac{z\, dz}{e^z + 1} = \Gamma(2)(1 - (1/2))\zeta(2) \tag{5.88}$$

To calculate the specific heat we calculate the energy using (5.83) with $f = \epsilon\mathcal{N}(\epsilon)$

$$\bar{E}(T) = \bar{E}(0) + (\mu(T) - \mu(0))\mathcal{N}(\mu(0))\mu(0) + 2(k_B T)^2 \left(\frac{3}{2}\right)\mathcal{N}(\mu(0))\frac{\pi^2}{12} \tag{5.89}$$

and using (5.87)

$$\bar{E}(T) = \bar{E}(0) + (k_B T)^2 \mathcal{N}(\mu(0))\frac{\pi^2}{6} \tag{5.90}$$

Here $\bar{E}$ has been expressed in terms of $N$ to leading order in $T$ ($\mu$ has been eliminated) so that we can compute the specific heat at constant volume and particle number directly from (5.90):

$$C_{V,N} = k_B \frac{\pi^2}{3} k_B T \mathcal{N}(\mu(0)) \tag{5.91}$$

(The factors associated with degeneracy are buried in $\mathcal{N}$ here so this formula is quite general.) As is well known, the violation of the equipartition theorem at low temperatures here was important historically in establishing that the ideal Fermi gas model was useful for modeling electrons in metals. Establishment of the reasons for the usefulness of the model in this strongly interacting system came later and we do not discuss it now.

## References

1. R. E. Dingle, *Applied Scientific Research* **B62** (1957) 25.
2. A. Einstein, *Sitzungsberichte der Preussischen Akademie der Wissenschaften* **1925** (1925) 3.
3. M. H. Anderson, J. R. Ensher, M. R. Matthews, C. E. Wieman and E. A. Cornell, *Science* **269** (1995) 198.
4. M.-O. Mewes, M. R. Andrews, N. J. van Druten, D. M. Kurn, D. S. Durfee and W. Ketterle, *Physical Review Letters* **77** (1996) 416.

## Problems

5.1 Carry out the kinetic theory estimates as described qualitatively in the text in order to estimate the rate of equilibration of a gas of density $\rho$, kinetic energy per particle $(3/2)k_B T$ and collision cross-section $\sigma$. Then confirm the condition under which the gas can be described as ideal by requiring that interaction energy be much smaller than the kinetic energy.

5.2 Find the specific heat $C_{V,\mu}$ for a perfect gas. Under what circumstances could the difference between $C_{V,N}$ and $C_{V,\mu}$ be observed experimentally? Try to describe an experiment in which this might be done.

5.3 Consider two hydrogen molecules with particle labels 1, 2, 3, 4 for the four protons. Illustrate the discussion of possible permutations by explicitly grouping the 24 permutations of the proton labels into sets corresponding to permutations of labels within molecules, permutations of entire sets of labels between molecules and permutations associated with different assignments of labels to molecules. Indicate the sign of each permutation and indicate a set of permutations which would suffice for describing the wave function (5.22) and one which would suffice for describing the semiclassical limit for the centers of mass of the molecules.

5.4 Work out the specific heat of a gas of deuterium molecules $D_2$. The nuclear spin of each nucleus is 1. Consider the case of equilibrium and the case of unequilibrated mixtures as discussed for $H_2$. To specify the rotational constant, let $\hbar^2/2Ik_BT = Bh/k_BT$ where $B = \hbar/2\pi I$ is a frequency. For deuterium $B = 0.912 \times 10^{12}\,\mathrm{s}^{-1}$ Make a graph of the specific heat as a function of temperature in each case, for the region between zero and room temperature.

5.5 Show that $E = (3/2)PV$ for a classical monatomic perfect gas.

5.6 Consider a system of noninteracting bosons with energies

$$\epsilon_{\vec{k}} = \hbar ck(1 - \alpha_1 k - \alpha_2 k^2)$$

Find an expansion of the form

$$C_V(T) = A_0T^3 + A_1T^4 + A_3T^5$$

for the specific heat at low temperatures and evaluate the coefficients in terms of the parameters given. Assume that the number is not fixed.

5.7 Show that the entropy (5.40) is in fact the number of ways of distributing particles among the one particle levels $\nu$ for a given set $n_\nu$. You must use a coarse graining in the scale of the quantum numbers $\nu$ and Stirling's approximation. Then show that you get (5.40) by maximizing this expression for $S$ at fixed energy and particle number.

5.8 Show that for $T < T_0$ the ratio of the smallest nonzero value of $\hbar^2k^2/2m$ to $\mu$ scales as $V^{-1/3}$ and make a clear statement of why this justifies the procedures used to describe the statistical mechanics of the Bose gas below $T_0$.

5.9 Show that, during an adiabatic process in the photon gas, $PV^3$ remains constant.

5.10 Consider an ideal gas of atoms obeying the following variant of the Pauli principle. Each solution of the one particle Schrödinger equation $\phi_\nu$ is allowed to appear $n_\nu = 0$, 1 or 2 times in the products which are used to describe the many body wave function, but not more. (You do not need to worry about the form of the wave functions, but only to assume that the many body energy eigenvalues are of the form $\sum_\nu n_\nu \epsilon_\nu$ with $n_\nu = 0$, 1 or 2.)

(a) Write down expressions for the grand canonical partition function $Z_{gc}$ and the thermodynamic potential $\Omega$ in terms of the temperature $T$, the chemical potential $\mu$ and the eigenenergies $\epsilon_\nu$ of the one particle Schrödinger equation.

(b) Use the result to write an expression for the average number of particles $\langle N \rangle$ and the average energy $\langle E \rangle$ in terms of the same quantities. Make a qualitatively correct sketch of the function $\langle n_\nu \rangle$ as a function of $\epsilon_\nu$ at low temperatures and, on the same graph, of the same function for the same values of $\langle N \rangle$ and the same one particle Hamiltonian for the case of fermions.

(c) Develop a low temperature expansion for the quantities $\langle N \rangle$ and $\langle E \rangle$, following the same general lines that were used for fermions in the text. Find explicit expressions for the first terms involving finite temperature corrections to the low temperature result, expressing the coefficients in terms of the quantities given and dimensionless integrals and the density of states of the eigenenergies $\mathcal{N}(\epsilon)$. (But do not try to evaluate the dimensionless integrals.) Give the low temperature specific heat in terms of the same quantities.

(d) Now consider, qualitatively only, the cases in which the allowed values of $n_\nu$ are $n_\nu = 0, 1, \ldots, n$ where $n$ is a finite number $n \geq 2$ . How do you expect the function $\langle n_\nu(\epsilon)\rangle$ to look as a function of $\epsilon$ at fixed $N$ and low (not zero) temperature? Make a graph like the one you drew for the last part of part (b), showing how you think the function will look for a series of increasing values of $n$, at a fixed low temperature.

5.11 Consider an ideal Bose gas of nonrelativistic particles of mass $m$ in a world of four spatial dimensions.

(a) Demonstrate that Bose condensation occurs and write an expression for the temperature $T_0$ below which it occurs, as a function of the number $N$ of particles, the four dimensional volume $V_4$ of the system, mass $m$ and Planck's constant.

(b) Write an expression which determines the chemical potential $\mu$ in terms of the variables named above when the temperature is just above the transition temperature $T_0$. Describe the behavior of $\mu$ at fixed $N$ as $T$ approaches $T_0$ from above in as much detail as you can.

(c) Find the specific heat at fixed $V$ and $N$ as a function of $T$ below $T_0$.

(d) Find an expression for the specific heat just above $T_0$ in terms of the analytical continuation of the expression found in part (c) plus a correction expressed in terms of the chemical potential. Using this result and the results of part (b) show that the specific heat again exhibits a cusp at $T_c$, as it did in three dimensions.

5.12 Find the second nonvanishing term in a series in powers of the temperature for the specific heat of an ideal Fermi gas.

5.13 Consider a large number $N$ of atoms of mass $m$ trapped in a spherical harmonic trap (potential $Kr^2/2$). Find the dependence of the Bose–Einstein condensation temperature $T_0$ on $N$ and the dependence of the condensate fraction $N_0$ on $T_0$, $N$ and $T$. (See Figure 5.3 and reference 4 for application.)

# 6

# Imperfect gases

Here we introduce interactions between particles, beginning with the classical case. In practice we will call a system an imperfect gas when it is sufficiently dilute so that an expansion of the pressure in a power series in the density converges reasonably quickly. This series is called the virial series and we will introduce it in this chapter. This definition of an imperfect gas thus can depend on the temperature. If the power series in the density does not converge we may refer loosely to the system as a liquid, as long as it does not exhibit long range order characteristic of various solids and liquid crystals. The experimental distinction between a gas and a liquid will be discussed more precisely in Chapter 10.

We will develop the virial series for a classical gas in two different, but equivalent, ways here. In the first method we develop a series for the partition function $Z$ using the grand canonical distribution. By making a partial summation of this series we get a series in the fugacity. In the second method we study a series for the free energy $F = -k_B T \ln Z$ and use the canonical ensemble. Though the two methods are equivalent, we discuss them both in order to provide an opportunity to introduce several concepts common in the statistical mechanical literature.

The classical virial series will clarify more precisely than we were able to do in the last two chapters the conditions under which a gas can be treated as perfect or ideal. It also makes systematic corrections for nonideal behavior possible as a series in the density.

At the end of this chapter we introduce quantum virial expansions and the Gross–Pitaevskii–Bogoliubov low temperature theory of a weakly interacting Bose gas well below its Bose–Einstein condensation temperature.

## Method I for the classical virial expansion

We begin with the classical Hamiltonian

$$H_N = \sum_{i=1}^{N} p_i^2/2m + \sum_{i<j} v_{ij} \tag{6.1}$$

where $v_{ij}$ is a *pairwise* potential energy of range much smaller than the size of the system. These are significant constraints both from the theoretical and the experimental point of view. Though the experimental systems of interest in non-relativistic statistical mechanics all interact, basically, via pairwise Coulomb interactions, this interaction is not of short range. Further, if one attempts to represent the interactions between atoms or molecules via effective atomic interactions (effectively "integrating out" the electronic degrees of freedom) then the resulting interatomic forces are often not pairwise but involve significant three and more body terms. Furthermore the requirement of pairwise, short range forces is quite essential theoretically. The development of cluster expansions for forces which involve three or more bodies at once is possible but substantially more complicated than what follows. In systems interacting via Coulomb interactions one can often show that screening makes the effective interactions short range but this is not a trivial exercise and we will not go into it in this chapter. In short, the constraints on the model are significant, but the systems to which they apply in good approximation are also quite abundant and the insights provided by the study are very valuable.

For concreteness, we mention here a common form for modeling the interaction potential between the atoms of a monatomic gas:

$$v_{ij} = 4\epsilon_{\text{LJ}} \left[ \left( \frac{\sigma}{r_{ij}} \right)^{12} - \left( \frac{\sigma}{r_{ij}} \right)^{6} \right] \tag{6.2}$$

This is called the Lennard-Jones interaction potential. By use of the two parameters $\epsilon$ and $\sigma$ it can be made to match the results of first principles calculations between many closed shell atoms moderately well. The physics of the interaction is quite clear: the short range repulsion describes the effects of the fact that the shells of the atoms are closed, resulting in a large energetic penalty for close approach, since the electrons of each atom cannot occupy low lying levels of its neighbor. The long range attraction in the Lennard-Jones interaction is of the form expected for the van der Waals interaction. Typical and important for our purpose, in addition to the fact that this interaction is widely used, is the fact that it is extremely divergent as $r_{ij} \to 0$. The potential is not integrable and has no integrable moments up to very high order.

In this chapter we will only consider thermodynamic quantities and consider the grand partition function written

$$Z_{gc} = \sum_{N=1}^{\infty} z^N Z_N \tag{6.3}$$

where $z = e^{\beta\mu}$ is the fugacity and $Z_N$ is the canonical partition function

$$Z_N = \frac{1}{N!h^{3N}} \int d^N\vec{p}\, d^N\vec{r}\, e^{-\beta H_N} \tag{6.4}$$

The integrals on momenta can be done at once as in the last chapter giving

$$Z_N = \frac{1}{N!\lambda^{3N}} \int d^N\vec{r}\, e^{-\beta \sum_{i<j} v_{ij}} \tag{6.5}$$

We wish to produce an expansion of this quantity in the density in a way that takes implicit account of the fact that the interactions, though strong, are of short range, so that at low densities, the particles only spend a small fraction of the time within range of the forces and the leading term in the expansion is the one appropriate to a perfect gas. For this purpose it is immediately clear that an expansion in the potential energy $v$ would not work well at all. Any formulation which expands the exponent in (6.5) will end up with integrals of $v(r_{ij})$ which are very large or divergent for the kinds of system of interest here. Even if such infinities could be controlled they would not express the physics of rare collisions of which we wish to take account in the expansion. Instead one considers the quantity

$$f_{ij} = e^{-\beta v_{ij}} - 1 \tag{6.6}$$

This has the attractive feature that it is not divergent as $r_{ij} \to 0$, even if, as is often the case, $v_{ij} \to \infty$ as $r_{ij} \to 0$. Furthermore, its integral on a volume element is finite for all reasonable potential functions and is of the order of the volume of a sphere over which the two atoms in question interact. Thus an expansion of the thermodynamic quantities in terms of such integrals times the density would express the fact that the volume per particle is large compared to the range of interaction of the particle. With this motivation we write the partition function in terms of the quantities $f_{ij}$. Note that we can write

$$Z_N = \frac{1}{N!\lambda^{3N}} \int d^N\vec{r} \prod_{i<j}(1 + f_{ij}) \tag{6.7}$$

The first few terms in such an arrangement of the integrand have the form

$$\prod_{i<j}(1+f_{ij}) = 1 + \sum_{i<j} f_{ij} + \sum_{i<j,k<l,i\neq l \text{ if } j=k; i\neq k \text{ if } j=l} f_{ij} f_{kl} + \cdots \tag{6.8}$$

To deal systematically with such a series it is useful to introduce a diagrammatic notation. One represents every term in the series for $Z_N$ by a series of $N$ circles. Each circle represents one (vector) coordinate over which the integrand must be integrated and its label goes inside the circle. To take account of the factors $f_{ij}$, one connects the two circles labelled $i$ and $j$ in the diagram by a line. Because the product on the left hand side of (6.8) contains each pair only once, this means that each pair of circles is directly connected by at most one line, though each circle may be connected to many different lines going to different circles. With these conventions, the first term in (6.8) is represented simply by $N$ circles:

(6.9)

The second term in (6.8) is

$$\sum_{i<j} f_{ij} =$$

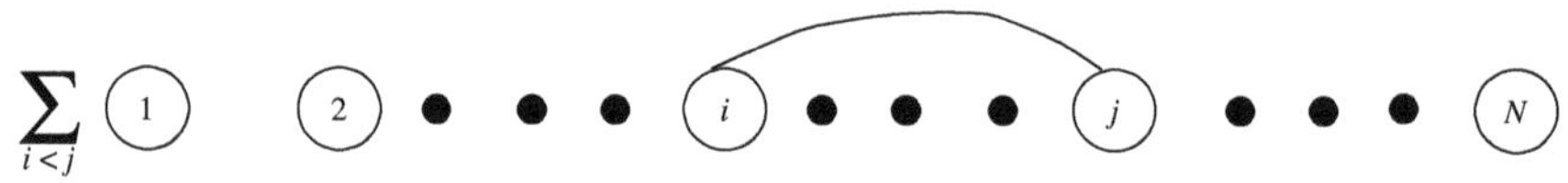


(6.10)

(We will often rearrange diagrams so that circles connected by lines are adjacent to one another.) It is clear that integrating with respect to coordinates associated in a diagram with circles not connected to any lines is easy. One just multiplies by a factor of the volume $V$ of the system for each such unconnected circle. We will express the partition function in terms of the *linked* parts of such diagrams, which consist of parts which contain no unconnected circles and which in addition are entirely connected in the following sense. We formally define a *linked l-cluster* as the integrand of a term in the series for $Z_l$ every circle of which is attached to at least one line in such a way that the diagram cannot be separated without cutting the line. The only linked 2-cluster is

$f_{12} =$ 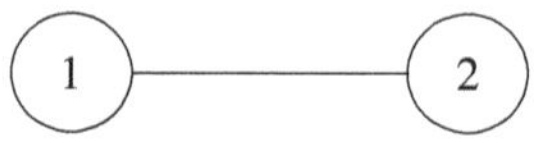

The linked 3-clusters are the following

$f_{12}f_{13} =$

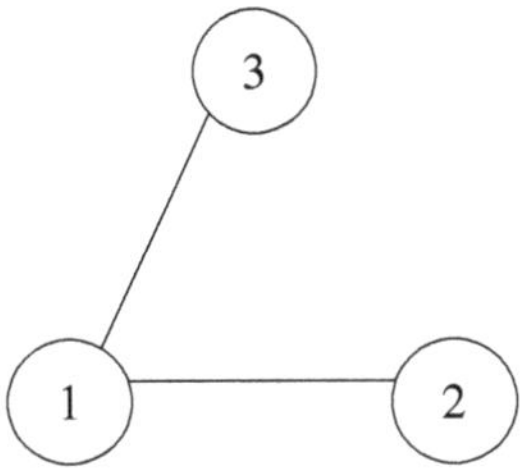


$f_{13}f_{32} =$

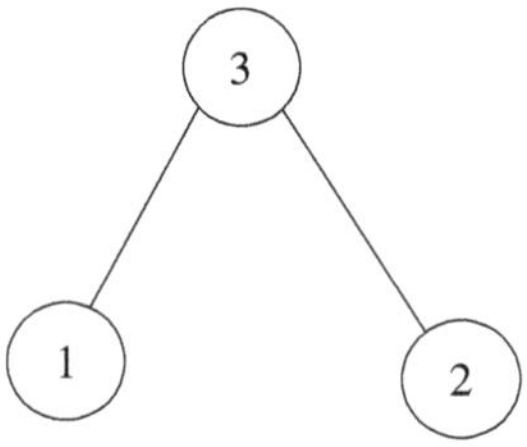


$f_{12}f_{23} =$

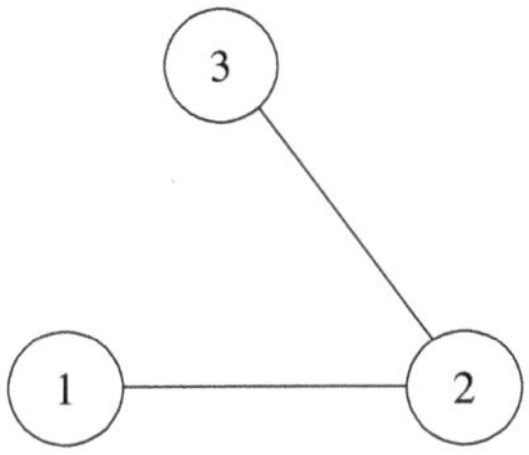


$f_{12}f_{13}f_{23} =$

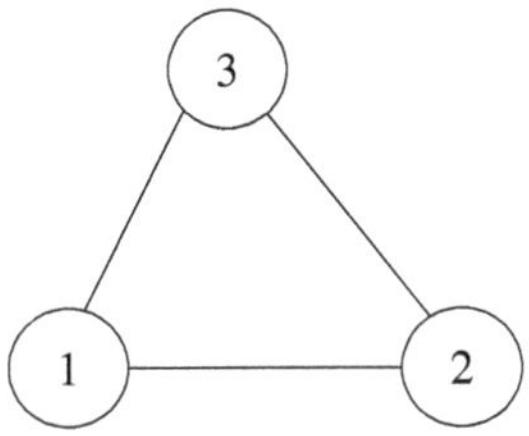


whereas

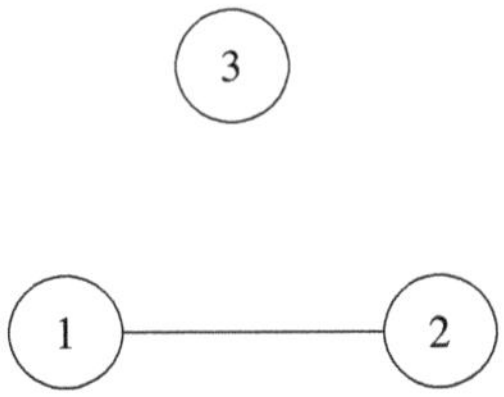


is an example of an unlinked 3-cluster. The linked $l$-clusters can only be integrated if we insert an explicit form for the potential. It turns out to be possible to express the series of $Z_N$ in terms of an appropriately defined sum of integrals of integrands corresponding to linked $l$-clusters. We begin by defining $b_l$ as

$$b_l = \frac{1}{\lambda^{3l-3} l! V} \int d^l \vec{r} \quad \{\text{the sum of all distinct linked } l\text{-clusters}\} \tag{6.11}$$

where we define the distinct $l$-clusters as those linked $l$-clusters which are found by drawing links between labelled points in all possible ways without having any line go between two points more than once. For example

$$b_2 = \frac{1}{2\lambda^3 V} \int d\vec{r}_1\, d\vec{r}_2$$

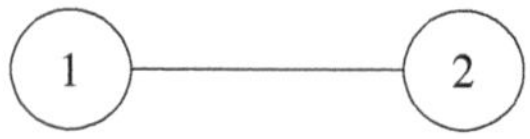


$$= \frac{1}{2\lambda^3 V} \int d\vec{r}_1\, d\vec{r}_2\, f_{12}$$

$$b_3 = \frac{1}{6\lambda^6 V} \int d\vec{r}_1\, d\vec{r}_2\, d\vec{r}_3 \tag{6.12}$$

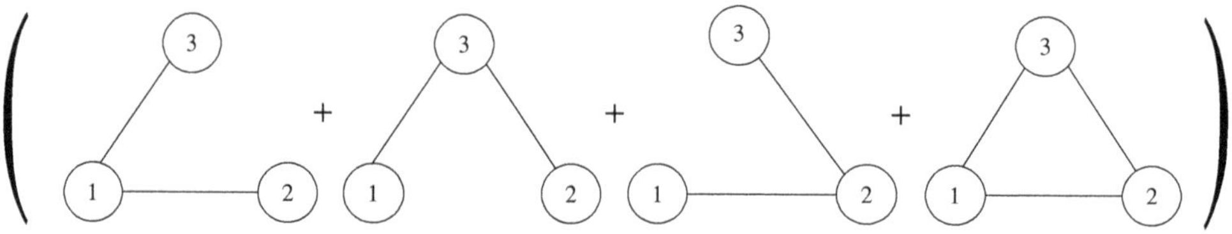


$$= \frac{1}{6V\lambda^6} \int d\vec{r}_1\, d\vec{r}_2\, d\vec{r}_3 (f_{12} f_{23} f_{13} + 3 f_{12} f_{13}) \tag{6.13}$$

Every term in $Z_N$ will involve the integral of the product of the integrands represented by a certain number of linked clusters. (We count single circles as linked 1-clusters.) We consider terms in which there are $m_l$ linked $l$-clusters. Because there are $N$ coordinates involved in the integrals for $Z_N$ we have the restriction $\sum_{l=1}^{N} l m_l = N$ on allowed sets $\{m_l\}$. Suppose that, in the term we are considering, there are $\eta_{l_1}$ linked $l$-clusters of one distinct type, $\eta_{l_2}$ linked $l$-clusters of a second type, ..., and $\eta_{N_l}$ of the last type. We have that

$$\sum_{i=1}^{N_l} \eta_{l_i} = m_l \tag{6.14}$$

where $N_l$ is the number of types of linked $l$-clusters ($N_1 = 1,\ N_2 = 1,\ N_3 = 4, \ldots$). How many terms in $Z_N$ will give the integral of this particular product of linked $l$-clusters? We can put the coordinate labels onto the diagram in $N!$ ways and will get the same answer every time after integration. Not all these permuted diagrams will give new contributions to $Z_N$. In particular, in order to get the correct series for $Z_N$ the following two kinds of permutations of the particle labels in the diagrams should be excluded.

1. Those permutations which involve permutations of labels within a linked $l$-cluster give contributions which will be counted later when a sum on $\eta_{l_i}$ of all possible types of $m_l$ $l$-clusters is done and should not be included at this stage. This is because we are regarding diagrams corresponding to permutations of labels of coordinates within a linked $l$-cluster as distinct types if the number of lines is less than the maximum of $B_{l,2} \equiv l!/2!(l-2)!$. For example

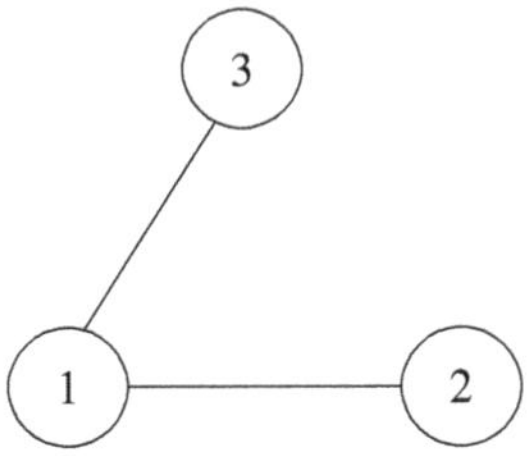


and

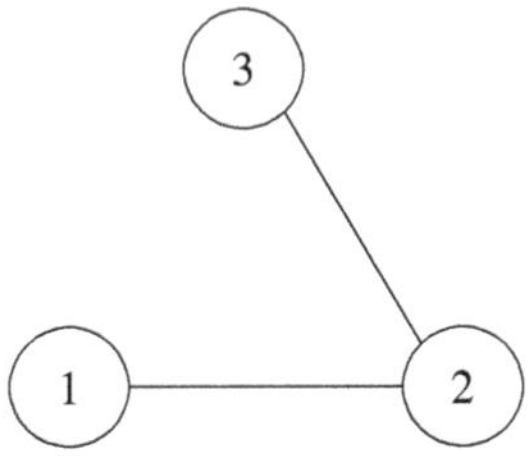


are regarded as distinct, even though the permutation $123 \to 213$ takes one into the other. For each placement of the coordinate labels which should be counted, there are $\Pi_l (l!)^{m_l} - 1$ more, due to the excluded permutations, which should not. Therefore the number $N!$ of permutations should be reduced to $N!/\Pi_l (l!)^{m_l}$ due to this first exclusion.

2. Those permutations which exchange all the labels in one linked $l$-cluster with all the labels in an identical linked $l$-cluster do not give a new contribution to $Z_N$ and should not be counted. For example

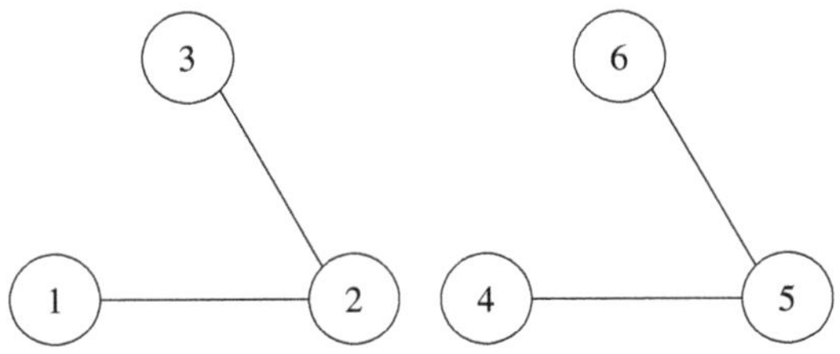


$$= f_{12} f_{23} f_{45} f_{56}$$

would be counted again as

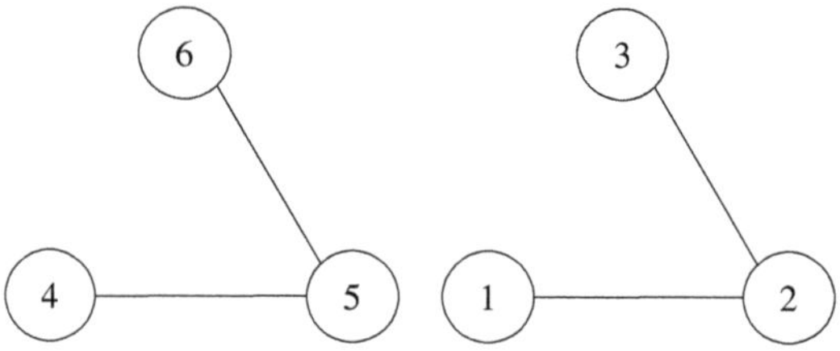


$$= f_{45} f_{56} f_{12} f_{23}$$

if the permutation 123456 → 456123 were counted. There are $\eta_{l_i}!$ ways to permute the clusters of type $l_i$ so this reduces the number of permutations by a further factor $1/\prod_i \eta_{l_i}!$.

These two exceptions thus imply that the number of permutations of coordinates which give identical contributions to $Z_N$ which should be counted from the terms characterized by the sets $\{m_l\}$ of linked $l$-clusters of types characterized by the types $\{\eta_{l_i}\}$ is

$$\frac{N!}{\prod_l (l!)^{m_l} \prod_l \prod_i^{N_l} \eta_{l_i}!} \tag{6.15}$$

We denote the integral of the linked $l$-cluster of type $l_i$ by $D_{l,i}$ so that

$$\lambda^{3l-3} l! b_l V = \sum_{i=1}^{N_l} D_{l,i} \tag{6.16}$$

For example labelling the four distinct 3-clusters by

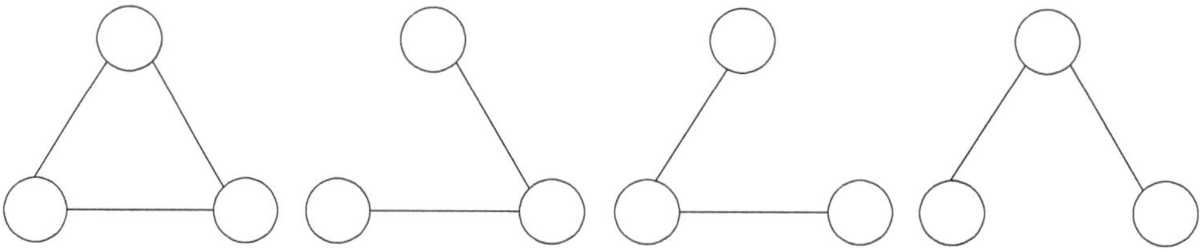

$i =$ 1 2 3 4

we have

$$D_{3,1} = \int \mathrm{d}\vec{r}_1\, \mathrm{d}\vec{r}_2\, \mathrm{d}\vec{r}_3\, f_{12} f_{13} f_{23} \tag{6.17}$$

Then the contribution of a term characterized by $\{m_l\}$ and $\{\eta_{l_i}\}$ together with all the identical terms which contribute to $Z_N$ is

$$\frac{N!}{\prod_l (l!)^{m_l} \prod_l \prod_i^{N_l} \eta_{l_i}!} \prod_{l,i} D_{l,i}^{\eta_{l_i}} \tag{6.18}$$

and

$$Z_N = \frac{1}{N!\lambda^{3N}} \sum_{\{m_l\}\text{ such that }\sum_l lm_l=N} \; \sum_{\{\eta_{l_i}\}\text{ such that }\sum_i \eta_{l_i}=m_l} \frac{N!}{\prod_l (l!)^{m_l} \prod_l \prod_i^{N_l} \eta_{l_i}!} \prod_{l,i} D_{l,i}^{\eta_{l_i}} \tag{6.19}$$

The inside sum on the $\{\eta_{l,i}\}$ is done as follows:

$$\begin{aligned} &\prod_l \frac{1}{m_l!} \sum_{\{\eta_{l_i}\}} \frac{m_l! \prod_i D_{l,i}^{\eta_{l_i}}}{\prod_i \eta_{l_i}!} \\ &= \prod_l \frac{1}{m_l!} \left( \sum_i D_{l,i} \right)^{m_l} = \prod_l \frac{\lambda^{(3l-3)m_l}}{m_l!} (l! b_l V)^{m_l} \end{aligned} \tag{6.20}$$

Here we have used the binomial theorem and also the definition (6.16). Now we put this back in the expression (6.19) for $Z_N$ and use the constraint $\sum_l m_l l = N$:

$$\begin{aligned} Z_N &= \frac{1}{\lambda^{3N}} \sum_{\{m_l\}\text{ such that }\sum_l lm_l=N} \prod_l \frac{\lambda^{(3l-3)m_l}}{m_l!} (b_l V)^{m_l} \\ &= \sum_{\{m_l\}\text{ such that }\sum_l lm_l=N} \prod_l \frac{1}{m_l!} \left( \frac{b_l V}{\lambda^3} \right)^{m_l} \end{aligned} \tag{6.21}$$

The sum on $\{m_l\}$ becomes possible when this is put back in the expression (6.3) so that the constraint disappears:

$$\begin{aligned} Z_{gc} &= \sum_N z^N Z_N = \sum_N \sum_{\{m_l\}\text{ such that }\sum_l lm_l=N} \prod_l z^{lm_l} \left( \frac{b_l V}{\lambda^3} \right)^{m_l} \frac{1}{m_l!} \\ &= \sum_{\{m_l\}} \prod_l \left( \frac{b_l V z^l}{\lambda^3} \right)^{m_l} \frac{1}{m_l!} \end{aligned} \tag{6.22}$$

where the last sum is not constrained. The sum and product can be interchanged giving

$$\begin{aligned} Z_{gc} &= \prod_l \sum_{m_l=0}^{\infty} \left( \frac{b_l V z^l}{\lambda^3} \right)^{m_l} \frac{1}{m_l!} \\ &= \exp\left( \sum_l \frac{z^l b_l V}{\lambda^3} \right) \end{aligned} \tag{6.23}$$

Thus the thermodynamic potential $\Omega$ is

$$\Omega = -k_B T \ln Z_{gc} = -k_B T \sum_l \frac{z^l b_l V}{\lambda^3} \tag{6.24}$$

This result has many nice features. It is properly proportional to the volume. (It is easy to show that the $b_l$ are independent of the volume.) Without summing the expression for $Z_{gc}$ on $m_l$ to all orders in $z$ this correct dependence on the volume would not have been obtained. Notice, for example, the tangle that would be obtained with respect to the factors $V$ if one had tried to keep just a few terms in the sums in (6.23), supposing that they were of decreasing size, and then had tried to take the ln of the result.

It is now quite straightforward to obtain an expansion of the pressure in terms of powers of the density. We use

$$\Omega = -PV = -k_B T \ln Z_{gc} \tag{6.25}$$

and

$$\bar{N} = \left(\frac{\partial \Omega}{\partial \mu}\right)_{T,V} \tag{6.26}$$

giving

$$\frac{P}{k_B T} = \frac{1}{\lambda^3} \sum_l z^l b_l \tag{6.27}$$

and

$$\frac{\bar{N}}{V} = \rho = \frac{1}{\lambda^3} \sum_{l=1}^{\infty} l z^l b_l \tag{6.28}$$

To get an expansion of $P$ in terms of $\rho$ one must eliminate $z$ between the last two equations. It is convenient to do this order by order by writing a series in $\rho$ for the pressure as

$$\frac{P}{k_B T} = \sum_{l=1}^{\infty} a_l \lambda^{3l-3} \rho^l \tag{6.29}$$

thus defining the coefficients $a_l$ (which can be easily shown to be dimensionless). Then one inserts the expressions (6.27) and (6.28) into (6.29) and solves order by order in powers of $z$ for the $a_l$ in terms of the $b_l$:

$$\frac{1}{\lambda^3} \sum_{l=1}^{\infty} z^l b_l = \sum_{l=1}^{\infty} a_l \lambda^{3l-3} \left( \frac{1}{\lambda^3} \sum_{l'=1}^{\infty} l' z^{l'} b_{l'} \right)^l \tag{6.30}$$

The $a_l$ can now be calculated term by term by equating powers of $z$ on both sides of this equation. The first few $a_l$ are

$$a_1 = b_1 = 1 \tag{6.31}$$

$$a_2 = -b_2 \tag{6.32}$$

$$a_3 = 4b_2^2 - 2b_3 \tag{6.33}$$

$$a_4 = -20b_2^3 + 18b_2 b_3 - 3b_4 \tag{6.34}$$

One sees from (6.29) and the value $a_1 = 1$ that the leading term gives the perfect gas equation of state. Thus we have a careful demonstration that an interacting gas will act like a noninteracting one at low enough densities and pressures. Another common form of (6.29) is

$$\frac{P}{k_B T} = \sum_l B_l \rho^l \tag{6.35}$$

in which the coefficients $B_l$ are related to the coefficients $a_l$ of (6.29) by $B_l = a_l \lambda^{3l-3}$. The last form is usually called the virial expansion and the $B_l$ are called virial coefficients.

## Method II for the virial expansion: irreducible linked clusters

### *Qualitative discussion*

In this section we obtain the virial expansion from a slightly different point of view. There are two reasons for doing this. In the first place it provides occasion for introducing several common and useful concepts in the use of series expansions in statistical physics. Equally important, it provides a systematic method of ensuring that the computational labor associated with finding the integrals which give the coefficients in the density expansion is minimized. We work with the canonical distribution function and find an expansion directly for $\ln Z_N = -F/k_B T$ instead of $Z_N$. You might guess that $\ln Z_N$ would be a better object to study than $Z_N$ from the fact that the series for $Z_{gc}$ had to be summed on $N$ in order to give a well behaved result in the last section.

Another hint concerning the usefulness of another approach is supplied by a closer look at the virial expansion of the last section. We have

$$\begin{aligned} a_3 &= 4b_2^2 - 2b_3 \\ &= 4\left(\frac{1}{2V\lambda^3}\int d\vec{r}_1\, d\vec{r}_2\, f_{12}\right)^2 - 2\left\{\frac{1}{6V\lambda^6}\int d\vec{r}_1\, d\vec{r}_2\, d\vec{r}_3\, \{f_{12}f_{32}f_{31} + 3f_{12}f_{13}\}\right\} \end{aligned} \tag{6.36}$$

Now we perform a transformation to center of mass coordinates in each of the integrals. This means using the coordinates

$$\vec{R} = \frac{\vec{r}_1 + \vec{r}_2}{2} \qquad \vec{r}_{12} = \vec{r}_1 - \vec{r}_2 \tag{6.37}$$

on the first integral and

$$\vec{R} = \frac{\vec{r}_1 + \vec{r}_2 + \vec{r}_3}{3} \qquad \vec{r}_{12} = \vec{r}_1 - \vec{r}_2 \qquad \vec{r}_{13} = \vec{r}_1 - \vec{r}_3 \tag{6.38}$$

on the last integral. The result is

$$a_3 = \frac{1}{\lambda^6}\left(\int d\vec{r}_{12}\, f_{12}\right)^2 - \frac{1}{3\lambda^6}\int d\vec{r}_{12}\, d\vec{r}_{13}\, f_{12} f_{13} f(\vec{r}_{13} - \vec{r}_{12})$$
$$- \frac{1}{\lambda^6}\int d\vec{r}_{12}\, f_{12}\int d\vec{r}_{13} f_{13}$$
$$= \frac{-1}{3\lambda^6}\int d\vec{r}_{12}\, d\vec{r}_{13}\, f_{12} f_{13} f(\vec{r}_{13} - \vec{r}_{12}) \tag{6.39}$$

In other words, the only surviving term is the one with the diagrammatic description

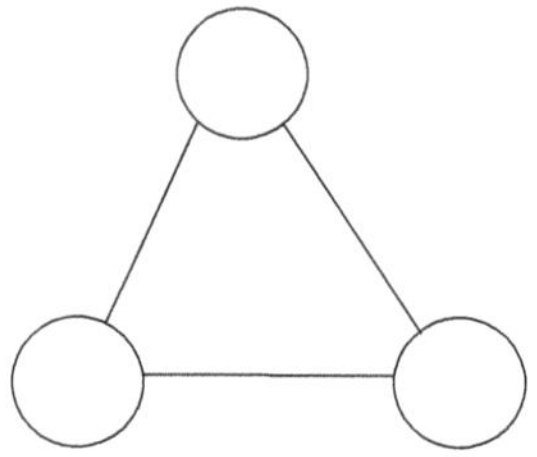

There is a general property which distinguishes this from the diagram

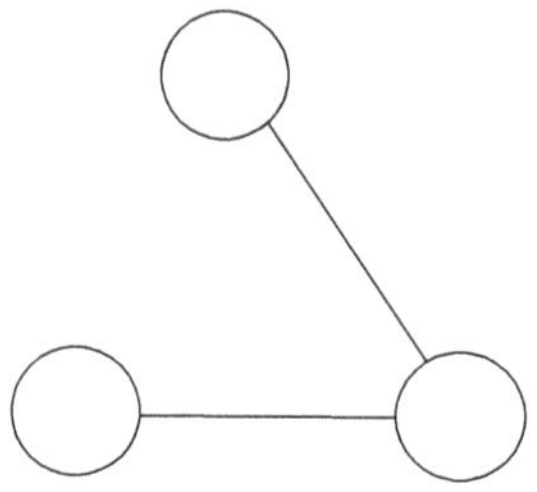

which dropped out. In terms of the integral,

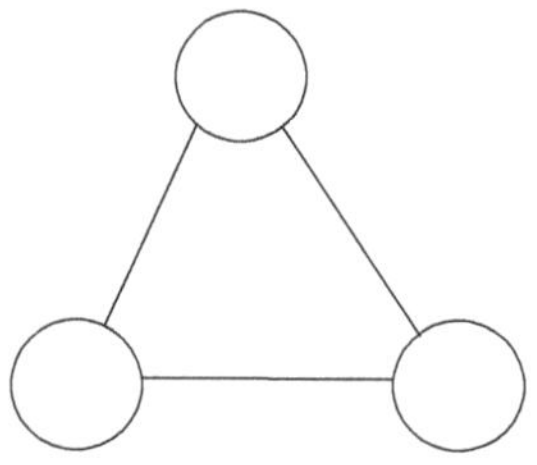

cannot be written as a product of integrals of the diagram

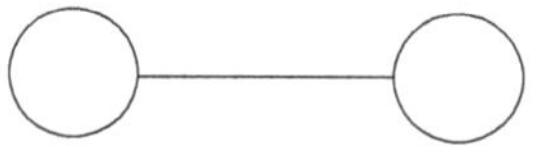

as

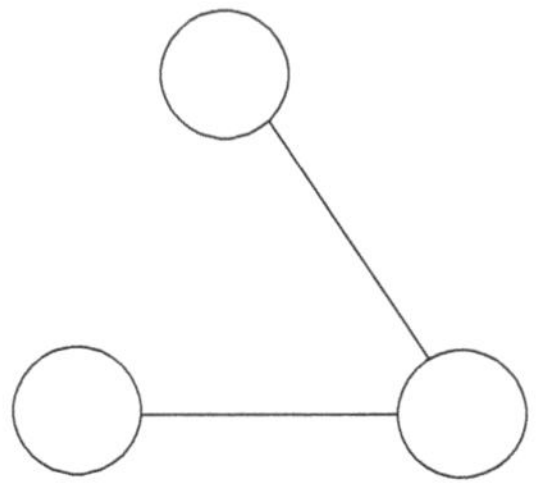

was (by the change to center of mass coordinates). In diagrammatic terms,

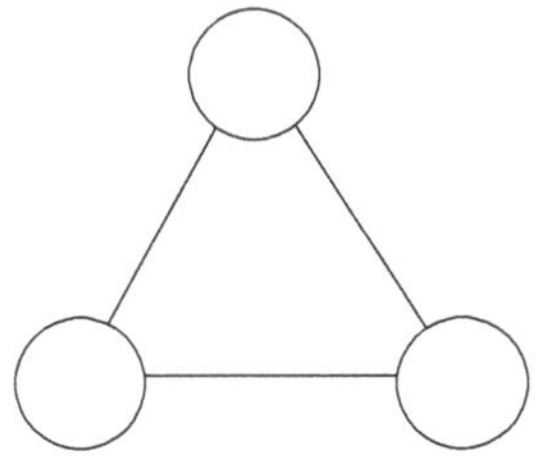

differs from

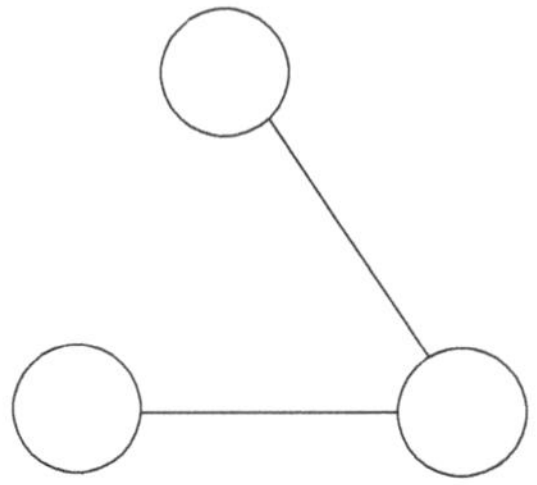

in having no circle which is connected to only one line. Diagrams with this property are called irreducible and can be shown not to be factorizable in general. In the course of developing this second method for evaluating the virial expansion, we will find that, in general, only irreducible diagrams contribute. Thus the simplifications associated with going to the center of mass coordinates in each cluster are automatically taken into account and no redundant calculations of cancelling terms are done.

Since we intend to expand $\ln Z_N$ directly, some mathematical preliminaries on the relationship of the expansion of a function to the expansion of its logarithm will be considered first.

### *Cumulant expansions*

The use of cumulants goes beyond the problem at hand so the discussion will be rather detailed. For clarity, we begin with an outline of the subsequent discussion.

(a) Cumulants for a function $M(\xi)$ of one variable.
(b) Cumulants for a function of one variable of the form $M(\xi) = \langle e^{\xi X}\rangle_X$ in which $\langle\ldots\rangle_X$ denotes averaging over a variable X.
(c) Cumulants for a function of several variables.
(d) Generalization to the "leveled" exponential function.

**(a) Cumulants for a function $M(\xi)$ of one variable** Suppose that a function $M(\xi)$ has the expansion

$$M(\xi) = \sum_{n=0}^{\infty} \frac{\mu_n}{n!}\xi^n \tag{6.40}$$

which defines expansion coefficients $\mu_n$. We consider the corresponding expansion of its ln: $K(\xi) \equiv \ln M(\xi)$. We will assume that $M(\xi = 0) = 1$. Then the constant term in the series for $K(\xi)$ is zero, giving

$$K(\xi) = \sum_{n=1}^{\infty} \frac{1}{n!}\xi^n \kappa_n \tag{6.41}$$

defining the coefficients $\kappa_n$. The $\mu_n$ are called moments of $M$ (for reasons which will become more evident below) and the $\kappa_n$ are called cumulants of $M$. To find the relationship between cumulants and moments one inserts (6.40) and (6.41) into the relationship $K(\xi) = \ln M(\xi)$ and expands the logarithm in powers of $\xi$. Then equating powers of $\xi$ on both sides of the equation gives the $\kappa_n$ in terms of the $\mu_n$. The first few $\kappa_n$ are

$$\kappa_1 = \mu_1 \tag{6.42}$$

$$\kappa_2 = \mu_2 - \mu_1^2 \tag{6.43}$$

$$\kappa_3 = \mu_3 - 3\mu_1\mu_2 + 2\mu_1^3 \tag{6.44}$$

$$\kappa_4 = \mu_4 - 3\mu_2^2 + 12\mu_1^2\mu_2 - 4\mu_1\mu_3 - 6\mu_1^4 \tag{6.45}$$

General expressions are available in the literature.[1]

**(b) Cumulants of a function $M(\xi) = \langle e^{\xi X}\rangle$** We consider the function $M(\xi) = \langle e^{\xi X}\rangle$ in which $X$ is a variable and a linear averaging procedure $\langle\ldots\rangle$ over the variable $X$ is specified. "Linear" means that for any two functions $A(X)$ and $B(X)$ and any two constants $\alpha$ and $\beta$, $\langle\alpha A(X) + \beta B(X)\rangle = \alpha\langle A(X)\rangle + \beta\langle B(X)\rangle$. In the application to the virial expansion below, the average will correspond to integration over all the spatial coordinates and division by the appropriate power of the volume: $\langle\ldots\rangle = \frac{1}{V^N}\int \mathrm{d}^N\vec{r}\cdots$. We will never take this average to be the thermal average.

As before we define $K(\xi) = \ln M(\xi)$. Expanding in powers of $\xi$ we then get (6.40) again, defining the quantities $\mu_n$. But we get another expression for the $\mu_n$ by expanding the right hand side of the definition $M(\xi) = \langle \mathrm{e}^{\xi X}\rangle$ as a function of $\xi$, using the assumed linearity of the average and equating the result term by term in powers of $\xi$ to the right hand side of (6.40). This gives

$$\mu_n = \langle X^n\rangle \tag{6.46}$$

This shows why we chose to call the $\mu_n$ "moments" in the last subsection. We again expand $K(\xi)$ in powers of $\xi$ defining $\kappa_n$ as in (6.41). By analogy with (6.46) we can write

$$\kappa_n \equiv \langle X^n\rangle_{\mathrm{c}} \tag{6.47}$$

defining an operation $\langle\ldots\rangle_{\mathrm{c}}$, called a cumulant average on $X^n$. But this averaging procedure is *not* linear. Combining equations (6.45), (6.46) and (6.47) we have

$$\langle X\rangle_{\mathrm{c}} = \langle X\rangle \tag{6.48}$$

$$\langle X^2\rangle_{\mathrm{c}} = \langle X^2\rangle - \langle X\rangle^2 \tag{6.49}$$

$$\langle X^3\rangle_{\mathrm{c}} = \langle X^3\rangle - 3\langle X\rangle\langle X^2\rangle + 2\langle X\rangle^3 \tag{6.50}$$

and so forth. Notice that if $\langle X^n\rangle = \langle X\rangle^n$ then $\langle X^n\rangle_{\mathrm{c}} = 0$. That is, if the averages of powers factor into the powers of the averages then the corresponding cumulant averages are zero. This turns out to be a general property.

**(c) Cumulants of an averaged exponential function of several variables** Next we generalize to several (say $N$) variables. We have

$$M(\xi_1,\ldots,\xi_N) \equiv \left\langle \mathrm{e}^{\sum_i^N \xi_i X_i}\right\rangle \tag{6.51}$$

a function of several variables. Define moments by the analogue of (6.40):

$$M(\xi_1,\xi_2,\ldots,\xi_M) = \sum_{\nu_1=1}^{\infty}\cdots\sum_{\nu_N=1}^{\infty}\frac{1}{\nu_1!}\cdots\frac{1}{\nu_n!}\mu_{\nu_1,\ldots,\nu_N}\xi_1^{\nu_1}\cdots\xi_N^{\nu_N} \tag{6.52}$$

Then by use of the linearity of the averaging procedure we have

$$\mu_{\nu_1,\ldots,\nu_N} = \left\langle X_1^{\nu_1}\cdots X_N^{\nu_N}\right\rangle \tag{6.53}$$

As before we define cumulants $\kappa_{\nu_1\ldots\nu_N}$ through

$$K(\xi_1,\ldots,\xi_N) \equiv \ln M(\xi_1,\ldots,\xi_N) = \sum_{\nu_1=1}^{\infty}\cdots\sum_{\nu_N=1}^{\infty}\frac{1}{\nu_1!}\cdots\frac{1}{\nu_N!}\kappa_{\nu_1\cdots\nu_N}\xi_1^{\nu_1}\cdots\xi_N^{\nu_N} \tag{6.54}$$

For these we introduce a notation like (6.47)

$$\kappa_{\nu_1\ldots\nu_N} = \left\langle X_1^{\nu_1}\cdots X_N^{\nu_N}\right\rangle_{\mathrm{c}} \tag{6.55}$$

By inserting (6.52) and (6.54) into $K(\xi_1,\ldots,\xi_N) \equiv \ln M(\xi_1,\ldots,\xi_N)$ and using (6.55) we obtain expressions for the many variable "cumulant averages" $\langle X_1^{\nu_1}\cdots X_N^{\nu_N}\rangle_{\mathrm{c}}$. The first two are

$$\langle X_i\rangle_{\mathrm{c}} = \langle X_i\rangle \tag{6.56}$$

$$\langle X_i X_j\rangle_{\mathrm{c}} = \langle X_i X_j\rangle - \langle X_i\rangle\langle X_j\rangle \tag{6.57}$$

**(d) Cumulants of "leveled" exponential functions** To obtain the virial expansion we will be interested in evaluating the quantity (see (6.19))

$$\left\langle \prod_{i<j}(1+f_{ij})\right\rangle \tag{6.58}$$

in which the average $\langle\ldots\rangle = \frac{1}{V^N}\int \mathrm{d}^N\vec{r}\cdots$. If we regard the $f_{ij}$ as playing the role of the variables $X_i$ of the last section, then we can introduce variables $\xi_{ij}$ and try to apply the formalism of the last section but we have a problem because we find the function $\prod_{i<j}(1+\xi_{ij}f_{ij})$ inside the average instead of $\exp(\sum_{i<j}\xi_{ij}X_{ij})$. We can adapt the formalism by noting that $\exp(\sum_i \xi_i X_i)$ contains all the terms to be found in $\prod_i(1+\xi_i X_i)$ but also contains many other terms in which the $X_i$ appear to higher powers than 1. Thus it is natural to introduce a "leveled exponential" function by the definition

$$\exp_{\mathrm{L}}\left(\sum_i \xi_i X_i\right) = \prod_i (1+\xi_i X_i) \tag{6.59}$$

This is called a leveled exponential because if we introduce a "leveling" operator $L_{\mathrm{op}}$ by the definition

$$L_{\mathrm{op}}\left(X_1^{\nu_1}\cdots X_N^{\nu_N}\right) = \begin{cases} X_1^{\nu_1}\cdots X_N^{\nu_N} & \text{if all } \nu_i = 0 \text{ or } 1\\ 0 & \text{otherwise}\end{cases} \tag{6.60}$$

then (assuming that $L_{\mathrm{op}}$ is linear)

$$\begin{aligned} L_{\mathrm{op}}\left(\exp\left(\sum_i \xi_i X_i\right)\right) &= L_{\mathrm{op}}\left\{\sum_{\nu_1=0}^{\infty}\cdots\sum_{\nu_N=0}^{\infty}\frac{\xi_1^{\nu_1}\cdots\xi_N^{\nu_N}}{\nu_1!\cdots\nu_N!}X_1^{\nu_1}\cdots X_N^{\nu_N}\right\}\\ &= \sum_{\nu_1=0}^{\infty}\cdots\sum_{\nu_N=0}^{\infty}\frac{\xi_1^{\nu_1}\cdots\xi_N^{\nu_N}}{\nu_1!\cdots\nu_N!}L_{\mathrm{op}}\left(X_1^{\nu_1}\cdots X_N^{\nu_N}\right)\\ &= \sum_{\nu_1=0}^{1}\cdots\sum_{\nu_N=0}^{1}\xi_1^{\nu_1}\cdots\xi_N^{\nu_N}\left(X_1^{\nu_1}\cdots X_N^{\nu_N}\right) \equiv \exp_{\mathrm{L}}\left(\sum_i \xi_i X_i\right) \end{aligned} \tag{6.61}$$

Thus $\exp_{\mathrm{L}}$ is reasonably called the leveled exponent.

We now average the function $\exp_{\rm L}(\sum_i \xi_i X_i)$ over the $X_i$ in the same manner as before to obtain a function $M(\xi_1, \ldots, \xi_N)$:

$$M(\xi_1, \ldots, \xi_N) = \left\langle \exp_{\rm L}\left(\sum_i \xi_i X_i\right)\right\rangle = \left\langle \prod_i (1 + \xi_i X_i)\right\rangle \tag{6.62}$$

so that if we write the general definition of the moment as before

$$M(\xi_1, \ldots, \xi_N) = \sum_{\nu_1=0}^{\infty} \cdots \sum_{\nu_N=0}^{\infty} \frac{\xi_1^{\nu_1} \cdots \xi_N^{\nu_N}}{\nu_1! \cdots \nu_N!} \mu_{\nu_1 \ldots \nu_N} \tag{6.63}$$

then

$$\mu_{\nu_1 \ldots \nu_N} = \begin{cases} \langle X_1^{\nu_1} \cdots X_N^{\nu_N} \rangle & \text{all } \nu_i = 0 \text{ or } 1 \\ 0 & \text{otherwise} \end{cases} \tag{6.64}$$

or equivalently

$$\mu_{\nu_1 \ldots \nu_N} = \langle L_{\rm op}(X_1^{\nu_1} \cdots X_N^{\nu_N}) \rangle \tag{6.65}$$

We let

$$K(\xi_1, \ldots, \xi_N) = \ln M(\xi_1, \ldots, \xi_N) \tag{6.66}$$

and

$$K(\xi_1, \ldots, \xi_N) = \sum_{\nu_1=1}^{\infty} \cdots \sum_{\nu_N=1}^{\infty} \frac{\xi_1^{\nu_1} \cdots \xi_N^{\nu_N}}{\nu_1! \cdots \nu_N!} \kappa_{\nu_1 \ldots \nu_N} \tag{6.67}$$

define cumulants as before. We denote $\kappa_{\nu_1 \ldots \nu_N}$ by

$$\kappa_{\nu_1 \ldots \nu_N} = \langle X_1^{\nu_1} \cdots X_N^{\nu_N} \rangle_{\rm L,c} \tag{6.68}$$

where the c is to remind us that this is a cumulant and the L that $M(\xi_1, \ldots, \xi_N)$ is defined in terms of a leveled exponent. The $\kappa_{\nu_1 \ldots \nu_N}$ in this case are obtained from (6.57) by inserting $L_{\rm op}$ in front of the powers of $X_i$ appearing inside averages on the right hand side. Thus

$$\langle X_i \rangle_{\rm L,c} = \langle X_i \rangle \tag{6.69}$$

$$\langle X_i^2 \rangle_{\rm L,c} = \langle L_{\rm op}(X_i^2) \rangle - \langle L_{\rm op}(X_i) \rangle^2 = -\langle X_i \rangle^2 \tag{6.70}$$

$$\langle X_i X_j \rangle_{\rm L,c} \underset{(i \neq j)}{=} \langle L_{\rm op}(X_i X_j) \rangle - \langle L_{\rm op}(X_i) \rangle \langle L_{\rm op}(X_j) \rangle = \langle X_i X_j \rangle - \langle X_i \rangle \langle X_j \rangle \tag{6.71}$$

$$\begin{aligned}\langle X_i X_j X_k \rangle_{\rm L,c} \underset{(i \neq j \neq k)}{=} & \langle X_i X_j X_k \rangle - \langle X_i X_j \rangle \langle X_k \rangle - \langle X_i X_k \rangle \langle X_j \rangle \\ & - \langle X_j X_k \rangle \langle X_i \rangle + 2 \langle X_i \rangle \langle X_j \rangle \langle X_k \rangle \end{aligned} \tag{6.72}$$

### Application of cumulants to the expansion of the free energy

Now we revisit the problem of a cluster expansion of the free energy of an imperfect gas, working in the canonical distribution and utilizing the concepts of cumulants just introduced. The model is the same as the model (6.1) considered in the first treatment of the virial expansion

$$H_N = \sum_{i=1}^{N} p_i^2/2m + \sum_{i<j} v_{ij}$$

which, as before, leads to the partion function (6.19):

$$Z_N = \frac{1}{N!\lambda^{3N}} \int \mathrm{d}^N\vec{r} \prod_{i<j}(1 + f_{ij})$$

where $f_{ij} = \mathrm{e}^{-\beta v_{ij}} - 1$ as before. Now to make contact with the definitions of cumulants defined in the preceding section we define a volume average (*not* a thermal average) as

$$\langle \ldots \rangle_V = \frac{1}{V^N} \int \mathrm{d}^N\vec{r}(\ldots) \tag{6.73}$$

then the partition function can be rewritten as

$$Z_N = \frac{V^N}{N!\lambda^{3N}} \left\langle \prod_{i<j}(1 + f_{ij}) \right\rangle_V = \frac{V^N}{N!\lambda^{3N}} \left\langle \exp_{\mathrm{L}} \left( \sum_{i<j} f_{ij} \right) \right\rangle_V \tag{6.74}$$

using the definition of the leveled exponent $\exp_{\mathrm{L}}$ given in the preceding section. The free energy is

$$F = k_{\mathrm{B}}T \ln \frac{\lambda^{3N} N!}{V^N} - k_{\mathrm{B}}T \ln \left\langle \exp_{\mathrm{L}} \left( \sum_{i<j} f_{ij} \right) \right\rangle_V \tag{6.75}$$

The first term is the free energy $F_0$ of the perfect gas. Writing $F = F_0 + F_{\mathrm{int}}$ we have

$$F_{\mathrm{int}} = -k_{\mathrm{B}}T \ln \left\langle \exp_{\mathrm{L}} \left( \sum_{i<j} f_{ij} \right) \right\rangle_V \tag{6.76}$$

A direct connection with the formalism of the preceding section is obtained by defining a function of $N(N-1)/2$ variables (one for each pair of particles) $\xi_{ij}$ as

$$f_{\mathrm{int}}(\xi_1, \ldots, \xi_{N(N-1)/2}) = -k_{\mathrm{B}}T \ln \left\langle \exp_{\mathrm{L}} \left( \sum_{i<j} \xi_{ij} f_{ij} \right) \right\rangle_V \tag{6.77}$$

Then

$$F_{\text{int}} = f_{\text{int}}(\xi_1, \ldots, \xi_{N(N-1)/2})|_{\text{all } \xi_{ij}=1} \tag{6.78}$$

Then we have an exact parallel with the development of the preceding section and can write

$$F = -k_B T \sum_{\nu_{ij}} \prod_{i<j} \frac{\xi_{ij}^{\nu_{ij}}}{\nu_{ij}!} \left\langle \prod_{i<j} f_{ij}^{\nu_{ij}} \right\rangle_{V,L,C} \Bigg|_{\text{all } \xi_{ij}=1} = -k_B T \sum_{\nu_{ij}} \prod_{i<j} \frac{1}{\nu_{ij}!} \left\langle \prod_{i<j} f_{ij}^{\nu_{ij}} \right\rangle_{V,L,C} \tag{6.79}$$

where the indices on the sets of integers cover all $N(N-1)/2$ pairs for which $i < j$. Note that there is a correspondence between the terms in the linked cluster expansion for $Z_N$ discussed in the first part of this chapter and terms in (6.79) with all the $\nu_{ij} = 0$ or 1. There are also some extra terms in (6.79), for which at least one of the $\nu_{ij} > 1$.

We wish to make the following points about (6.79).

(a) For terms for which all $\nu_{ij} = 0$ or 1 in (6.79), the leveled exponent and cumulant character of the average in (6.79) means that cancellations of the sort discussed after (6.36) above occur automatically in (6.79) and no redundant calculation occurs using it.
(b) There is a diagrammatic characterization of the surviving terms in (6.79) for which all $\nu_{ij} = 0$ or 1 which makes it straightforward to sort out which linked $l$-clusters contribute at each level. The surviving linked $l$-clusters will be called irreducible linked $l$-clusters.
(c) By considering the thermodynamic limit one can show that once one has confined attention to the surviving terms (associated with irreducible linked $l$-clusters) then the actual computation of integrals can take place as if the average is an ordinary volume average $\langle \ldots \rangle_V$.

In summary these three points mean that the calculation of the free energy proceeds by calculation only of the irreducible linked $l$-clusters (which we have not yet defined in detail) but otherwise is rather similar to the calculation of the partition function as was done using linked (not irreducible) $l$-clusters in method I. We will now discuss points (a)–(c) in more detail. There will be no formal proofs but we will try to make each point convincing by example and in some cases by informal proof.

(a) Cancellations like those discussed after (6.36) are automatically taken into account by (6.79) so no redundant calculation of integrals occurs. We illustrate this by consideration of the example discussed after (6.36). First recall that the cumulant average (leveled or not) always vanishes if the corresponding ordinary average factors into products of ordinary averages. Now consider a term of form $\langle f_{ij} f_{jk} \rangle_V$ represented by the diagram

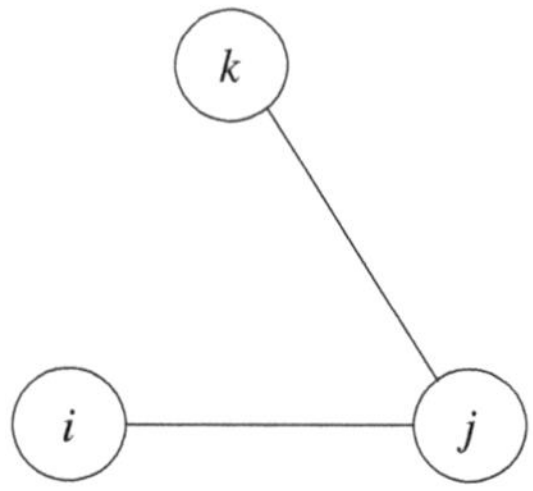


Writing this out

$$\langle f_{ij} f_{jk} \rangle_V = \frac{1}{V^3} \int \mathrm{d}\vec{r}_i \, \mathrm{d}\vec{r}_j \, \mathrm{d}\vec{r}_k \; f_{ij} f_{jk} \tag{6.80}$$

We change the variables in the integral to

$$\vec{r}_{ij} = \vec{r}_i - \vec{r}_j \tag{6.81}$$

$$\vec{r}_{jk} = \vec{r}_j - \vec{r}_k \tag{6.82}$$

$$\vec{r} = \frac{\vec{r}_i + \vec{r}_j + \vec{r}_k}{3} \tag{6.83}$$

(which is the center of mass transformation) giving

$$\begin{aligned} \langle f_{ij} f_{jk} \rangle_V &= \frac{1}{V^2} \left( \int \mathrm{d}\vec{r}_{ij} \; f_{ij} \right) \left( \int \mathrm{d}\vec{r}_{jk} \; f_{jk} \right) \\ &= \frac{1}{V^4} \int \mathrm{d}\vec{r}_i \, \mathrm{d}\vec{r}_j \; f_{ij} \int \mathrm{d}\vec{r}_j \, \mathrm{d}\vec{r}_k \; f_{jk} = \langle f_{ij} \rangle_V \langle f_{jk} \rangle_V \end{aligned} \tag{6.84}$$

Thus this average factors into the product of two averages and it follows that $\langle f_{ij} f_{jk} \rangle_{V,L,C} = 0$. Thus this diagram, which we showed in the section after (6.36) did not ultimately contribute to the virial expansion, never enters (6.79). The general property which this example represents is that clusters for which the averages factor do not contribute. It is easy to show, for example, by going through the same exercise for the term represented by the diagram

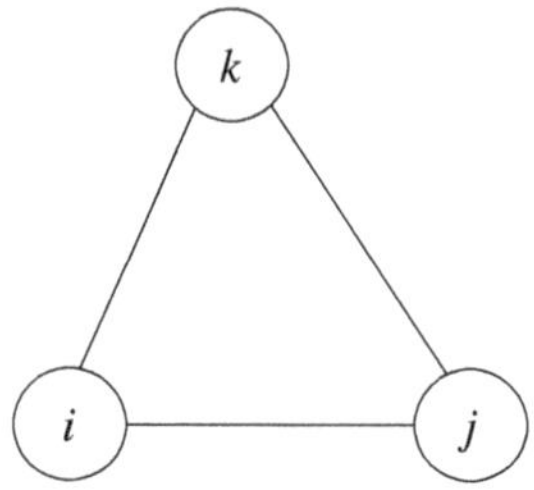


that the average of this diagram does *not* factor.

(b) For the diagrammatic description of the surviving terms, note that, on the basis of the example just discussed, a term will not survive in (6.79) just because its diagram is a linked $l$-cluster. The general requirement turns out to be that the diagrams have the property that they cannot be separated into two parts by cutting all the lines connected to just one circle. (Such diagrams are called *stars* by G. E. Uhlenbeck and G. W. Ford.[2]) We will call all such diagrams *irreducible* linked clusters. We claim that an average $\langle f_{i_1 i_2} f_{i_3 i_4} \cdots \rangle_V$ can be factored if and only if it is represented by a diagram which is not irreducible. (We say a cluster is *reducible* if it is not irreducible.)

We supply an informal proof. An unlinked cluster can obviously be factored. A reducible linked $l$-cluster has at least one circle with the property that if all the lines to it are cut, then the diagram separates into two disconnected parts. Let the label on such a circle be $i$:

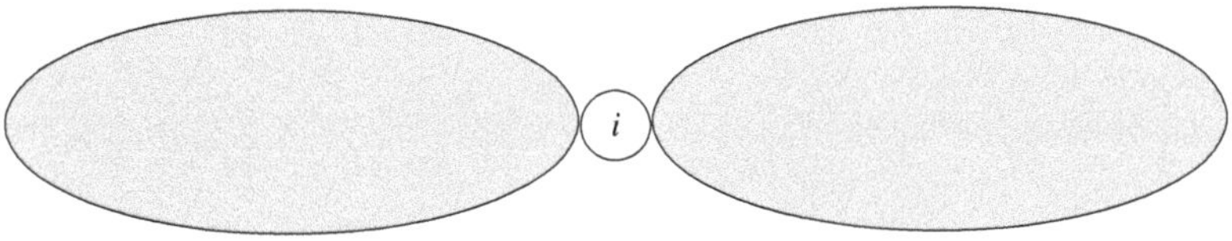


The corresponding expression is

$$\left\langle \prod_{j_l} f_{i j_l} \prod_{j_r} f_{i j_r} \prod f \text{ not involving } i \right\rangle_V \tag{6.85}$$

where the $j_l$ and $j_r$ label the coordinates connected to $i$ from the left and right respectively. We introduce change of variables to the new variables

$$\vec{r}_i \qquad \text{and} \qquad \vec{r}_{ji} = \vec{r}_j - \vec{r}_i \qquad \text{for} \qquad j \neq i. \tag{6.86}$$

Using this coordinate system one may do the integral on $r_i$ giving a factor $V$. The integrals on the difference coordinates then factor into two factors, one on the coordinates appearing to the right and the other on the coordinates appearing to the left. Thus the reducible diagrams (or those which are not stars) factor as claimed. To prove the converse is more complicated. We refer the reader to Uhlenbeck and Ford for a general proof. However, one can easily see that the above approach does not result in factorization if the left and right parts of the diagram are connected at some other point besides the coordinate $i$. The minimal such connection is a single bond like this:

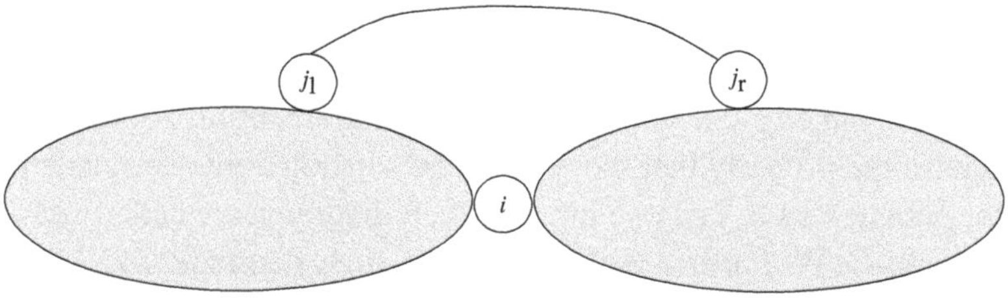


Now the same coordinate transformation will not result in a factorization because the integral will contain a factor $f(\vec{r}_{j_r i} - \vec{r}_{j_l i})$ which prevents the factorization.

(c) The cumulant averages of terms corresponding to irreducible linked $l$-clusters can be taken to be ordinary volume averages in the thermodynamic limit. That is, the correction terms associated with the cumulant can be dropped. We will illustrate how this works through the example

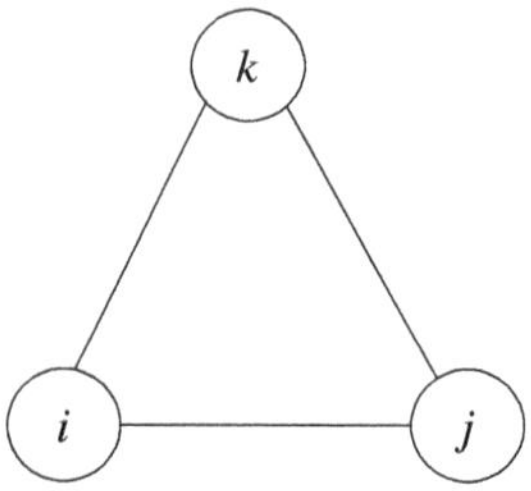


The leveled cumulant average is just the cumulant average which can be written as

$$\begin{aligned}\langle f_{ij} f_{jk} f_{ki} \rangle_{V,L,C} = \langle f_{ij} f_{jk} f_{ki} \rangle_V - \langle f_{ij} f_{jk} \rangle_V \langle f_{ki} \rangle_V - \langle f_{ij} \rangle_V \langle f_{jk} f_{ki} \rangle_V \\ - \langle f_{ij} f_{ki} \rangle_V \langle f_{jk} \rangle_V + 2 \langle f_{ij} f_{jk} f_{ki} \rangle_V \end{aligned} \tag{6.87}$$

Compare the second term on the right hand side to the first:

$$\frac{\langle f_{ij} f_{jk} \rangle_V \langle f_{ki} \rangle_V}{\langle f_{ij} f_{jk} f_{ki} \rangle_V} = \frac{\frac{1}{V^5} \int d\vec{r}_i \, d\vec{r}_j \, d\vec{r}_k \, f_{ij} f_{jk} \int d\vec{r}_k \, d\vec{r}_i \, f_{ki}}{\frac{1}{V^3} \int d\vec{r}_i \, d\vec{r}_j \, d\vec{r}_k \, f_{ij} f_{jk} f_{ki}} \tag{6.88}$$

Changing the variable in each average as before to eliminate the center of mass gives

$$\frac{\langle f_{ij} f_{jk} \rangle_V \langle f_{ki} \rangle_V}{\langle f_{ij} f_{jk} f_{ki} \rangle_V} = \frac{1}{V} \frac{\left( \int d\vec{r}_{ij} \, f_{ij} \right)^3}{\int d\vec{r}_{ij} \, d\vec{r}_{jk} \, f_{ij} f_{jk} f_{ki}} \tag{6.89}$$

The remaining integrals are all less than or equal to a finite volume within which the range of the interaction is confined. Let this volume, which is independent of the volume of the system $V$, be $v_r$. Then

$$\left| \lim_{V\to\infty} \frac{\langle f_{ij} f_{jk} \rangle_V \langle f_{ki} \rangle_V}{\langle f_{ij} f_{jk} f_{ki} \rangle_V} \right| \leq \lim_{V\to\infty} \frac{v_r}{V} = 0 \tag{6.90}$$

It is not hard to generalize this argument and thus see that all the correction terms in the cumulant average of irreducible diagrams can be dropped.

Now we rewrite (6.79) using these results

$$F_{\text{int}} = -k_{\text{B}} T \sum_{l=1}^{\infty} \sum_{\substack{\text{all irreducible linked } l\text{-clusters} \\ \text{for the } N\text{-particle system}}} \langle \text{integrand corresponding to the irreducible linked } l\text{-cluster} \rangle_V \tag{6.91}$$

Notice that, by the definition of irreducibility, we do not find products of $m_l$ or $\eta_{l,i}$ factors in these terms as we did in method I because only one cluster appears only once in each term. To turn this into a density expansion we define

$$V^{-l} l! \beta_l = \left\langle \sum_{\substack{\text{linked irreducible } l+1 \text{ clusters} \\ \text{for the } l+1 \text{ particle system}}} \text{integrand corresponding to the cluster} \right\rangle_V \tag{6.92}$$

It is easy to show that $\beta_l$ is independent of $V$. To illustrate this definition, note that we must count each irreducible graph for the $l + 1$ particle system as distinct if it represents a distinct set of connections. Thus there are three types of irreducible linked 4-clusters of the type

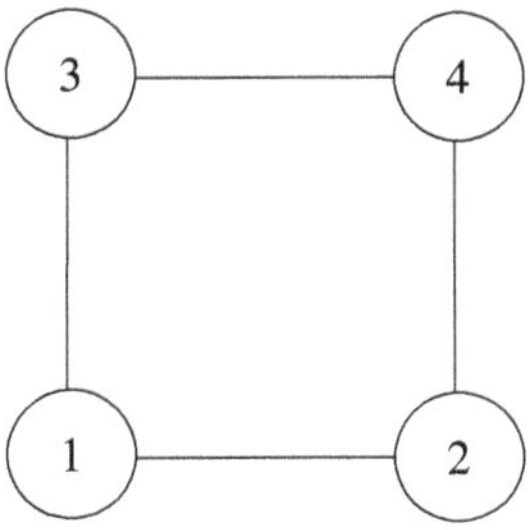


and six of the type

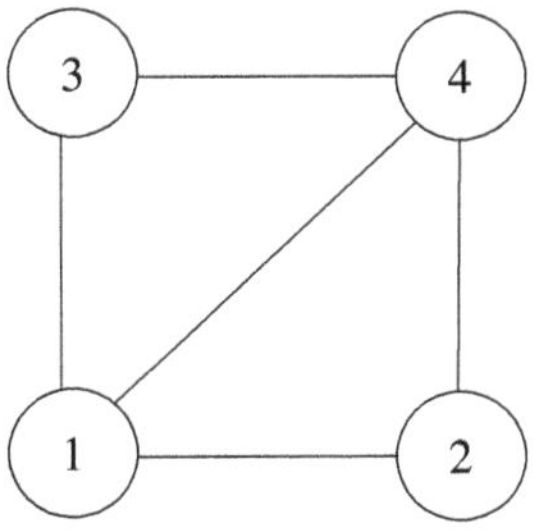

By counting in this way in the definition of $\beta_l$ we can express $F_{\text{int}}$ in terms of the $\beta_l$. There are $N!/(l+1)!(N-(l+1))!$ ways to pick $l+1$ circles from $N$ circles, thus

$$F_{\text{int}} = -k_{\text{B}}T \sum_{l+1}^{\infty} \frac{N!l!\beta_l}{(l+1)!V^l(N-(l+1))!} \tag{6.93}$$

In the thermodynamic limit we can reexpress this using Stirling's approximation

$$\frac{N!}{(N-(l+1))!} \to \frac{N^N}{N^{N-(l+1)}} = N^{l+1} \tag{6.94}$$

so that

$$F_{\text{int}} \to -Nk_{\text{B}}T \sum_{l=1}^{\infty} \frac{\beta_l \rho^l}{(l+1)} \tag{6.95}$$

in which $\rho = N/V$. Thus we get a density expansion directly in terms only of irreducible linked $l$-clusters.

### Cluster expansion for a quantum imperfect gas (extension of method I)

We now describe the development analogous to method I for the quantum case. There is less work on this aspect, partly because the determination of the $l$th cluster coefficient requires a complete solution of the $l$-body quantum mechanical problem. Nevertheless the development is of significant pedagogical value as well as practical use because it establishes the limits of the semiclassical approximation more precisely than we were able to do in Chapter 4. We first write the partition function

$$\begin{aligned} Z_N = \text{Tr}\mathrm{e}^{-\beta H} &= \sum_{\alpha} \int \mathrm{d}^N\vec{r}\,\psi_{\alpha}^*(\vec{r}_1,\ldots,\vec{r}_N)\,\mathrm{e}^{-\beta H}\psi_{\alpha}(\vec{r}_1,\ldots,\vec{r}_N) \\ &= \int \mathrm{d}^N\vec{r} \sum_{\alpha} \psi_{\alpha}^*(\vec{r}_1,\ldots,\vec{r}_N)\,\mathrm{e}^{-\beta H}\psi_{\alpha}(\vec{r}_1,\ldots,\vec{r}_N) \\ &\equiv \frac{1}{\lambda^{3N}N!} \int \mathrm{d}^N\vec{r}\; W_N(\vec{r}_1,\ldots,\vec{r}_N) \end{aligned} \tag{6.96}$$

Here $\alpha$ labels the functions of a complete basis for the Hilbert space and the interchange of summation and integration is assumed to be valid. The last line defines the function $W_N(\vec{r}_1,\ldots,\vec{r}_N)$, which is a function of the coordinates and the temperature.

We now use $W_N$ to define quantities which behave as the integrands corresponding to linked clusters did in the classical case. In fact, the quantities we define yield the classical coefficients in the appropriate limit. Note that in the classical case, the integrands corresponding to linked clusters vanish as any pair of the coordinates approaches infinity. For example the 3-cluster

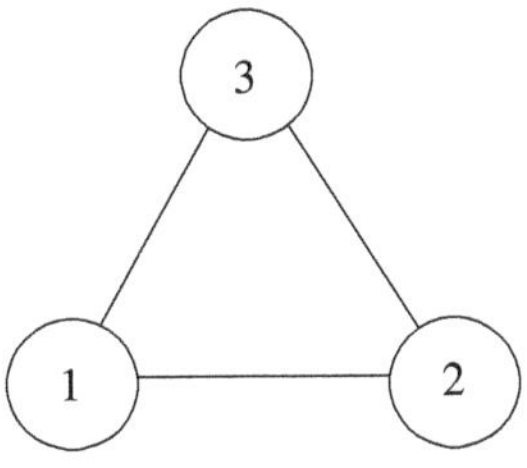


$$= f_{12} f_{23} f_{31} \underset{r_{ij}\to\infty}{\longrightarrow} 0$$

for any of the pairs $ij = 12, 23, 31$. To find the appropriate analogue in the quantum case, we consider $W_l(\vec{r}_1, \ldots, \vec{r}_l)$ and subtract from it any part which does not vanish as the separation between all pairs of coordinates becomes infinite. The remaining function is called $U_l(\vec{r}_1, \ldots, \vec{r}_l)$ and will play the role of the integrand of the linked $l$-cluster. To perform the subtractions systematically one begins with the smallest $l$ and works up. For $l = 2$, one has

$$W_2 \underset{r_{12}\to\infty}{\longrightarrow} 1 = W_1(\vec{r}_1)W_1(\vec{r}_2) = U_1(\vec{r}_1)U_1(\vec{r}_2) \tag{6.97}$$

This is intuitively plausible but it is worthwhile to work out the details (Problem 6.5). So we define

$$U_2(\vec{r}_1, \vec{r}_2) = W_2(\vec{r}_1, \vec{r}_2) - U_1(\vec{r}_1)U_1(\vec{r}_2) \tag{6.98}$$

which may be rewritten

$$W_2(\vec{r}_1, \vec{r}_2) = U_1(\vec{r}_1)U_1(\vec{r}_2) + U_2(\vec{r}_1, \vec{r}_2) \tag{6.99}$$

$U_2(\vec{r}_1, \vec{r}_2)$ now approaches zero as $r_{12} \to \infty$ and plays a role like $f_{12}$ in the classical case. We may think of the last equation as follows. The first term on the right is what results when we take all the particles in the cluster to infinity and the second term is what is left.

Next consider $l = 3$. As in the preceding case, we first separate out the result of taking all three particles to infinity:

$$W_3(\vec{r}_1, \vec{r}_2, \vec{r}_3) = U_1(\vec{r}_1)U_1(\vec{r}_2)U_1(\vec{r}_3) + W_3'(\vec{r}_1, \vec{r}_2, \vec{r}_3) \tag{6.100}$$

Now in $W_3'(\vec{r}_1, \vec{r}_2, \vec{r}_3)$ consider the result of taking $\vec{r}_1$ to a place infinitely far away from the other two particle coordinates. This must yield $U_2(\vec{r}_2, \vec{r}_3)U_1(\vec{r}_1)$ since the result of taking all three particles far away from each other has already been taken into account. There are two other ways to take one particle to infinity while leaving the other two in place. Including these in our expression for $W_3(\vec{r}_1, \vec{r}_2, \vec{r}_3)$ we obtain

$$W_3(\vec{r}_1, \vec{r}_2, \vec{r}_3) = U_1(\vec{r}_1)U_1(\vec{r}_2)U_1(\vec{r}_3) + U_2(\vec{r}_2, \vec{r}_3)U_1(\vec{r}_1) + U_2(\vec{r}_1, \vec{r}_3)U_1(\vec{r}_2) + U_2(\vec{r}_2, \vec{r}_1)U_1(\vec{r}_3) + U_3(\vec{r}_1, \vec{r}_2, \vec{r}_3) \tag{6.101}$$

Since all possible ways of separating the particles have been taken into account in the first four terms, the remainder, termed $U_3(\vec{r}_1, \vec{r}_2, \vec{r}_3)$ vanishes as any particle is removed to a large distance. This function plays the role of the sum of integrands associated with linked 3-clusters in the classical virial expansion.

It should now be possible to understand the general expression for $W_l$. We consider all possible groupings of the $l$ coordinates into groups (analogous to clusters in the previous, classical, discussion) each containing $l'$ coordinates. Let there be $m_{l'}$ groups of $l'$ coordinates with the constraint that $\sum_{l'} m_{l'} l' = l$. Then

$$W_l(\vec{r}_1, \ldots, \vec{r}_l) = \sum_{\{m'_l\}} \sum_{\mathcal{P}}{}' U_1(\mathcal{P}(1)) \cdots U_1(\mathcal{P}(m_1))U_2(\mathcal{P}(m_1+1), \mathcal{P}(m_1+2)) \cdots U_2(\mathcal{P}(m_1+2m_2-1), \mathcal{P}(m_1+2m_2)) \cdots U_l\left(\mathcal{P}\left(\sum_{l'=1}^{l-1} l' m_{l'} + 1\right), \ldots, \mathcal{P}(l)\right) \tag{6.102}$$

The prime on the sum over permutations means that permutations in which the coordinates of a group are permuted among themselves, or in which whole sets associated with different groups are interchanged, are to be excluded. These are exactly the constraints on allowed permutations which we imposed in the discussion of the counting of linked clusters in method I in the classical case and they lead, as in that case, to $l!/\prod_{l'}(l'!)^{m'_l} m_{l'}!$ contributions of the same value, for a given set $\{m_{l'}\}$, after integration. Following the analogy with the classical case we define

$$b_l = \frac{1}{l!\lambda^{3l-3}V} \int \mathrm{d}^l \vec{r}\, U_l(\vec{r}_1, \ldots, \vec{r}_l) \tag{6.103}$$

Using this definition and equations (6.96) and (6.102) we find a general expression for $Z_N$

$$Z_N = \frac{1}{\lambda^{3N} N!} \sum_{m_l}{}' \frac{N!}{\prod_l (l!)^{m_l} m_l!} \prod_l (l!\lambda^{3l-3} V b_l)^{m_l} \tag{6.104}$$

This is formally identical to the corresponding expression in the classical case. Exactly as in that case it leads to an expression for the grand canonical partition function which takes a compact form:

$$Z_{\mathrm{gc}} = \exp\left(\sum_{l=1}^{\infty} z^l b_l V/\lambda^3\right) \tag{6.105}$$

The only difference is that the $b_l$ are defined differently. Clearly, the thermodynamic arguments leading to fugacity expansions for the density and the pressure are the same so that

$$\frac{N}{V} = \frac{1}{\lambda^3}\sum_{l=1}^{\infty} l z^l b_l \tag{6.106}$$

and

$$\frac{P}{k_B T} = \frac{1}{\lambda^3}\sum_{l=1}^{\infty} b_l z^l \tag{6.107}$$

Thus the first terms give the perfect gas law, and we may find quantum corrections equation of state by evaluating the $b_l$. The derivation of the virial expansion for $P/k_B T$ as an expansion in the density proceeds exactly as in the classical case so that equation (6.29) is unchanged with the coefficients defined in terms of the $b_l$ by equations (6.32)–(6.34) as in the classical case, except that the $b_l$ in equations (6.32)–(6.34) are calculated differently (using (6.103)).

Next, we consider applications of these general results to the perfect quantum gas and to an interacting gas. In the case of a perfect quantum gas, it is not surprising that we can find all of the $b_l$, since the exact solution is known. With interactions, we will only present results for $b_2$.

For the perfect quantum gas, one could proceed by evaluating the $U_l$ explicitly, using the known symmetrized or antisymmetrized product wave functions. It turns out to be easier, and must obviously be equivalent, to expand the exact form of the thermodynamic potential $\Omega$ in a series in the fugacity and compare term by term with (6.107) to obtain the $b_l$. The exact form (equation (5.36))

$$\Omega = \pm k_B T \sum_{\nu} \ln\left(1 \mp e^{(\mu-\epsilon_\nu)\beta}\right)$$

Thus since $\Omega = -PV$ we find

$$\frac{P}{k_B T} = \mp\frac{1}{V}\sum_{\nu} \ln\left(1 \mp z\, e^{-\epsilon_\nu \beta}\right) \tag{6.108}$$

Now taking the Hamiltonian to be the kinetic energy only and using periodic boundary conditions in a cubic box as before we obtain

$$\begin{aligned}\frac{P}{k_B T} &= \mp \frac{4\pi}{(2\pi)^3\hbar^3}\int_0^{\infty} p^2 \mathrm{d}p \ln\left(1 \mp z\, e^{-p^2/2mk_B T}\right)\\ &= \mp \frac{4}{\sqrt{\pi}\lambda^3}\int_0^{\infty} x^2 \ln\left(1 \mp z\, e^{-x^2}\right)\mathrm{d}x \end{aligned} \tag{6.109}$$

Now expand the integrand in $z$ using

$$\ln(1 \mp y) = \begin{cases} -\sum_{l=1}^{\infty} \frac{y^l}{l} \\ \sum_{l=1}^{\infty} (-)^{l+1} \frac{y^l}{l} \end{cases} \tag{6.110}$$

This gives

$$\frac{P}{k_B T} = \frac{4}{\sqrt{\pi}\lambda^3} \begin{cases} \sum_{l=1}^{\infty} \frac{z^l}{l} \int_0^{\infty} e^{-x^2 l} x^2 \, dx \\ \sum_{l=1}^{\infty} (-)^{l+1} \frac{z^l}{l} \int_0^{\infty} e^{-x^2 l} x^2 \, dx \end{cases} \tag{6.111}$$

The integral is

$$\int_0^{\infty} e^{-x^2 l} x^2 \, dx = \frac{1}{4l} \sqrt{\frac{\pi}{l}} \tag{6.112}$$

so that

$$\frac{P}{k_B T} = \frac{1}{\lambda^3} \sum_{l=1}^{\infty} \frac{(\pm)^{l+1}}{l^{5/2}} z^l \tag{6.113}$$

or by comparison with (6.107)

$$b_l^{(0)} = \frac{(\pm)^{l+1}}{l^{5/2}} \tag{6.114}$$

where the superscript "(0)" is added to distinguish this from the values of $b_l$ in the interacting case. This result can be used to estimate more quantitatively the range of validity of the assumptions leading to the semiclassical perfect gas description of the gas.

Next consider the calculation of $b_2$ in the interacting case. As noted, a calculation of $b_l$ in the interacting case requires solution of the $l$-body quantum problem and this is analytically intractable beyond $l = 2$. Even at the $l = 2$ level, $b_2$ will clearly depend on the detailed nature of the potential. We can, however, reexpress $b_2$ in terms of a few features of the solution to the two body problem, namely the bound state energies and the scattering phase shifts. We present that development here. We consider the difference between $b_2$ and the value $b_2^{(0)}$ which it has when there are no interactions (the latter was calculated in (6.114)). One has

$$\begin{aligned} b_2 - b_2^{(0)} &= \frac{1}{2\lambda^3 V} \int d^2\vec{r} \, (U_2(\vec{r}_1, \vec{r}_2) - U_2^{(0)}(\vec{r}_1, \vec{r}_2)) \\ &= \frac{1}{2\lambda^3 V} \int d^2\vec{r} \, (W_2(\vec{r}_1, \vec{r}_2) - W_2^{(0)}(\vec{r}_1, \vec{r}_2)) \end{aligned} \tag{6.115}$$

because $W_2(\vec{r}_1, \vec{r}_2) = U_2(\vec{r}_1, \vec{r}_2) - U_1(\vec{r}_1)U_1(\vec{r}_2)$ and $U_1$ is the same for the interacting and noninteracting cases. We write the two body problem as

$$H_2 \psi_\alpha(\vec{r}_1, \vec{r}_2) = E_\alpha \psi_\alpha(\vec{r}_1, \vec{r}_2) \tag{6.116}$$

in which

$$H_2 = \frac{-\hbar^2}{2m}(\nabla_1^2 + \nabla_2^2) + v(|\vec{r}_1 - \vec{r}_2|) \tag{6.117}$$

in which we have assumed that the interaction potential is spherically symmetric. This is fine for monatomic gases, but not at all appropriate for molecular gases. We are assuming that the energy is independent of any electronic or nuclear spin degrees of freedom. We make a center of mass transformation defining

$$\vec{R} = \frac{1}{2}(\vec{r}_1 + \vec{r}_2) \qquad \vec{r} = \vec{r}_1 - \vec{r}_2 \tag{6.118}$$

so that the Hamiltonian is

$$H_2 = \frac{-\hbar^2}{4m}\nabla_{\vec{R}}^2 - \frac{\hbar^2}{m}\nabla_{\vec{r}}^2 + v(r) \tag{6.119}$$

The eigenvalue problem (6.116) then separates:

$$\psi_\alpha = \Psi_{\vec{P}}(\vec{R})\psi_n(\vec{r}) \tag{6.120}$$

where

$$\Psi_{\vec{P}}(\vec{R}) = \frac{1}{\sqrt{V}} e^{\frac{i\vec{P}\cdot\vec{R}}{\hbar}} \tag{6.121}$$

and the wave function $\psi_n(\vec{r})$ is determined by

$$\left(\frac{-\hbar^2}{m}\nabla_{\vec{r}}^2 + v(r)\right)\psi_n(\vec{r}) = \left(E_{n,\vec{P}} - \frac{P^2}{4m}\right)\psi_n(\vec{r}) = \epsilon_n\psi_n(\vec{r}) \tag{6.122}$$

where we define $\epsilon_n = (E_{n,\vec{P}} - P^2/4m)$ and write $\alpha = n, \vec{P}$. In terms of the $\psi_n(\vec{r})$, $W_2(\vec{r}_1, \vec{r}_2)$ is written

$$\begin{aligned}
W_2(\vec{r}_1, \vec{r}_2) &= 2\lambda^6 \sum_{\vec{P}} \sum_n e^{-\beta \vec{P}^2/4m} e^{-\beta\epsilon_n} \frac{1}{V} |\psi_n(\vec{r})|^2 \\
&= \frac{2\lambda^6}{(2\pi\hbar)^3} \int d^3\vec{P} e^{-\beta\vec{P}^2/4m} \sum_n e^{-\beta\epsilon_n} |\psi_n(\vec{r})|^2 \\
&= 4\sqrt{2}\lambda^3 \sum_n e^{-\beta\epsilon_n} |\psi_n(\vec{r})|^2
\end{aligned} \tag{6.123}$$

Using this we rewrite (6.115) as

$$\begin{aligned}
b_2 - b_2^{(0)} &= \frac{1}{2\lambda^3 V} \int d\vec{r}_1\, d\vec{r}_2\, (W_2(\vec{r}_1, \vec{r}_2)) - W_2^{(0)}(\vec{r}_1, \vec{r}_2)) \\
&= \frac{2^{3/2}}{V} \int d\vec{R} \int d\vec{r} \left\{ \sum_n e^{-\beta\epsilon_n} |\psi_n(\vec{r})|^2 - \sum_n e^{-\beta\epsilon_n^{(0)}} |\psi_n^{(0)}|^2 \right\} \\
&= 2^{3/2} \sum_n (e^{-\beta\epsilon_n} - e^{-\beta\epsilon_n^{(0)}})
\end{aligned} \tag{6.124}$$

In the last line we used the assumed normalization of the wave functions $\psi_n$. It is useful to separate off the bound state part of the eigenvalue spectrum of the interacting two body problem. (The noninteracting problem obviously does not have any bound states.) We denote these bound state energy eigenvalues by $\epsilon_{n_{\rm B}}$ and have

$$b_2 - b_2^{(0)} = 2^{3/2} \sum_{n_{\rm B}} {\rm e}^{-\beta\epsilon_{n_{\rm B}}} + \sum_k \left({\rm e}^{-\beta\epsilon_k} - {\rm e}^{-\beta\epsilon_k^{(0)}}\right) \tag{6.125}$$

in which the unbound states in the continuum are labelled $k$. (Note that the dependence on $V$ has disappeared here so that we can take the limit $V \to \infty$ and work with a continuum of unbound states.) To evaluate the sum on continuum states, note that in both the interacting and the noninteracting cases, the spectrum of continuum energies runs from 0 to $\infty$ but that the number of states per unit energy interval will be different in the interacting and noninteracting cases. (This is clear because, for example, the interaction can pull some states out of the continuum into the bound state spectrum.) In each case we write the energy as $\epsilon_k = \hbar^2 k^2/m$ and have

$$b_2 - b_2^{(0)} = 2^{3/2} \left\{ \sum_{n_{\rm B}} {\rm e}^{-\beta\epsilon_{n_{\rm B}}} + \int {\rm d}k \left(g_k - g_k^{(0)}\right) {\rm e}^{-\beta\hbar^2 k^2/m} \right\} \tag{6.126}$$

where $g_k {\rm d}k$ is the number of continuum states with $k$ between $k$ and $k + {\rm d}k$ in the interacting case and $g_k^{(0)} {\rm d}k$ is the corresponding quantity in the noninteracting case. We can express the corresponding difference in densities of states which appears in (6.126) in terms of the phase shifts of the two body scattering problem.

To do this, note that the solutions to the two body scattering problem are characterized by angular momentum quantum numbers $L$ and $m$ and take the form

$$\psi_{Lmk} = \text{constant} \times Y_L^m \frac{u_{Lmk}(r)}{r} \tag{6.127}$$

In the limit of large $r$

$$u_{Lmk}(r) \to \sin(kr + L\pi/2 + \eta_L(k)) \tag{6.128}$$

where $\eta_L(k)$ is the phase shift. (The allowed values of $L$ will depend on whether we are dealing with fermions or bosons and on whether there are any internal degrees of freedom. In the case that we are dealing with no internal degrees of freedom such as spin, the wave function $\psi_{Lmk}$ must be even under inversion for bosons and odd under inversion for fermions. Therefore in this case of spinless particles, we are confined to even $L$ for bosons and odd $L$ for fermions. We leave more complicated cases for exercises.)

To find the density of states consider a sphere of large radius $R$ around the origin in $\vec{r}$ space. On that sphere, the asymptotic form (6.128) will apply and a complete set of scattering states is obtained by requiring that the logarithmic derivative of the wave function take some fixed value on that sphere. (It turns out not to matter precisely what that value is as long as it is fixed.) Thus, the states of the continuum are obtained by requiring that

$$\tan(kR + L\pi/2 + \eta_l(k)) = A \tag{6.129}$$

where $A$ is some fixed value. But the solutions to (6.129) are

$$kR + L\pi/2 + \eta_l(k) = n\pi + \delta \tag{6.130}$$

where $n$ is any integer and $\delta$ is a fixed number. The interval $\Delta k$ between states is the change in $k$ required to take a value on the left hand side which satisfies (6.130) for integer $n$ on the right hand side to a value on the left hand side which satisfies (6.130) for the integer $n+1$ on the right hand side. That is

$$(k + \Delta k)R + L\pi/2 + \eta_L(k + \Delta k) = (n+1)\pi + \delta \tag{6.131}$$

We take the difference between the last two equations and expand $\eta_L(k + \Delta k)$ assuming $\Delta k/k \ll 1$:

$$\Delta k\left(R + \frac{\mathrm{d}\eta_L}{\mathrm{d}k}\right) = \pi \tag{6.132}$$

The density of states $g_k$ (for each $l$) is clearly $1/\Delta k$ and the noninteracting density $g_k^{(0)}$ is the same without the phase shift. Thus finally

$$g_k - g_k^{(0)} = \frac{1}{\pi}\frac{\mathrm{d}\eta_L}{\mathrm{d}k} \tag{6.133}$$

and the virial coefficient can be written as

$$b_2 - b_2^{(0)} = 2^{3/2}\left\{\sum_{n_\mathrm{B}} \mathrm{e}^{-\beta\epsilon_{n_\mathrm{B}}} + \int_0^\infty \mathrm{d}k \sum_L{}' \frac{2l+1}{\pi}\frac{\mathrm{d}\eta_L}{\mathrm{d}k}\mathrm{e}^{\frac{-\hbar^2k^2\beta}{m}}\right\} \tag{6.134}$$

Here we have used the fact that there are $2L+1$ values of $m$ associated with each $L$. The prime on the sum on $L$ is to remind us to take into account restrictions on the sum on $L$ arising from the statistics of the particles (fermions or bosons) as discussed above. In the case that there are internal degrees of freedom, the sums on even and odd values of $L$ could be weighted differently, but both might be present.

## Gross–Pitaevskii–Bogoliubov theory of the low temperature weakly interacting Bose gas

The quantum cluster expansion is useful at temperatures high enough so that $\lambda^3\rho \ll 1$ and, roughly speaking, $\rho$ times the collision cross-section of atoms to the $3/2$ power is small. The first condition is clearly not met below the Bose–Einstein condensation temperature which is roughly set by the condition $\lambda^3\rho \approx 1$. To deal with this low temperature situation in which the interactions are weak, one proceeds by starting with a ground state of the same form as the noninteracting ground state (which is Bose condensed) and finding states associated with small fluctuations around it arising from the presence of interactions. These low energy states are the only relevant ones at low temperatures and one can use them to find the partition function and thermodynamic functions. To formulate this approach it is very convenient to use a second quantized representation of the Hamiltonian. We give a brief description here and refer to other texts for more details. One may think of second quantization as a way of representing the Hamiltonian (and other relevant operators) in terms of its matrix elements in a basis of symmetrized product wave functions

$$\langle \vec{r} \mid \{n_\nu\}\rangle = \langle \vec{r}_1, \ldots, \vec{r}_N \mid \vec{n}_{\nu_1}, \ldots, \rangle = \frac{1}{\sqrt{\eta(n_{\nu_1}, \ldots, )}} \sum_{\mathcal{P}}{}' (\pm)^{\mathcal{P}} \prod_{i=1}^{N} \phi_{\nu_i}(\vec{r}_{\mathcal{P}(i)}) \tag{6.135}$$

as described in Chapter 4 except that here we allow for the possibility that the one particle part of the Hamiltonian includes potential as well as kinetic energy:

$$H = \sum_i H_1(i) + V \tag{6.136}$$

in which

$$H_1(i) = -\hbar^2\nabla_i^2/2m + V_1(\vec{r}_i) \tag{6.137}$$

and

$$H_1\phi_\nu = \epsilon_\nu\phi_\nu \tag{6.138}$$

In boson systems, some of the $\nu_i$ can be identical, with the factor $\eta$ defined to take account of this:

$$\eta(\nu_1, \ldots) = \frac{N!}{n_{\nu_1}!n_{\nu_2}!\cdots} \tag{6.139}$$

In the denominator, each set of quantum numbers appears only once and $n_{\nu_i}$ is the number of times $\nu_i$ appears in the wave function (6.135). It turns out that the matrix elements of operators in this basis can be described by the following prescription. (For the rest of this section we confine attention to the boson case.) Define operators $a_\nu, a_\nu^\dagger$ such that

$$[a_\nu, a_{\nu'}^\dagger] = \delta_{\nu,\nu'} \tag{6.140}$$

(as for the raising and lowering operators of a harmonic oscillator, of which this formalism is a generalization). Following the same logic used for harmonic oscillator raising and lowering operators, the eigenstates $|n_\nu\rangle$ of the operator $a_\nu^\dagger a_\nu$ have integer eigenvalues $n_\nu = 0, 1, 2, \ldots$ and can be written

$$|n_\nu\rangle \equiv \frac{(a_\nu^\dagger)^{n_\nu}}{\sqrt{n_\nu!}}|0\rangle \tag{6.141}$$

in which $a_\nu|0\rangle = 0$. One constructs a basis involving all the $\nu$ by taking the direct product of all the states in (6.141):

$$|\{n_\nu\}\rangle = \prod_\nu \frac{(a_\nu^\dagger)^{n_\nu}}{\sqrt{n_\nu!}}|0\rangle \tag{6.142}$$

Now the idea is to construct a Hamiltonian in terms of the operators $a_\nu, a_\nu^\dagger$ which when evaluated in the abstract basis $|n_\nu\rangle$ has the same matrix elements as the original Hamiltonian had when evaluated in the basis (6.135). (These states are characterized by the same sets of quantum numbers $\{n_\nu\}$.) First consider the part of the Hamiltonian without particle interactions:

$$H_0 = \sum_i H_1(i) \tag{6.143}$$

It is not hard to show that

$$\int \mathrm{d}^N\vec{r}\,\langle\{n_\nu\}|\vec{r}_1, \ldots, \vec{r}_N\rangle H_0\langle\vec{r}_1, \ldots, \vec{r}_N \mid \{n_\nu\}\rangle = \sum_\nu n_\nu\epsilon_\nu \tag{6.144}$$

so that the equivalent Hamiltonian for use in the abstract ("second quantized") basis is

$$H_0' = \sum_\nu \epsilon_\nu a_\nu^\dagger a_\nu \tag{6.145}$$

A careful analysis of the matrix elements shows that if the interaction term has the form

$$V = \sum_{i<j} v(\vec{r}_{ij}) \tag{6.146}$$

then the corresponding operator is

$$V' = \sum_{\nu_1,\nu_2,\nu_3,\nu_4} v_{\nu_1,\nu_2,\nu_3,\nu_4} a_{\nu_1}^\dagger a_{\nu_2}^\dagger a_{\nu_3} a_{\nu_4} \tag{6.147}$$

in which

$$v_{\nu_1,\nu_2,\nu_3,\nu_4} = \int \mathrm{d}\vec{r}_1\,\mathrm{d}\vec{r}_2\,\phi_{\nu_1}^*(\vec{r}_1)\phi_{\nu_2}^*(\vec{r}_2)v(\vec{r}_{12})\phi_{\nu_3}(\vec{r}_2)\phi_{\nu_4}(\vec{r}_1) \tag{6.148}$$

For thinking about approximations it is useful to recast this Hamiltonian in terms of a wave operator $\psi(\vec{r})$ defined so that

$$\psi(\vec{r}) = \sum_{\nu} \phi_{\nu} a_{\nu} \tag{6.149}$$

Using completeness of the functions $\phi_{\nu}$ one shows without difficulty that

$$[\psi(\vec{r}), \psi^{\dagger}(\vec{r}')] = \delta(\vec{r} - \vec{r}') \tag{6.150}$$

and that

$$H' = H_0' + V' = \int d\vec{r} \psi^{\dagger}(\vec{r}) H_1 \psi(\vec{r}) + (1/2) \int d\vec{r}_1 \int d\vec{r}_2 \psi^{\dagger}(\vec{r}_1) \psi^{\dagger}(\vec{r}_2) v(\vec{r}_{12}) \psi(\vec{r}_2) \psi(\vec{r}_1) \tag{6.151}$$

Now we consider a weakly interacting Bose gas below its condensation temperature. In the noninteracting case, the system is characterized by the basis (6.135) (which are eigenstates in that case) and the density matrix at finite temperature is diagonal with a very large amplitude for states which have a macroscopically large number of factors $\phi_{\nu_0}(\vec{r})$ associated with the ground state of the one particle Hamiltonian $H_1$. We postulate that when the interactions are weak a similar situation prevails in that only admixtures of states in the basis (6.135) with a macroscopically large number of factors $\phi_{\nu_0}(\vec{r})$ play a significant role in the relevant eigenstates. (However we do not assume that $\phi_{\nu_0}$ has the same form as it did in the noninteracting case.) Now consider the wave operator $\psi(\vec{r}) = \sum_{\nu} \phi_{\nu} a_{\nu}$. Acting on the restricted set of states postulated to be relevant at low temperatures, the summand $\phi_{\nu_0} a_{\nu_0}$ will always produce terms with one less factor $\phi_{\nu_0}$ times $\sqrt{N_0}$ where $N_0 \gg 1$ is the macroscopically large number of factors $\phi_{\nu_0}$ in the basis function on which it is acting. Thus this single term in the summand, when acting on the restricted basis, will yield large numbers and will give results much larger than the other terms in the summand. Similarly, the dominant term in $\psi^{\dagger}(\vec{r})$ will yield terms with one more factor $\phi_{\nu_0}$ times $\sqrt{N_0 + 1}$, but since $N_0 \gg 1$ this factor is nearly the same as the one obtained for $\psi(\vec{r})$. Guided by these arguments one constructs a systematic low temperature approximation by writing

$$\psi(\vec{r}) = \sqrt{N_0} \phi_0(\vec{r}) + \lambda \delta \psi(\vec{r}) \tag{6.152}$$

where $N_0$ is defined to be $\langle a_{\nu_0}^{\dagger} a_{\nu_0} \rangle$, the expectation value of the number of factors $\phi_{\nu_0}$ evaluated at zero temperature. $\lambda$ is a counting device, set equal to 1 at the end of the calculation. $\phi_0(\vec{r})$ is assumed to be a function (not changing the number of particles) and is not assumed to be the same as $\phi_{\nu_0}$. Rather, $\phi_0(\vec{r})$ is determined self consistently by minimization of the energy at zero temperature. $\delta\psi(\vec{r})$ is assumed within the restricted basis to yield smaller results than the first term but its operator character is retained. One can then construct a systematic calculational scheme by inserting (6.152) into (6.151) and solving the resulting problem order by order in increasing powers of $\lambda$. At zeroth order in $\lambda$, the Hamiltonian becomes

$$N_0 \int d\vec{r}\, \phi_0^*(r) H_1 \phi_0(r) + (N_0^2/2) \int d\vec{r}\, d\vec{r}'\, \phi_0^*(r)\phi_0^*(r') v(\vec{r}-\vec{r}')\phi_0(r')\phi_0(r) \tag{6.153}$$

This is just a number in the approximation used, so its expectation value is the same as its value. $\phi_0(r)$ is determined by minimizing it with respect to variations in $\phi_0^*$ subject to the constraint that the total number of particles (equal to $N_0$ at zero order in $\lambda$) remain fixed. The constraint is imposed by subtracting the term $\mu N_0 \int d\vec{r} \phi_0^*(r)\phi_0(r)$ where $\mu$ is a Lagrange multiplier equal to the chemical potential, prior to minimization. The resulting equation for $\phi_0(r)$,

$$H_1\phi_0(r) + N_0 \int d\vec{r}'|\phi_0(r')|^2 v(\vec{r}-\vec{r}')\phi_0(r) = \mu\phi_0(r) \tag{6.154}$$

is called the Gross–Pitaevskii equation. It has nearly the form of a Schrödinger equation but it is nonlinear and the eigenvalue is the chemical potential, not the energy of any particle. Its form is quite typical of mean field theories for the behavior of low temperature phases of condensed matter systems. (A comparison with Hartree theory of electrons is also instructive but we will not dwell on that here.) In the case that $H_1$ is simply the kinetic energy and the particles are confined to a large rectangular box with periodic boundary conditions, any plane wave will solve the Gross–Pitaevskii equation. The preferred solution is then the one with lowest energy and this will always be the solution $\phi_0(\vec{r}) = 1/\sqrt{\Omega}$ where $\Omega$ is the system volume giving a chemical potential $\mu = (N_0/\Omega)\int d\vec{r}' v(\vec{r}-\vec{r}')$. Notice that for most realistic interatomic potentials such as the Lennard-Jones potential (equation (6.2)) this integral is *divergent* so the approach fails utterly. This difficulty makes the Gross–Pitaevskii equation a hopeless starting point for the study of superfluid $^4$He. (This kind of problem also motivated the introduction of cluster expansions as mentioned earlier in this chapter.) However, in the special case of a dilute gas which has been kept in a metastable vapor phase at very low temperatures (which is precisely the situation in the alkali and atomic hydrogen vapors in which Bose–Einstein condensation has been observed) one can get around this difficulty as follows. Because the energies of the particles are low, the divergent part of the potential is never probed in collisions and one can use an effective interaction which reproduces the low energy features of the collision cross-section. This scattering length approximation, discussed in many elementary quantum mechanics books, takes the form

$$v(\vec{r}-\vec{r}') = (4\pi\hbar^2 a_s/m)\delta(\vec{r}-\vec{r}\,') \tag{6.155}$$

where $a_s$ is a parameter called the scattering length which characterizes the low energy collisions. Using this, the chemical potential for a gas in a rectangular box

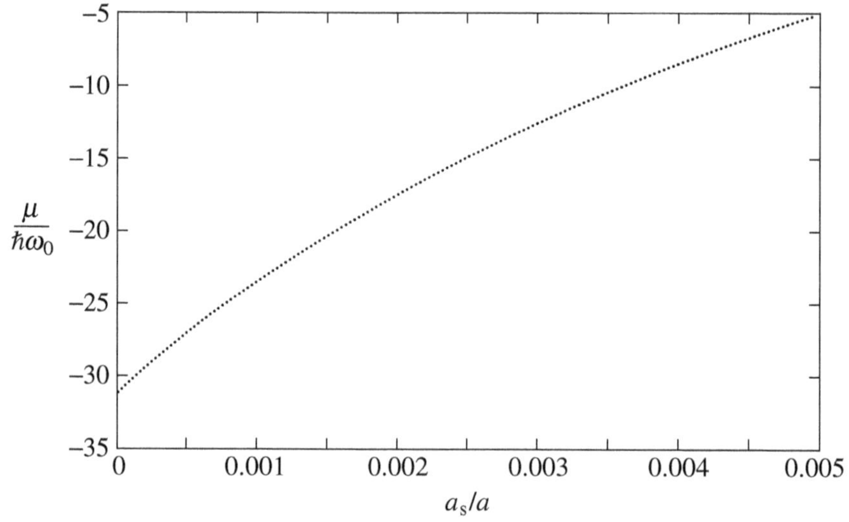


Figure 6.1 Chemical potential as a function of scattering length for a gas in a spherical well of radius $a$ and depth 38 $\hbar\omega_0$ where $\omega_0 = \hbar/2ma^2$. From reference 3.

is finite and of reasonable magnitude. In the real dilute systems in which Bose–Einstein condensation has been observed, the confining potential for the atoms does not consist of a set of impenetrable walls forming a large rectangular box however. Instead, $H_1$ includes both the kinetic energy and a harmonic potential which, in fact, is cut off at large distances from the origin. In such cases one must solve the Gross–Pitaevskii equation numerically. For example, in Figure 6.1 we show values of the chemical potential as a function of scattering length for a spherical well of finite depth.

The next order in $\lambda$ in the Hamiltonian is $\lambda^2$ because the linear term in $\lambda$ has the same coefficient which was required to be zero by the Gross–Pitaevskii equation. The term of order $\lambda^2$ is (with $\lambda$ set to 1)

$$\begin{aligned} H_2' = & \int d\vec{r}\, \delta\psi^\dagger(r) H_1 \delta\psi(r) + (N_0) \int d\vec{r}\, d\vec{r}\,' |\phi_0(r)|^2 \delta\psi(r')^\dagger v(\vec{r} - \vec{r}\,')\delta\psi(r') \\ & + (N_0) \int d\vec{r}\, d\vec{r}\,'\, \phi_0(r)^* \phi_0(r') \delta\psi^\dagger(\vec{r}\,') v(\vec{r} - \vec{r}\,') \delta\psi(\vec{r}) \\ & + (N_0/2) \int d\vec{r}\, d\vec{r}\,'\, \phi_0(\vec{r})^* \phi_0(\vec{r}\,')^*(\vec{r}\,') v(\vec{r} - \vec{r}\,') \delta\psi(r') \delta\psi(r) \\ & + (N_0/2) \int d\vec{r}\, d\vec{r}\,'\, \delta\psi(\vec{r})^\dagger \delta\psi(r)^\dagger v(\vec{r} - \vec{r}\,') \phi_0(\vec{r}) \phi_0(\vec{r}\,') \end{aligned} \tag{6.156}$$

This is a second quantized operator in which the operators $\delta\psi^\dagger(r)$ and $\delta\psi(\vec{r})$ may be regarded as adding or removing particles from the $\phi_0$ condensate state in the product wave functions which make up the underlying basis. To use this to find states corresponding to the small amplitude excitations of the system about the Gross–Pitaevskii solution one writes the Heisenberg equation of motion

$$i\hbar\frac{\partial\delta\psi}{\partial t} = [\delta\psi, H_2] = \left[\frac{-\hbar^2\nabla^2}{2m} + V_1 - \mu + 2gN_0 \mid \phi_0(\vec{r}) \mid^2\right]\delta\psi(\vec{r})$$
$$+ N_0 g \mid \phi_0(\vec{r}) \mid^2 \delta\psi^\dagger(\vec{r}) \tag{6.157}$$

in which the scattering length approximation has been used and $g = 4\pi\hbar^2 a_s/m$. Because both $\delta\psi$ and $\delta\psi^\dagger$ appear here, the eigenstates will correspond to acting on the ground, Gross–Pitaevskii, state with a linear combination of terms involving $\delta\psi$ and $\delta\psi^\dagger$. By appropriate normalization, such a linear combination can be required to obey Bose commutation relations. Let the required linear combinations be denoted $b_\mu^\dagger$, which creates an excited state, and $b_\mu$ which destroys one. By inversion of the linear relation, $\delta\psi$ can be written as a linear combination of the $b_\mu$ and $b_\mu^\dagger$:

$$\delta\psi(\vec{r}, t) = \sum_\mu \left[U_\mu(\vec{r})b_\mu\, e^{-i\omega_\mu t} + V_\mu(\vec{r})b_\mu^\dagger\, e^{i\omega_\mu t}\right] \tag{6.158}$$

Inserting this in (6.157) and taking $[\ldots, b_\nu^\dagger]$ and $[b_\nu, \ldots]$ of the result gives

$$\left[-\frac{\hbar^2}{2m}\nabla^2 + V_1 - \mu + 2N_0 g\phi_0^2 - \hbar\omega_\nu\right]U_\nu(\vec{r}) + N_0 g\phi_0^2 V_\nu = 0 \tag{6.159}$$

$$\left[-\frac{\hbar^2}{2m}\nabla^2 + V_1 - \mu + 2N_0 g\phi_0^2 + \hbar\omega_\nu\right]V_\nu(\vec{r}) + N_0 g\phi_0^2 U_\nu = 0 \tag{6.160}$$

which are sometimes called the Bogoliubov equations. It is not too hard to show that in terms of the eigenvalues $\omega_\nu$ $H_2$ becomes

$$H_2 = \sum_\nu \hbar\omega_\nu b_\nu^\dagger b_\nu \tag{6.161}$$

so that at this level of approximation one can regard the excitations created by the ladder like creation operators $b_\nu^\dagger$ as noninteracting bosons. However, because they are derived from $\delta\psi$ which does not include the part of the wave function associated with the factors $\phi_0$, there is no reason to expect the number of boson like excitations associated with the $b_\nu^\dagger$ to be conserved. The low temperature statistical mechanics associated with (6.161) is that of nonconserved boson excitations with no chemical potential (or equivalently with zero chemical potential). Thus the energy at low temperatures is

$$U = \sum_\nu \frac{\hbar\omega_\nu}{e^{\hbar\omega_\nu\beta} - 1} \tag{6.162}$$

and the specific heat associated with these excitations is easily obtained once the spectrum $\omega_\nu$ is known. The spectrum is obtained by solution of the equations (6.160) and this depends in turn on the boundary conditions and on the potential $V_1$. In the case of a rectangular box with periodic boundary conditions, $V_1$ is zero and the equations are solved by plane waves in the form $U_\nu(\vec{r}) = u_{\vec{k}} e^{i\vec{k}\cdot\vec{r}}, V_\nu(\vec{r}) = v_{\vec{k}} e^{-i\vec{k}\cdot\vec{r}}$. Putting these forms into the equations (6.160) gives ($V_0 \equiv N_0 g/\Omega$, $\epsilon_{\vec{k}} \equiv \hbar^2 k^2/2m$)

$$(\epsilon_{\vec{k}} - \mu + 2V_0 - \hbar\omega_{\vec{k}})u_{\vec{k}} + V_0 v^*_{\vec{k}} = 0 \tag{6.163}$$

$$V_0 u_{\vec{k}} + (\epsilon_{\vec{k}} - \mu + 2V_0 + \hbar\omega_{\vec{k}})v^*_{\vec{k}} = 0 \tag{6.164}$$

The corresponding secular equation is

$$\begin{vmatrix} \epsilon_{\vec{k}} - \mu + 2V_0 - \hbar\omega_{\vec{k}} & V_0 \\ V_0 & \epsilon_{\vec{k}} - \mu + 2V_0 + \hbar\omega_{\vec{k}} \end{vmatrix} = 0 \tag{6.165}$$

which has solutions

$$\hbar\omega_{\vec{k}} = \sqrt{\epsilon^2_{\vec{k}} + 2\epsilon_{\vec{k}} V_0} \tag{6.166}$$

where we used the value $\mu = N_0 g/\Omega$ for the chemical potential as obtained from the expression from the Gross–Pitaevskii equation above, specialized to this case. The interesting thing about (6.166) is that as $|\vec{k}| \to 0$, $\omega_{\vec{k}} \to c|\vec{k}|$ with $c = \sqrt{N_0 g/\Omega m}$ so sound like modes are found at long wavelength. These are very far from ordinary hydrodynamic sound waves however. For other cases, the Bogoliubov equations have also been solved numerically.[3]

## References

1. S. A. Rice and P. Gray, *The Statistical Mechanics of Simple Liquids*, New York: Wiley Interscience, 1965, p. 25.
2. G. E. Uhlenbeck and G. W. Ford, *Lectures in Statistical Mechanics*, Providence, RI: American Mathematical Society, 1963, p. 29.
3. A. Wynveen, A. Setty, A. Howard, J. W. Halley and C. E. Campbell, *Physical Review A* **62** (2000) 023602.

## Problems

6.1 Consider a classical gas of particles in $d$ spatial dimensions which interact through the two body potential

$$v(r) = \epsilon \left(\frac{b}{r}\right)^n$$

where $\epsilon$ is a positive number with the dimensions of energy and $n$ is a positive integer. $b$ is a parameter with the dimensions of length.

(a) Reduce the expression for the sum $b_2$ of the integrals of all the distinct 2-clusters to a dimensionless integral times powers of $b$, $\epsilon$ and the temperature.

(b) By analysis of the integral, show that the virial series cannot be convergent unless $d < n$.

(c) Using the expression found in (a) find the equation of state for this gas including the first correction to the ideal gas equation of state arising from $b_2$. By drawing qualitative graphs of some isotherms in the $P$–$V$ plane, indicate how the equation of state differs for different values of $n$ at fixed $d$.

6.2 Consider a model semiclassical imperfect gas in which the atoms are interacting through the two body potential

$$v(r) = \begin{cases} \to \infty & r < \sigma \\ -V_0 & \sigma < r < 2\sigma \end{cases}$$

and 0 for $r > 2\sigma$.

(a) Calculate the value of the linked cluster integral $b_2$ as a function of the temperature, the mass of the particles and the parameters in the potential.

(b) Evaluate the pressure as a function of the density $\rho$ in a series to second order in the density. At what temperature does the predicted equation of state change qualitatively? Sketch a graph of the pressure as a function of $1/\rho$ for temperatures above, equal to and below this temperature.

(c) Find the fugacity $z = \mathrm{e}^{\beta\mu}$ as a function of the density $\rho$ to second order in $\rho$ when $\langle N \rangle$, the number of particles, is fixed.

(d) Using the result of (c), write the Helmholtz free energy $F$ in terms of the density $\rho$, showing the ideal gas result, plus the first correction due to the interactions.

(e) Using (d), calculate an expression for the entropy $S$ and the specific heat $C_v$ to the same order. Make a qualitative graph of the specific heat as a function of $T$, comparing it with the ideal gas result.

6.3 For the Lennard-Jones interaction, calculate the sum of linked 2- and 3-clusters using parameters for argon of $\epsilon_{\mathrm{LJ}} = 125$ K and $\sigma = 3.45$ Å. Evaluate your expressions numerically for a range of temperatures from 0.1 of room temperature to twice room temperature. Using your evaluation of the linked 2- and 3-clusters for argon, calculate and plot isotherms for $P$ as a function of $1/\rho$ for argon in the same temperature range for which you found $b_2$ and $b_3$. Compare the results with the ideal gas equation of state and with the van der Waals equation of state. Indicate the regions in which the first three terms of the virial expansion should be adequate. The van der Waals equation of state is of the form

$$P_{\text{van der Waals}} = N_{\mathrm{A}} k_{\mathrm{B}} T/(V N_{\mathrm{A}}/N - b) - a/(V N_{\mathrm{A}}/N)^2$$

where $N_{\mathrm{A}}$ is Avogadro's number. $V N_{\mathrm{A}}/N$ is the volume per mole of the gas or fluid and is sometimes denoted $\tilde{V}$. $a$ and $b$ are constants. One gets the values for argon by using the values of $P, T$ and $\tilde{V}$ at the critical point at which

$(\partial P/\partial V)_T = 0$ and $(\partial^2 P/\partial V^2)_T = 0$. The critical values for the van der Waals equation of state are $P_c = a/27b^2$, $\tilde{V}_c = 3b$ and $T_c = (8/27)(a/N_A k_B)$. Experimentally for argon, $\tilde{V}_c = 75.2\,\text{cm}^3/\text{mol}$ and $P_c = 48$ atm yielding $b = 25.07\,\text{cm}^3/\text{mol}$, $a = 10828.8\,\text{atm cm}^3/\text{mol}$.

6.4 Show that there are respectively three and six distinct contributions from irreducible 4-clusters of the following two types respectively, as claimed in the text. To make your proof convincing, show the distinct contributions explicitly in each case.

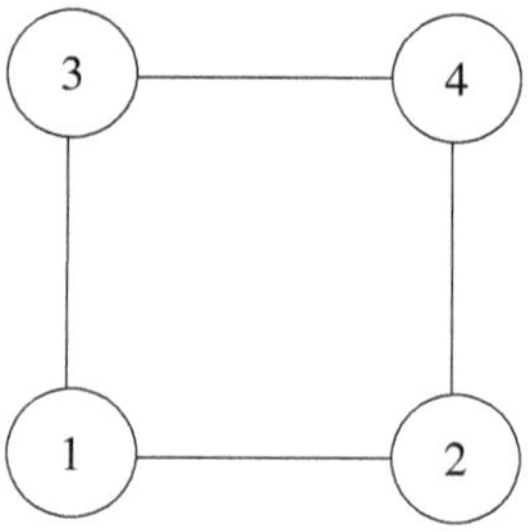


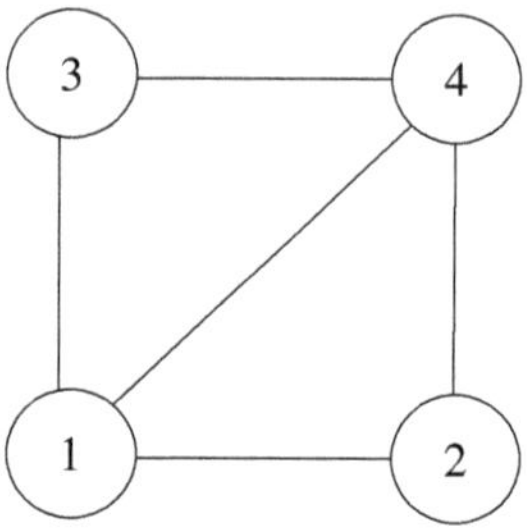


Are there any more irreducible 4-clusters? If so, draw them and show how many contributions there are from them.

6.5 (a) Find an explicit form for $U_1$ in the quantum cluster expansion and evaluate it for the case of periodic boundary conditions. (b) Show explicitly that $W_2(1,2) \rightarrow U_1(1)U_1(2)$ in the same case.

# 7

# Statistical mechanics of liquids

Here we briefly review some aspects of the statistical mechanics of liquids. The distinction between a liquid and a gas is not sharp except in the neighborhood of the transition between them which we will discuss in Chapter 10. As a working definition we will consider a system to be a liquid if it lacks the geometrical structure associated, for example, with crystals and for which the density expansions discussed in the last chapter do not converge. This distinction can be made somewhat sharper when we have discussed the relevant correlation functions and phase diagrams. It is to be noted that for atomic and molecular systems which can be treated classically and for which the two body interactions contain a hard core at short distances, there will always be a region of the thermodynamic phase space (for example in the $PT$ diagram) for which the system will behave as a liquid according to this definition.

We begin the discussion by defining correlation functions which are very useful for characterizing the structure of liquids and also for making measurements and formulating theories to describe them. The considerations here apply equally well to the imperfect gases discussed in the last chapter, but they are particularly useful and necessary for the discussion of liquids. We next describe experimental techniques which directly measure some of these correlation functions. Finally we briefly describe two distinct theoretical approaches to the description of liquids: analytical formulations based on approximate summations of series like those described in the last chapter, and numerical simulation. Numerical simulation is quite a general method but it is particularly useful for liquids where the complexity and approximate nature of analytical theories limits their usefulness. In most of this chapter the considerations are classical though similar ideas are useful in the study of quantum fluids, which is the subject of the following chapter.

**Definitions of $n$-particle distribution functions**

We consider an $N$-particle classical system characterized by a $3N$ dimensional coordinate space $\mathbf{R}$. (Generalization to any dimension is trivial for these definitions.) We define the $n$-particle distribution function $n_n(\vec{r}_1, \ldots, \vec{r}_n)$ as

$$n_n(\vec{r}_1, \ldots, \vec{r}_n)\mathrm{d}\vec{r}_1 \cdots \mathrm{d}\vec{r}_n$$
$$= \{\text{Probability that there is an atom in } \mathrm{d}\vec{r}_1\text{, an atom in } \mathrm{d}\vec{r}_2, \ldots\text{, in } \mathrm{d}\vec{r}_n \tag{7.1}$$

In the canonical distribution the probability of finding the atoms labelled $1, \ldots, n$ in the volume element $\mathrm{d}\vec{r}_1 \cdots \mathrm{d}\vec{r}_n$ is

$$\frac{\int \mathrm{d}\vec{r}_{n+1} \cdots \mathrm{d}\vec{r}_N \, \mathrm{e}^{-\beta U(\mathbf{R})}}{\int \mathrm{d}\vec{r}_1 \cdots \mathrm{d}\vec{r}_N \, \mathrm{e}^{-\beta U(\mathbf{R})}} \tag{7.2}$$

However, the definition of $n_n(\vec{r}_1, \ldots, \vec{r}_n)$ calls for the probability that *any* $n$ atoms lie in the relevant volume elements. Therefore, to obtain an expression for $n_n(\vec{r}_1, \ldots, \vec{r}_n)$ we must multiply (7.2) by the number of ways of selecting $n$ atoms from $N$, counting all the permutations of the $n$ atoms as distinct. Thus

$$n_n(\vec{r}_1, \ldots, \vec{r}_n) = \frac{N!}{(N-n)!} \times \frac{\int \mathrm{d}\vec{r}_{n+1} \cdots \mathrm{d}\vec{r}_N \, \mathrm{e}^{-\beta U(\mathbf{R})}}{\int \mathrm{d}\vec{r}_1 \cdots \mathrm{d}\vec{r}_N \, \mathrm{e}^{-\beta U(\mathbf{R})}} \tag{7.3}$$

It should be obvious that $n_n(\vec{r}_1, \ldots, \vec{r}_n)$ can only be defined in the canonical ensemble when $n \leq N$. In the grand canonical ensemble the corresponding expression is obtained as follows. The distribution function is

$$\frac{1}{\lambda^{3N} N!} \mathrm{e}^{\beta N \mu - \beta U(\mathbf{R})} \tag{7.4}$$

The normalization of probabilities requires that we divide by this quantity summed over all $N$. By the preceding argument, the numerator is

$$\frac{N!}{(N-n)!} \frac{1}{\lambda^{3N} N!} \int \mathrm{d}\vec{r}_{n+1} \cdots \mathrm{d}\vec{r}_N \, \mathrm{e}^{\beta N \mu - \beta U(\mathbf{R})} \tag{7.5}$$

summed over all $N \geq n$. Thus, in the grand canonical distribution

$$n_n(\vec{r}_1, \ldots, \vec{r}_n) = \frac{\sum_{N=n}^{\infty} \frac{z^N}{\lambda^{3N}(N-n)!} \int \mathrm{d}\vec{r}_{n+1} \cdots \mathrm{d}\vec{r}_N \, \mathrm{e}^{-\beta U(\mathbf{R})}}{\sum_{N=1}^{\infty} \frac{z^N}{\lambda^{3N} N!} \int \mathrm{d}\vec{r}_1 \cdots \mathrm{d}\vec{r}_N \, \mathrm{e}^{-\beta U(\mathbf{R})}} \tag{7.6}$$

For most of the rest of the discussion we focus attention on $n_2(\vec{r}_1, \vec{r}_2)$. This is usually expressed in terms of the radial distribution function $g(\vec{r}_1, \vec{r}_2)$ which is defined for homogeneous systems as

$$g(\vec{r}_1, \vec{r}_2) = n_2(\vec{r}_1, \vec{r}_2)/\rho^2 \tag{7.7}$$

where $\rho$ is the mean density $N/V$. Note that $g(\vec{r}_1, \vec{r}_2)\rho \mathrm{d}\vec{r}_2$ can be interpreted as the probability that there is a particle in $\mathrm{d}\vec{r}_2$ at $\vec{r}_2$ given that there is a particle in $\mathrm{d}\vec{r}_1$ at $\vec{r}_1$.

With these definitions, and in the thermodynamic limit for a homogeneous system, $\lim_{|r_1 - \vec{r}_2| \to \infty} g(\vec{r}_1, \vec{r}_2) = 1$. In the case that the function $U(\mathbf{R})$ is a sum of pair potentials $U(\mathbf{R}) = \sum_{i<j} v(\vec{r}_i - \vec{r}_j)$ one can express all the thermodynamic properties of the system in terms of $g(r)$ as we now show. Generally (in the canonical distribution) the average energy is

$$E = -\frac{\partial}{\partial \beta} \ln Z_N = -\frac{1}{Z_N} \frac{\partial}{\partial \beta} Z_N \tag{7.8}$$

Thence

$$= \frac{\frac{\partial}{\partial \beta}\left(\frac{1}{N!\lambda^{3N}} \int \mathrm{d}^N \vec{r}\, \mathrm{e}^{-\beta \sum_{i<j} v_{ij}}\right)}{\frac{1}{N!\lambda^{3N}} \int \mathrm{d}^N \vec{r}\, \mathrm{e}^{-\beta \sum_{i<j} v_{ij}}} \tag{7.9}$$

Taking the derivative of the thermal wavelength with respect to $\beta$ one obtains

$$E = \frac{3}{2} N k_{\mathrm{B}} T + \frac{\int \mathrm{d}^N \vec{r} \sum_{i<j} v_{ij}\, \mathrm{e}^{-\beta \sum_{i<j} v_{ij}}}{\int \mathrm{d}^N \vec{r}\, \mathrm{e}^{-\beta \sum_{i<j} v_{ij}}} \tag{7.10}$$

The numerator of the second term can be rewritten as

$$\frac{N(N-1)}{2} \int \mathrm{d}\vec{r}_3 \cdots \mathrm{d}\vec{r}_N\, \mathrm{d}\vec{r}_1\, \mathrm{d}\vec{r}_2\, v(\vec{r}_1, \vec{r}_2)\, \mathrm{e}^{-\beta \sum_{i<j} v_{ij}} \tag{7.11}$$

so that the second term is

$$\frac{\frac{1}{2} \int \mathrm{d}\vec{r}_1\, \mathrm{d}\vec{r}_2 \frac{N!}{(N-2)!} \int \mathrm{d}\vec{r}_3 \cdots \mathrm{d}\vec{r}_N\, v(\vec{r}_1, \vec{r}_2)\, \mathrm{e}^{-\beta \sum_{i<j} v_{ij}}}{\int \mathrm{d}^N r\, \mathrm{e}^{-\beta \sum_{i<j} v_{ij}}}$$
$$= \frac{1}{2} \int \mathrm{d}\vec{r}_1\, \mathrm{d}\vec{r}_2\, v(\vec{r}_1, \vec{r}_2) n_2(\vec{r}_1, \vec{r}_2) = \frac{N\rho}{2} \int \mathrm{d}\vec{r}\, v(r) g(r) \tag{7.12}$$

Thus the internal energy is

$$E = N\left(\frac{3}{2} k_{\mathrm{B}} T + \frac{\rho}{2} \int \mathrm{d}\vec{r} v(r) g(r)\right) \tag{7.13}$$

To calculate the pressure, it is useful first to write the free energy as

$$F = -k_{\mathrm{B}} T \ln Z_N = -k_{\mathrm{B}} T \ln\left(\frac{1}{N!\lambda^{3N}} \int \mathrm{d}^N \vec{r}\, \mathrm{e}^{-\beta \sum_{i<j} v_{ij}}\right) \tag{7.14}$$

The pressure is then

$$P = -\left(\frac{\partial F}{\partial V}\right)_{V,T}$$
$$= k_{\mathrm{B}} T \frac{\frac{\partial}{\partial V}\left\{\frac{1}{\lambda^{3N}} \int \mathrm{d}^N \vec{r}\, \mathrm{e}^{-\beta \sum_{ij} v_{ij}}\right\}}{\left\{\frac{1}{\lambda^{3N}} \int \mathrm{d}^N \vec{r}\, \mathrm{e}^{-\beta \sum_{i<j} v_{ij}}\right\}} \tag{7.15}$$

In the numerator, the volume appears in the limits of integration. To obtain a useful expression for the pressure, it is best to remove this dependence in the limits by changing the variables of integration to dimensionless ones

$$\vec{r}_i' = \frac{1}{V^{1/3}}\vec{r}_i \tag{7.16}$$

Then

$$P = k_B T \frac{\frac{\partial}{\partial V}\left\{\frac{1}{\lambda^{3N}} V^N \int \mathrm{d}^N \vec{r}' \mathrm{e}^{-\beta \sum_{ij} v_{ij}}\right\}}{\left\{\frac{1}{\lambda^{3N}} \int \mathrm{d}^N \vec{r}\, \mathrm{e}^{-\beta \sum_{i<j} v_{ij}}\right\}} = \frac{N k_B T}{V} - \frac{V^N \int \mathrm{d}^N r' \sum_{ij} \frac{\partial v_{ij}}{\partial V} \mathrm{e}^{-\beta \sum_{ij} v_{ij}}}{\int \mathrm{d}^N \vec{r}\, \mathrm{e}^{-\beta \sum_{i<j} v_{ij}}} \tag{7.17}$$

The derivative in the second term is written

$$\frac{\partial}{\partial V} v(\vec{r}_{ij}) = \frac{\partial}{\partial V} v\left(V^{1/3}\vec{r}_{ij}'\right) = \frac{\partial}{\partial V}\left(V^{1/3}\vec{r}_{ij}'\right)\frac{\partial v_{ij}}{\partial \vec{r}_{ij}} = \frac{1}{3V}\vec{r}_{ij} \cdot \frac{\partial v_{ij}}{\partial \vec{r}_{ij}} \tag{7.18}$$

Thus we obtain

$$P = \frac{1}{V}\left(N k_B T + \frac{1}{3}\left\langle \sum_{i<j} \vec{r}_{ij} \cdot \vec{F}_{ij} \right\rangle\right) \tag{7.19}$$

This form is quite useful in simulations. The last term in brackets on the right hand side is called the "internal virial." One can express the last term in terms of the radial distribution function giving:

$$P = \frac{N k_B T}{V} - \frac{1}{6V}\int \mathrm{d}\vec{r}_1\, \mathrm{d}\vec{r}_2\, n_2(\vec{r}_1, \vec{r}_2)\vec{r}_{12} \cdot \frac{\partial v}{\partial \vec{r}_{12}} = \frac{N k_B T}{V} - \frac{\rho^2}{6}\int g(\vec{r})\vec{r} \cdot \frac{\partial v}{\partial \vec{r}} \mathrm{d}\vec{r} \tag{7.20}$$

## Determination of $g(r)$ by neutron and x-ray scattering

Neutron and x-ray scattering provide information on the structure of liquids and other dense materials on basically the same range of lengths, from a few tenths of angstroms out to nearly micrometers (though the methods on the longer scales are a bit different). (On the other hand neutrons and x-rays provide information about dynamics on quite distinct time scales.) Here we provide some details concerning how neutron scattering is used to determine liquid structure and, in particular, $g(r)$. However, the technique and the analysis have very wide applicability to the study of condensed matter. In the basic experiment, neutrons are prepared in a monochromatic beam with an initial wave vector $\vec{k}_\mathrm{i}$, usually by Bragg scattering a thermal beam of neutrons from a crystal. This monochromatic beam is then directed at the sample to be studied and the scattered neutrons at wave vectors $\vec{k}_\mathrm{f}$ are counted. (For experimental details, one can consult numerous books on neutron scattering, or for a brief description relevant to the monatomic liquid example of interest here, see reference 1.)

From the data on the numbers and energies of neutrons scattered in various directions from the sample, one deduces a differential scattering cross-section $d^2\sigma/d\Omega_{\vec{k}_f}d\epsilon_{\vec{k}_f}$ from the experiment. Here $d\Omega_{\vec{k}_f}$ is an element of solid angle within which the direction of the wave vector of the scattered neutron is fixed and $\epsilon_{\vec{k}_f}$ is the energy of the scattered neutron. This cross-section provides information about both statics and dynamics of the sample. The part of the cross-section around zero energy transfer (when $\epsilon_{\vec{k}_f}$ is the same as the energy of the incoming beam) gives the structural information and is called quasielastic scattering.

To understand how $g(r)$ is determined from neutron scattering data, it is useful to begin with an expression for the inelastic cross-section for neutron scattering from a many body system. The Born approximation is suitable for this problem because the interaction between neutrons and nuclei of the atoms of the liquid is weak when the neutrons are at thermal energies. The Born approximation is equivalent to use of the Fermi golden rule for the scattering rate. In terms of the rate, the cross-section is

$$\frac{d^2\sigma}{d\Omega_{\vec{k}_f}d\epsilon_{\vec{k}_f}} = \frac{\text{rate}}{\text{incident flux}} \tag{7.21}$$

We choose to normalize neutron wave functions in a large volume $V$ and have, supposing that the potential describing the interaction between the neutron and the liquid is $\mathcal{V}(\vec{r})$, from the Fermi golden rule for the rate:

$$\frac{d^2\sigma}{d\Omega_{\vec{k}_f}d\epsilon_{\vec{k}_f}} = \frac{\sum_{\nu_f,\nu_i} \frac{2\pi}{\hbar}\frac{e^{-\beta E_{\nu_i}}}{Z} \mid \langle \vec{k}_i, \nu_i \mid \mathcal{V} \mid \vec{k}_f, \nu_f\rangle \mid^2 \delta\left(\epsilon_{\vec{k}_i} + E_{\nu_i} - E_{\nu_f} - \epsilon_{\vec{k}_f}\right)}{\frac{\hbar k_i}{mV}}$$
$$\times \frac{V}{(2\pi)^3}\frac{k_f^2 d\Omega_{\vec{k}_f}dk_f}{d\Omega_{k_f}d\epsilon_{\vec{k}_f}} \tag{7.22}$$

for the cross-section. We have summed over all possible final states $\nu_f$ of the many body system (the sample) from which the neutron is scattering and have averaged over all initial states $\nu_i$ of that many body system, assuming that the target of the neutron scattering experiment is in thermal equilibrium and can be described by the canonical ensemble. We wrote $\hbar k_i/mV$ for the incident flux. Using $\epsilon_{\vec{k}_f} = \hbar^2 k_f^2/2m$ and rearranging this gives

$$\frac{d^2\sigma}{d\Omega_{\vec{k}_f}d\epsilon_{\vec{k}_f}} = \frac{V^2m^2}{\hbar^4(2\pi)^2}\left(\frac{k_f}{k_i}\right)\sum_{\nu_f,\nu_i}\frac{e^{-\beta E_{\nu_i}}}{Z} \mid \langle \vec{k}_i, \nu_i|V|\vec{k}_f, \nu_f\rangle \mid^2 \delta\left(\epsilon_{\vec{k}_i} + E_{\nu_i} - E_{\nu_f} - \epsilon_{\vec{k}_f}\right) \tag{7.23}$$

We now make a series of manipulations, usually attributed to Van Hove, which show that this quantity is closely related to a correlation function of the potential $\mathcal{V}(\vec{r}, t)$ with itself (where $\mathcal{V}(\vec{r}, t)$ is the Heisenberg representation of the operator

$\mathcal{V}(\vec{r})$ with respect to the Hamiltonian describing the sample). The delta function is rewritten using the identity

$$\delta(E) = \frac{1}{2\pi}\int_{-\infty}^{\infty}\frac{\mathrm{d}t}{\hbar}\mathrm{e}^{-\mathrm{i}Et/\hbar} \tag{7.24}$$

to give

$$\frac{\mathrm{d}^2\sigma}{\mathrm{d}\Omega_{\vec{k}_\mathrm{f}}\mathrm{d}\epsilon_{\vec{k}_\mathrm{f}}} = \frac{m^2}{\hbar^5(2\pi)^3}\left(\frac{k_\mathrm{f}}{k_\mathrm{i}}\right)\sum_{\nu_\mathrm{i}}\frac{\mathrm{e}^{-\beta E_{\nu_\mathrm{i}}}}{Z}\sum_{\nu_\mathrm{f}}\int \mathrm{d}\vec{r}\int \mathrm{d}\vec{r}\,'\int \mathrm{d}t$$
$$\times\langle\nu_\mathrm{i} \mid \mathcal{V}(\vec{r}) \mid \nu_\mathrm{f}\rangle\langle\nu_\mathrm{f} \mid \mathrm{e}^{-\mathrm{i}Ht/\hbar}\mathcal{V}(\vec{r}\,')\mathrm{e}^{\mathrm{i}Ht/\hbar} \mid \nu_\mathrm{i}\rangle\mathrm{e}^{-\mathrm{i}(\vec{r}-\vec{r}\,')\cdot(\vec{k}_\mathrm{f}-\vec{k}_\mathrm{i})}\,\mathrm{e}^{\mathrm{i}(\epsilon_{\vec{k}_\mathrm{i}}-\epsilon_{\vec{k}_\mathrm{f}})t/\hbar} \tag{7.25}$$

The initial and final wave functions were assumed to be separable products: $\langle\vec{r} \mid \vec{k}_{\mathrm{i,f}}, \nu_{\mathrm{i,f}}\rangle = \frac{\mathrm{e}^{\mathrm{i}\vec{k}_{\mathrm{i,f}}\cdot\vec{r}}}{\sqrt{V}} \mid \nu_{\mathrm{i,f}}\rangle$. The dependence of the interaction potential $\mathcal{V}(\vec{r})$ on the neutron coordinate $\vec{r}$ has been explicitly displayed. The sum on final states $\nu_\mathrm{f}$ of the target can now be done using the fact that they are complete so that $\sum_{\nu_\mathrm{f}} \mid \nu_\mathrm{f}\rangle\langle\nu_\mathrm{f} \mid = 1$ with the result

$$\frac{\mathrm{d}^2\sigma}{\mathrm{d}\Omega_{\vec{k}_\mathrm{f}}\mathrm{d}\epsilon_{\vec{k}_\mathrm{f}}} = \frac{m^2}{\hbar^5(2\pi)^3}\left(\frac{k_\mathrm{f}}{k_\mathrm{i}}\right)\int \mathrm{d}\vec{r}\int \mathrm{d}\vec{r}\,'\int \mathrm{d}t\,\mathrm{e}^{-\mathrm{i}\omega t}\mathrm{e}^{-\mathrm{i}\vec{Q}\cdot(\vec{r}-\vec{r}\,')}\langle\mathcal{V}(\vec{r},0)\mathcal{V}(\vec{r}\,',t)\rangle \tag{7.26}$$

Here the Heisenberg representation $\mathcal{V}(\vec{r},t) = \mathrm{e}^{\mathrm{i}Ht/\hbar}\mathcal{V}(\vec{r})\mathrm{e}^{-\mathrm{i}Ht/\hbar}$ of the interaction potential with respect to the target Hamiltonian $H$ has been used. The variables $\vec{Q} = \vec{k}_\mathrm{i} - \vec{k}_\mathrm{f}$ and $\omega = (1/\hbar)(\epsilon_\mathrm{i} - \epsilon_\mathrm{f})$ have been defined. Up to this point the result is quite general and is useful in measurement of dynamical correlation functions. To measure static correlations functions we note that if the sample is an isotropic liquid then on general symmetry grounds the quantity

$$\int \mathrm{d}\vec{r}\int \mathrm{d}\vec{r}\,'\,\mathrm{e}^{-\mathrm{i}\vec{Q}\cdot(\vec{r}-\vec{r}\,')}\langle\mathcal{V}(\vec{r},0)\mathcal{V}(\vec{r}\,',t)\rangle \tag{7.27}$$

will depend only on the magnitude of the vector $\vec{Q}$. In that case, it is possible to integrate the cross-section $\mathrm{d}^2\sigma/\mathrm{d}\Omega_{\vec{k}_\mathrm{f}}\mathrm{d}\epsilon_{\vec{k}_\mathrm{f}}$ on the variable $\omega$ at fixed magnitude of $\vec{Q}$ using the experimental scattering data. Some details concerning how this is done appear in the Appendix in reference 1. We thus obtain

$$\hbar\int \mathrm{d}\omega\left(\frac{k_\mathrm{i}}{k_\mathrm{f}}\right)\frac{\mathrm{d}^2\sigma}{\mathrm{d}\Omega_{\vec{k}_\mathrm{f}}\mathrm{d}\epsilon_{\vec{k}_\mathrm{f}}} = \frac{m^2}{\hbar^4(2\pi)^2}\int \mathrm{d}\vec{r}\int \mathrm{d}\vec{r}\,'\,\mathrm{e}^{-\mathrm{i}\vec{Q}\cdot(\vec{r}-\vec{r}\,')}\langle V(\vec{r},0)V(\vec{r}\,',0)\rangle \tag{7.28}$$

At the low neutron energies of interest, the interaction potential can be written in scattering length approximation:

$$\mathcal{V}(\vec{r}) = \frac{2\pi\hbar^2}{m}\sum_{i=1}^{N} b_{i,\nu}\delta(\vec{r}-\vec{r}_i) \tag{7.29}$$

where the $\vec{r}_i$ are the positions of the atoms of the liquid. We include an index $\nu$ on the scattering length $b_{i,\nu}$ to account for the fact that the nucleus of the $i$th atom may in some cases be in more than one state, either because of various possible nuclear spin values for the nucleus (as in the case, for example, of hydrogen) or because of the existence of an isotopic mixture in the atoms of the fluid. In either case, the sum over initial states (the ensemble average) must include a sum over this index $\nu$. ($\mathcal{V}(\vec{r})$ is to be regarded as the diagonal matrix element of the full nuclear Hamiltonian, with respect to the degrees of freedom represented by this index $\nu$.) Then

$$\int \mathrm{d}\,\hbar\omega \left(\frac{k_\mathrm{i}}{k_\mathrm{f}}\right) \frac{\mathrm{d}^2\sigma}{\mathrm{d}\Omega_{\vec{k}_\mathrm{f}}\mathrm{d}\epsilon_{\vec{k}_\mathrm{f}}} = \left\langle \left| \int \mathrm{d}\vec{r}\, \mathrm{e}^{\mathrm{i}\vec{Q}\cdot\vec{r}} \sum_{i=1}^{N} b_{i,\nu}\delta(\vec{r}-\vec{r}_i) \right|^2 \right\rangle \tag{7.30}$$

It is useful to rewrite this expression by expressing $b_{i,\nu}$ in terms of the average $\bar{b} = \sum_\nu P_\nu b_{i,\nu}$ and the difference $\Delta b_{i,\nu} = b_{i,\nu} - \bar{b}$: $b_{i,\nu} = \Delta b_{i,\nu} + \bar{b}$. Here $P_\nu$ is the probability of the nuclear state $\nu$ and the average $\bar{b}$ will usually be independent of $i$. Equation (7.30) now becomes

$$\int \mathrm{d}\,\hbar\omega \left(\frac{k_\mathrm{i}}{k_\mathrm{f}}\right) \frac{\mathrm{d}^2\sigma}{\mathrm{d}\Omega_{\vec{k}_\mathrm{f}}\mathrm{d}\epsilon_{\vec{k}_\mathrm{f}}} = \sum_i \sum_\nu P_\nu(\bar{b}^2 + \Delta b_{i,\nu}^2) + \sum_{i\neq j} \left( \sum_{\nu,\nu'} P_\nu P_{\nu'} \Delta b_{i,\nu'} \Delta b_{j,\nu} \right.$$
$$\left. + \sum_\nu P_\nu \Delta b_{i,\nu}\bar{b} + \bar{b} \sum_{\nu'} P_{\nu''} \Delta b_{j,\nu'} + \bar{b}^2 \right) \langle \mathrm{e}^{\mathrm{i}\vec{Q}\cdot(\vec{r}_i - \vec{r}_j)} \rangle \tag{7.31}$$

The averages on the nuclear variables $\nu$ clearly give zero except in the case of the first and last terms with the result

$$\int \mathrm{d}\,\hbar\omega \left(\frac{k_\mathrm{i}}{k_\mathrm{f}}\right) \frac{\mathrm{d}^2\sigma}{\mathrm{d}\Omega_{\vec{k}_\mathrm{f}}\mathrm{d}\epsilon_{\vec{k}_\mathrm{f}}} = N(\bar{b}^2 + \Delta\bar{b}^2) + \bar{b}^2 \sum_{i,j} \langle \mathrm{e}^{\mathrm{i}\vec{Q}\cdot(\vec{r}_i - \vec{r}_j)} \rangle \tag{7.32}$$

The second term is called the coherent scattering elastic cross-section and can be determined from $g(r)$ or, conversely, the experiment can be used to determine $g(r)$ from it. To see this we note that, in the case of a liquid, it follows from the definition of $n_2$ that

$$n_2(\vec{r}) = (\rho/N) \sum_{i\neq j} \langle \delta(\vec{r} - \vec{r}_{ij}) \rangle \tag{7.33}$$

so that

$$g(\vec{r}) = \frac{n_2}{\rho^2} = \frac{1}{N\rho} \sum_{i\neq j} \langle \delta(\vec{r} - \vec{r}_{ij}) \rangle \tag{7.34}$$

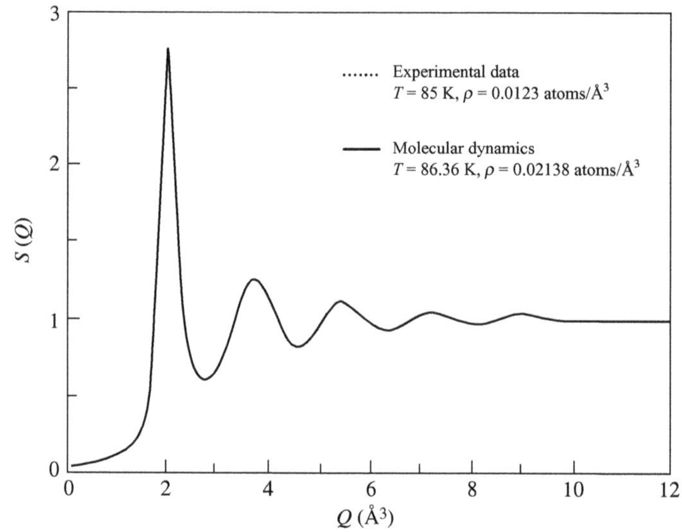


Figure 7.1 Experimentally determined $S(Q)$ for $^{36}$Ar at 85 K from reference 1. The solid line is the result of a molecular dynamics simulation, as described at the end of the chapter. It is indistinguishable from the experimental result.

Using this we rewrite the last term of the expression for the scattering cross-section as

$$\begin{aligned}\bar{b}^2 \sum_{i,j} \int d\vec{r} \langle \delta(\vec{r} - \vec{r}_{ij}) \rangle \, e^{\vec{Q}\cdot\vec{r}} &= N\bar{b}^2 \left[ \int d\vec{r} \, (\rho g(r) + \delta(\vec{r})) \, e^{i\vec{Q}\cdot\vec{r}} \right] \\ &= N\bar{b}^2 \left[ 1 + \rho \int d\vec{r} \, g(r) e^{i\vec{Q}\cdot\vec{r}} \right] \\ &= N\bar{b}^2 \left[ 1 + \rho \int d\vec{r} \, (g(r) - 1) \, e^{i\vec{Q}\cdot\vec{r}} + \rho \int d\vec{r} \, e^{i\vec{Q}\cdot\vec{r}} \right] \\ &= N\bar{b}^2 \left[ 1 + \rho \int d\vec{r} \, (g(r) - 1) \, e^{i\vec{Q}\cdot\vec{r}} + \rho \delta(\vec{Q})(2\pi)^3 a \right] \end{aligned} \tag{7.35}$$

The last two forms remove a singularity which occurs in forward scattering $\vec{Q} = 0$ and which is quite easily avoided when the experiment concerns wavelengths of the order of atomic separations. It is customary to define the expression

$$S(\vec{Q}) = 1 + \rho \int d\vec{r} (g(r) - 1) \, e^{i\vec{Q}\cdot\vec{r}} \tag{7.36}$$

By inversion of these relations one obtains $g(r)$ from experiment. Some results for simple fluids are shown in Figures 7.1 and 7.2. They are easily intepreted in terms of local structure in the fluid.

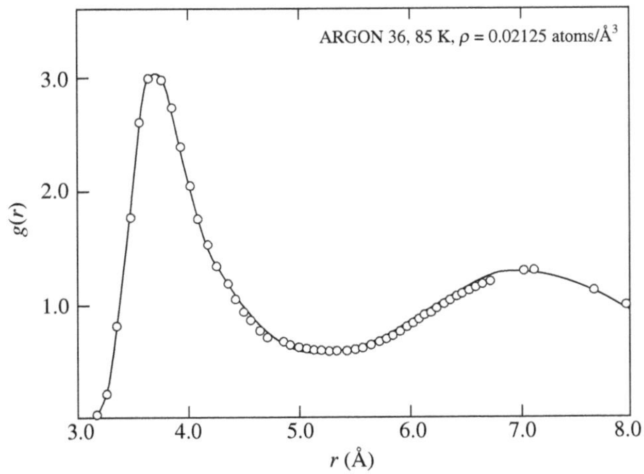


Figure 7.2 Radial distribution function for argon at 85 K determined from the experimentally derived data in the preceding figure and compared with results of molecular dynamics simulation as described later in the text.

## BBGKY hierarchy

Though $g(\vec{r})$ gives the thermodynamics of a fluid interacting through pair interactions and $g(r)$ can be measured directly, we have not yet explained how to determine $g(r)$ from the underlying Hamiltonian which describes the microscopic model. As a practical matter, this is often done these days by simulating the motion of several hundred to a few thousand particles computationally and this often proves quite adequate for determining $g(r)$ uniquely from the underlying model. The methods for doing such simulations are discussed below. Whether such calculations complete the program of many body physics for classical fluids is to some extent a matter of taste. The unique solution can be found to arbitrary accuracy but the lack of a comprehensible analytical theory may lead to some loss of insight. Furthermore, because such simulations only became possible quite recently, there is a long history of distinguished efforts to produce analytical theories permitting the calculation of $g(r)$ in closed form. These theories are of interest not only because of their usefulness to the theory of classical fluids, but also because the same methods have proved very useful in the study of quantum fluids, where brute force simulation is still not nearly as useful a technique as it is in the study of classical fluids. Here we begin by describing a hierarchy which is the basis for approximations leading to such approximate closed form theories for $g(r)$. Then we describe some of the leading approximate closed form theories briefly.

We may define a "potential of mean force" $\phi(r)$ by the equation

$$\phi(\vec{r}) = -k_B T \ln g(\vec{r}) \tag{7.37}$$

or

$$g(\vec{r}) = \mathrm{e}^{-\phi(\vec{r})/k_{\mathrm{B}}T} \tag{7.38}$$

The origin of the name is understood by taking the gradient of the last expression with respect to its argument, renamed $\vec{r}_{12}$ for reasons that should shortly be evident:

$$\begin{aligned} -\nabla_{\vec{r}_{12}}\phi(\vec{r}_{12}) &= \frac{k_{\mathrm{B}}T}{g(\vec{r}_{12})}\nabla_{\vec{r}_{12}}g(\vec{r}_{12}) = \frac{k_{\mathrm{B}}T}{n_2(\vec{r}_{12})}\nabla_{\vec{r}_1}n_2(\vec{r}_{12}) \\ &= \frac{\int \mathrm{d}\vec{r}_3 \cdots \int \mathrm{d}\vec{r}_N \left(\frac{-\partial}{\partial \vec{r}_1}\sum_{i<j} v_{ij}\right) \mathrm{e}^{-\beta\sum_{i<j} v_{ij}}}{\int \mathrm{d}\vec{r}_3 \cdots \int \mathrm{d}\vec{r}_N \, \mathrm{e}^{-\beta\sum_{i<j} v_{ij}}} \end{aligned} \tag{7.39}$$

This is seen to be the average force on an atom fixed at $\vec{r}_1$ when another atom is fixed at $\vec{r}_2$ and all the other atoms are free to move as if in equilibrium. We obtain a useful expression by separating out the term containing $\nabla_{\vec{r}_1}v_{12}$ in the numerator with the result

$$-\nabla_{\vec{r}_1}\phi(\vec{r}_{12}) = -\nabla_{\vec{r}_1}v(\vec{r}_{12}) - \sum_{j\neq 1,2}\left\langle \nabla_{\vec{r}_1}v_{1j}\right\rangle_T^{(1,2)} \tag{7.40}$$

The labels $\langle \ldots \rangle_T^{(1,2)}$ on the average indicate that a thermal average is to be taken over all particles except the first and second, which are fixed. We rewrite the last term as

$$\int \mathrm{d}\vec{r}_3 P(\vec{r}_3 \mid \vec{r}_1, \vec{r}_2)(-\nabla_{\vec{r}_1}v(\vec{r}_3, \vec{r}_1)) \tag{7.41}$$

where

$$P(\vec{r}_3 \mid \vec{r}_1, \vec{r}_2)\mathrm{d}\vec{r}_3 = \begin{cases} \text{Probability of finding an atom within } \mathrm{d}\vec{r}_3 \\ \text{of } \vec{r}_3 \text{ given that that there are atoms at } \vec{r}_1, \vec{r}_2 \end{cases} \tag{7.42}$$

By the rules relating conditional probabilities:

$$n_3(\vec{r}_1, \vec{r}_2, \vec{r}_3) = P(\vec{r}_3 \mid \vec{r}_1, \vec{r}_2)n_2(\vec{r}_1, \vec{r}_2) \tag{7.43}$$

so that the expression (7.41) can be rewritten as

$$\int \mathrm{d}\vec{r}_3 \frac{n_3(\vec{r}_1, \vec{r}_2, \vec{r}_3)}{n_2(\vec{r}_1, \vec{r}_2)}(-\nabla_{\vec{r}_1}v(\vec{r}_3, \vec{r}_1)) \tag{7.44}$$

Thus combining (7.44) with (7.39) and (7.40) we have

$$k_{\mathrm{B}}T\nabla_{\vec{r}_1}\ln g(\vec{r}_{12}) = -\nabla_{\vec{r}_1}v(\vec{r}_{12}) - \frac{1}{\rho^2}\int \mathrm{d}\vec{r}_3 \frac{n_3(\vec{r}_1, \vec{r}_2, \vec{r}_3)}{g(\vec{r}_{12})}\nabla_{\vec{r}_1}v(\vec{r}_{13}) \tag{7.45}$$

This is an equation relating $g(\vec{r})$ to itself but one easily sees that it is not closed because of the presence of the factor $n_3(\vec{r}_1, \vec{r}_2, \vec{r}_3)$ in the integrand of the second term. By writing an equation for $n_3$ following similar procedures we could get a similar equation for it involving $n_4$, etc. This leads to an exact hierarchy which, however, turns out to be more difficult to solve numerically than the original equations of motion of the fluid and is therefore not very useful. On the other hand (7.45) is very useful for motivating approximate theories giving closed form equations for $g(\vec{r})$.

Some other useful forms of (7.45) can be obtained in the case (of main interest here) in which $g(\vec{r})$ and $v(\vec{r})$ are functions only of the magnitude of the vector $\vec{r}$. Then by projecting (7.45) onto the direction $\hat{r}_{12}$ we obtain

$$\frac{\partial \ln g(r_{12})}{\partial r_{12}} = -\frac{\partial}{\partial r_{12}}\left(\frac{v(r_{12}) + W(r_{12})}{k_B T}\right) \tag{7.46}$$

in which

$$\frac{\partial W}{\partial r_{12}} = \frac{1}{\rho^2}\int d\vec{r}_3 \frac{n_3(\vec{r}_1, \vec{r}_2, \vec{r}_3)}{g(\vec{r}_1, \vec{r}_2)}(\hat{r}_{12}\cdot\hat{r}_{13})\frac{\partial v(r_{13})}{\partial r_{13}} \tag{7.47}$$

The last two equations can be integrated using the fact that both $v(r)$ and $W(r)$ vanish at large $r$ to give

$$g(r) = e^{-(v(r)+W(r))/k_B T} \tag{7.48}$$

in which

$$W(r) = +\int_r^\infty dr_{12}\frac{1}{\rho^2}\int d\vec{r}_3 \frac{n_3(\vec{r}_1, \vec{r}_2, \vec{r}_3)}{g(r_{12})}(\hat{r}_{12}\cdot\hat{r}_{13})\frac{\partial v(r_{13})}{\partial r_{13}} \tag{7.49}$$

These equations have a very physical interpretation in terms of a direct or "bare" force on particle 1 due to particle 2 and an indirect effect of 1 on 2, represented by $W$ which arises because if 2 interacts with 3 which in turn interacts with 1 then the presence of 1 can affect the likelihood of finding an atom at 2.

## Approximate closed form equations for $g(\vec{r})$

The simplest of the closed form equations to describe is the Yvon–Born–Green approximation which is obtained by replacing the three particle distribution function $n_3(\vec{r}_1, \vec{r}_2, \vec{r}_3)$ which occurs in the integrands of the equations of the last section by the "Kirkwood superposition approximation":

$$n_3(\vec{r}_1, \vec{r}_2, \vec{r}_3) \approx \frac{1}{\rho^3}n_2(\vec{r}_1, \vec{r}_2)n_2(\vec{r}_1, \vec{r}_3)n_2(\vec{r}_2, \vec{r}_3) \tag{7.50}$$

This approximation has the attractive feature of giving zero for $n_3$ when any pair of the arguments $\vec{r}_1, \vec{r}_2, \vec{r}_3$ is equal, as is physically required.

Using the Kirkwood superposition approximation in equations (7.48) and (7.49) we find

$$\ln g(r_{12}) = \frac{-v(r_{12})}{k_B T} + \frac{\rho}{k_B T}\int_{r_{12}}^\infty dr'_{12}\int d\vec{r}'_3\, g(r'_{13})g(r'_{23})\frac{\partial v(r'_{13})}{\partial r'_{13}}(\hat{r}'_{12}\cdot\hat{r}'_{13}) \tag{7.51}$$

This is one form of the Yvon–Born–Green equation for $g(r)$. It is a closed nonlinear equation for $g(r)$ which may be solved numerically with the boundary condition $g(r) \underset{r\to\infty}{\to} 1$ for a given potential $v(r)$. Because the Kirkwood approximation is uncontrolled (that is, we have no systematic way to improve it and estimate errors) the results must be checked against experiments and/or simulations.

There are two other popular approximate equations for $g(r)$, based on partial summations of infinite series of selected terms in the virial series for $g(r)$. The principles upon which such a virial series is obtained are the same as those used in the last chapter but we will not go through the derivation here. The interested reader is referred to Rice and Gray.[2] An early paper by Verlet[3] provides a clear account of the derivation of the hypernetted chain approximation. The result for the hypernetted chain approximation is that $W(r)$ of (7.49) may be written:

$$W(r_{12}) = -k_B T \rho \int d\vec{r}_3\, c(r_{13})(g(r_{32}) - 1) \tag{7.52}$$

in which the function $c(r)$ satisfies the integral equation:

$$c(r_{12}) = g(r_{12}) - 1 - \rho \int d\vec{r}_3\, (g(r_{32}) - 1)c(r_{13}) \tag{7.53}$$

The last equation is sometimes called the Ornstein–Zernike equation and $c(r)$ is called the direct correlation function. The third approximate formulation, called the Percus–Yevick approximation, consists in using the same expression for $W(r)$ but effectively assuming that it is small so that (7.48) becomes

$$g(r) \approx e^{-\beta v(r)}(1 - \beta W(r)) \tag{7.54}$$

## Molecular dynamics evaluation of liquid properties

In practice, results of approximations such as those described in the last section are often compared to results of direct numerical simulations of the Newtonian equations of motion for a small sample of the atomic or molecular constituents of the fluid, rather than to the results of experiment. This is because in the case of simple liquids, numerical simulation can now give essentially exact results for the short range equilibrium radial distribution functions of dense liquids away from critical points. Here we briefly review the techniques for making such simulations, commonly called molecular dynamics. The reader is referred to other texts (such as Allen and Tildesley[4]) for more details.

In such simulations, one works directly with the equations of motion for the atoms or molecules of the fluid. For simplicity we confine attention here to monatomic liquids and to systems interacting through short range interactions. The equations of motion are

$$m\frac{\mathrm{d}^2\vec{r}_i}{\mathrm{d}t^2} = -\nabla_i \sum_{j\neq i} v(\vec{r}_i - \vec{r}_j) \tag{7.55}$$

We describe the method for a system of $N$ particles. In practice $N$ can be made as large as $10^6$ with current computers but simulations with $\mathcal{O}(10^3)$ particles are more common. Techniques are available for fixing the local pressure and/or the local kinetic energy (obviously related to the temperature) during a simulation but a more common (and more realistic) method fixes the volume $V$ and energy of the system. To simulate a bulk fluid, it is useful to use periodic boundary conditions because this reduces the effects of the boundaries, which can extend significantly into the sample if other boundary conditions are used. Extension of the method to simulation of liquid–solid interfaces is straightforward.

In practice, one chooses a set of initial positions $\{\vec{r}_i(0)\}$. If we were dealing with the equations (7.55) directly, without discretizing the time, we would also require initial data on the velocities in order to specify a unique solution. In the present case we will find approximate numerical solutions which give solutions for the $\{\vec{r}_i\}$ at a series of discrete times $n\Delta t$ where $n$ is a positive integer and $\Delta t$ is called the time step. The appropriate additional initial data are then the values $\{\vec{r}_i(-\Delta t)\}$ of the positions at the time step previous to the initial one. From $\{\vec{r}_i(0)\}$ and $\{\vec{r}_i(-\Delta t)\}$, approximations to the initial velocities are easily constructed as we will shortly discuss. These values are sufficient if the simplest integration algorithm, which is the only one we will discuss here, is used. To describe this algorithm, known as the "Verlet" algorithm in the molecular dynamics literature, we expand $\vec{r}_i(t+\Delta t)$ and $\vec{r}_i(t-\Delta t)$ as Taylor series in $\Delta t$:

$$\vec{r}_i(t+\Delta t) = \vec{r}_i(t) + \frac{\mathrm{d}\vec{r}_i}{\mathrm{d}t}(t)\Delta t + \frac{1}{2}\frac{\mathrm{d}^2\vec{r}_i}{\mathrm{d}t^2}\Delta t^2 + \frac{1}{6}\frac{\mathrm{d}^3\vec{r}_i}{\mathrm{d}t^3}\Delta t^3 + \mathcal{O}(\Delta t^4) \tag{7.56}$$

$$\vec{r}_i(t-\Delta t) = \vec{r}_i(t) - \frac{\mathrm{d}\vec{r}_i}{\mathrm{d}t}(t)\Delta t + \frac{1}{2}\frac{\mathrm{d}^2\vec{r}_i}{\mathrm{d}t^2}(t)\Delta t^2 - \frac{1}{6}\frac{\mathrm{d}^3\vec{r}_i}{\mathrm{d}t^3}(t)\Delta t^3 + \mathcal{O}(\Delta t^4) \tag{7.57}$$

Adding these we easily obtain

$$\vec{r}_i(t+\Delta t) = 2\vec{r}_i(t) - \vec{r}_i(t-\Delta t) + \frac{\mathrm{d}^2\vec{r}_i}{\mathrm{d}t^2}(t)\Delta t^2 + \mathcal{O}(\Delta t^4) \tag{7.58}$$

so that by use of (7.55)

$$\vec{r}_i(t+\Delta t) = 2\vec{r}_i(t) - \vec{r}_i(t-\Delta t) + \frac{\vec{F}_i(t)}{m}\Delta t^2 + \mathcal{O}(\Delta t^4) \tag{7.59}$$

where

$$\vec{F}_i(t) = -\nabla_i \sum_{j\neq i} v(\vec{r}_i(t) - \vec{r}_j(t)) \tag{7.60}$$

is the force on the $i$th particle at time $t$. Equation (7.59) is used in the simulation to compute the values of the positions of the particles at times $(n+1)\Delta t$ from their values at the previous two time steps $(n-1)\Delta t$ and $n\Delta t$, starting with $n=0$

$$\vec{r}_i((n+1)\Delta t) = 2\vec{r}_i(n\Delta t) - \vec{r}_i((n-1)\Delta t) + \frac{\vec{F}_i(n\Delta t)}{m}\Delta t^2 + \mathcal{O}(\Delta t^4) \quad (7.61)$$

Obviously each such computation requires evaluating the force $\vec{F}_i$ on each particle at time $t = n\Delta t$. This evaluation of the forces after each new set of positions is calculated is the most time consuming part of the calculations in the simulation. Subtracting (7.56) from (7.57) we get an approximate expression for the velocity at time $t$:

$$\vec{r}_i((n+1)\Delta t) - \vec{r}_i((n-1)\Delta t) = 2\frac{d\vec{r}_i}{dt}(n\Delta t) + \mathcal{O}(\Delta t^3) \quad (7.62)$$

or

$$\vec{v}_i(n\Delta t) = \left(\frac{\vec{r}_i((n+1)\Delta t) - \vec{r}_i((n-1)\Delta t)}{2\Delta t}\right) + \mathcal{O}(\Delta t^3) \quad (7.63)$$

This is used to calculate the velocities at time $t$ after the time step to time $t = (n+1)\Delta t$ has been made.

For realistic simulations, the choice of forces is crucial in this technique. Here we confine attention to forces arising from the sum of pair potentials though, in principle, one can include many body forces at significant computational cost. Often pair potentials are determined by making all electron quantum mechanical calculations on pairs of atoms and fitting the results to simple forms. For the rare gas liquids which we will use as an example, such a fit can be made very satisfactorily with the Lennard-Jones form in equation (6.2). For systems in which the constituent classical entities carry a charge (as in molten salts for example) Coulomb interactions complicate the technique considerably, because it is not possible to cut off the Coulomb interactions at achievable box sizes and achieve realistic results. (See reference 4 for details on methods to handle the case of Coulomb interactions.)

The interpretation of periodic boundary conditions is straightforward with respect to the treatment of a particle which passes through a wall of the confining volume or box: it is simply reflected to the opposite side of the volume (Figure 7.3). To determine the forces on a particle, one assumes that the particle interacts with all the particles in the box *and* with all the particles in all the periodic images (Figure 7.4) of the box. In the case that Coulomb interactions exist this is quite a complicated task for which several methods are available.[4] However, we will confine attention here to short range forces. Then one may usually take the box size to be much larger than twice the range of the forces and confine attention only

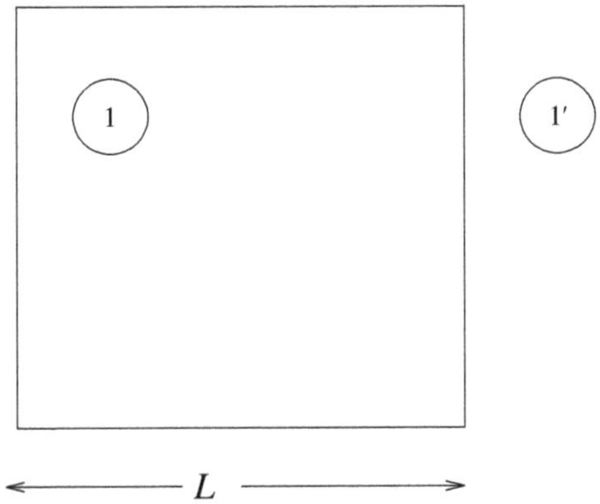


Figure 7.3 Illustration of implementation of boundary conditions. If particle 1 ends up at 1′ after the time step it is moved to 1.

to those images of the particles which are within half the box size of the particle whose force is being calculated (Figure 7.5). This is known as the minimum image convention. To implement these conditions suppose the confining volume is cubic for simplicity and that the length of its side is $L$. Suppose that new positions $\{x_i((n+1)\Delta t), y_i((n+1)\Delta t), z_i((n+1)\Delta t)\}$ have been determined from earlier positions using (7.61). There is nothing in (7.61) which prevents some of these positions from ending up outside the confining volume and they must be reflected back into the box before proceeding. This is done with the rule:

$$\begin{aligned} &\text{if } x_i((n+1)\Delta t) > L/2,\ x_i((n+1)\Delta t) - L \text{ replaces } x_i((n+1)\Delta t) \\ &\text{if } x_i((n+1)\Delta t) < -L/2,\ x_i((n+1)\Delta t) + L \text{ replaces } x_i((n+1)\Delta t) \end{aligned} \tag{7.64}$$

with similar replacements for $y_i((n+1)\Delta t)$, $z_i((n+1)\Delta t)$.

Now to compute the forces associated with these new positions in preparation for the next step one takes differences $\Delta\vec{r}_{ij}((n+1)\Delta t) \equiv \vec{r}_i((n+1)\Delta t) - \vec{r}_j((n+1)\Delta t)$ for use in the computation of the force $\vec{F}_{ij}((n+1)\Delta t)$ exerted by $j$ on $i$ and vice versa. To implement the minimum image convention, one can see quite easily that one must apply the same rule (7.64) to the differences $\Delta x_{ij}((n+1)\Delta t)$, $\Delta y_{ij}((n+1)\Delta t)$, $\Delta z_{ij}((n+1)\Delta t)$ before using them in computation of the force $\vec{F}_{ij}((n+1)\Delta t)$:

$$\begin{aligned} &\text{if}\Delta x_{ij}((n+1)\Delta t) > L/2,\ \Delta x_{ij}((n+1)\Delta t) - L \text{ replaces } \Delta x_{ij}((n+1)\Delta t) \\ &\text{if}\Delta x_{ij}((n+1)\Delta t) < -L/2,\ \Delta x_{ij}((n+1)\Delta t) + L \text{ replaces } \Delta x_{ij}((n+1)\Delta t) \end{aligned} \tag{7.65}$$

The result of the implementation of these procedures is a time series of positions for all the particles. Usually, not all this information is stored because it is very voluminous and not easy to interpret without further analysis. The quantities which

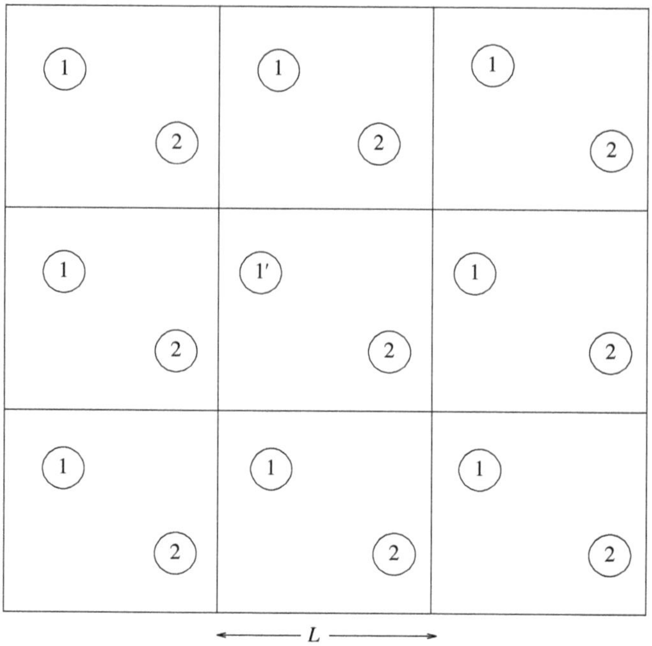


Figure 7.4 For models containing long range interactions one includes the interaction of 1′ with all the other particles in the simulation cell and with all the periodic images of all the particles in the cell.

are kept for analysis are of two types: those which are primarily used to establish the validity of the code and those which are used to compute quantities leading to predictions for the results of experimental measurements on the fluid. The most important variables of the first type are the total energy, the kinetic energy per particle and the local pressure, which is calculated using equation (7.19). We first discuss each of these briefly.

Because the system of equations (7.55) conserves energy, the total energy is exactly independent of time in an exact solution. In the numerical solution, the energy variations arising from numerical error must be much smaller than the temporal variations in the total kinetic and total potential energies if the results are to be physically meaningful. In practice, with the simple systems to be considered here one finds that in a correct code the energy variations are not larger than a part in $10^4$ and contain no detectable secular drift. This turns out to be a very sensitive diagnostic for errors in codes of this type. Of course the total energy is also useful in determining the equation of state of the model of the fluid.

Because we are simulating a classical system, the average kinetic energy per particle is equal to $\frac{3}{2}k_B T$ and hence fixes the effective temperature at which the simulation is carried out. (As mentioned earlier, it is possible[4] to add artificial terms

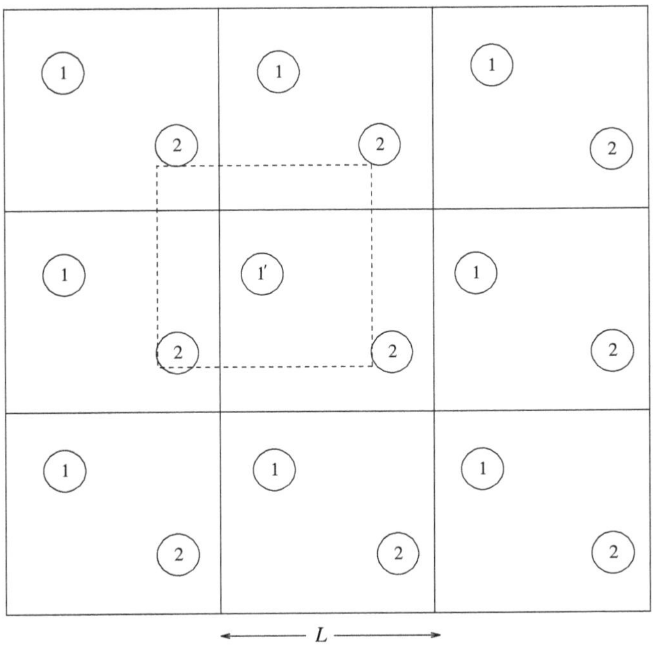


Figure 7.5 Minimum image convention. If the forces are of short range ($<L/2$) then only interactions of $1'$ with particles within a box (dashed lines) centered on $1'$ are kept.

to (7.55) which effectively fix the total kinetic energy instantaneously, but we will not discuss these here.) In practice, when a simulation of a new system is begun, the particles are not in positions which are close to mechanical equilibrium and they accelerate rapidly, resulting in a high average kinetic energy and hence a high effective temperature. To bring the kinetic energy down to a level consistent with the desired simulation temperature, one can use various algorithmic "thermostats" of which the simplest is the following. If the average kinetic energy per particle $\langle t \rangle$ differs from the desired average kinetic energy $\langle t \rangle_w = (3/2)k_B T$ by more than a fraction $\Delta f$, then rescale all the velocities by

$$\sqrt{\langle t \rangle_w / \langle t \rangle}\vec{v}_i(n\Delta t) \equiv f\vec{v}_i(n\Delta t) \rightarrow \vec{v}_i(n\Delta t) \tag{7.66}$$

Because the programs do not deal directly with velocities this is accomplished by changing the values of the positions at the *preceding* time step. This is done as follows. After calculating (preliminary) values of $\vec{r}_i((n+1)\Delta t)$ using (7.61), and (preliminary) velocities using (7.63), one calculates the ratio $f$. If $|(1-f)| < \Delta f$ one uses the $\vec{r}_i((n+1)\Delta t)$ as the values on the trajectory for the next step. However, if $|(1-f)| > \Delta f$, one modifies $\vec{r}_i((n+1)\Delta t)$ to $\vec{r}_i((n+1)\Delta t)'$ and (implicitly)

$\vec{r}_i((n-1)\Delta t)$ to $\vec{r}_i((n-1)\Delta t)'$ so that they satisfy

$$\vec{v}_i(n\Delta t)' = \left(\frac{\vec{r}_i((n+1)\Delta t) - \vec{r}_i((n-1)\Delta t)'}{2\Delta t}\right) = f\vec{v}_i(n\Delta t)$$
$$= f\left(\frac{\vec{r}_i((n+1)\Delta t) - \vec{r}_i((n-1)\Delta t)}{2\Delta t}\right)$$

and

$$\vec{r}_i((n+1)\Delta t)' = 2\vec{r}_i(n\Delta t) - \vec{r}_i((n-1)\Delta t)' + \frac{\vec{F}_i(n\Delta t)}{m}\Delta t^2$$

This amounts to changing the values of the positions at the preceding time step so that when the positions at the next time step are recalculated, the velocities will come out changed by the factor $f$. Solving the last two equations for $\vec{r}_i(n+1)\Delta t)'$, leaving the present positions $\vec{r}_i(n)\Delta t)$ fixed and eliminating the $\vec{r}_i((n-1)\Delta t)'$ one finds that the recalculated positions at the next step are

$$\vec{r}_i((n+1)\Delta t)' = \vec{r}_i(n\Delta t) + (f/2)(\vec{r}_i((n+1)\Delta t) - \vec{r}_i((n-1)\Delta t)) + \frac{\vec{F}_i(n\Delta t)}{2m}\Delta t^2 \tag{7.67}$$

(When $f = 1$, $\vec{r}_i((n+1)\Delta t)' = \vec{r}_i((n+1)\Delta t)$. One could also explicitly calculate $\vec{r}_i((n-1)\Delta t)'$ but it is not needed.)

With an appropriate choice of $\Delta f$ one can leave this algorithm in place throughout the simulation, and it will almost never be activated once a configuration approximating equilibrium is achieved. The correct choice of $\Delta f$ for this purpose is a somewhat subtle problem. The mean square deviation of the kinetic energy in a system described by a canonical ensemble is quite easily computed to be

$$\frac{\sqrt{\langle (KE)^2\rangle - \langle KE\rangle^2}}{\langle KE\rangle} = \sqrt{\frac{2}{3N}} \tag{7.68}$$

However, this is an ensemble dependent result. In a simulation, this ratio remains proportional to $1/\sqrt{N}$ but the coefficient is different.[5] It is shown in reference 5 that, in the microcanonical ensemble

$$\frac{\sqrt{\langle (KE)^2\rangle - \langle KE\rangle^2}}{\langle KE\rangle} = \sqrt{\frac{2}{3N}(1 - 3Nk_B/2C_V)} \tag{7.69}$$

Thus in principle the fluctuations cannot be estimated until the specific heat is known, and in fact the last relation can be used to compute the specific heat. On the other hand, one can get a very reasonable (but uncontrolled) estimate for the expected fluctuations by using the value for $N$ three dimensional harmonic oscillators of $C_V = 3Nk_B$ giving

$$\frac{\sqrt{\langle (KE)^2\rangle - \langle KE\rangle^2}}{\langle KE\rangle} \approx \sqrt{\frac{1}{3N}} \tag{7.70}$$

and using a $\delta f$ which is twice this.

To determine the pressure, one uses (7.19). For a system containing only short range forces this is straightforward, but with Coulomb interactions more care is required.[4] We do not discuss the latter case here. As for the temperature, if the initial conditions are chosen badly, the simulation may result in a very high pressure which must be adjusted by adjusting the volume. Algorithms also exist which (rather artificially) continuously hold the pressure fixed during the simulation.

In addition to thermodynamic quantities such as the energy, pressure and temperature, one can collect microscopic information about the simulated equilibrium state. Because of its central role in describing this state as explained earlier in this chapter, the radial distribution function $g(r)$ is of particular interest in this regard. For this purpose, we note that an alternative expression for $g(r)$ in an isotropic fluid is

$$g(r)\rho 4\pi r^2\,\mathrm{d}r = \left\langle \frac{1}{N}\sum_i \sum_{i\neq j} \mathrm{d}N(r <\mid \vec{r}_i - \vec{r}_j \mid < r + \mathrm{d}r)\right\rangle \tag{7.71}$$

in which $\mathrm{d}N$ is the number of particles $j$ which are found within the indicated distances from particle $i$. Thus to evaluate $g(r)$ on a finite grid of values of $r = i_g \Delta r$, $i_g = 0, \ldots, NG$ we use

$$g(i_g) = \left\langle \frac{V}{N^2 4\pi \Delta r^3 i_g^2}\sum_i \sum_{j\neq i} \Delta N(i_g \Delta r <\mid \vec{r}_i - \vec{r}_j \mid < (i_g+1)\Delta r)\right\rangle \tag{7.72}$$

Results of implementing these procedures for argon liquid were compared with experiment in Figures 7.1 and 7.2.

## References


1. J. L Yarnell, M. J. Katz, R. G. Wenzel and S. H. Koenig, *Physical Review A* **7** (1973) 2130.
2. S. A. Rice and P. Gray, *The Statistical Mechanics of Simple Liquids: an Introduction to the Theory of Equilibrium and non-Equilibrium Phenomena*, New York: Interscience, 1965.
3. L. Verlet, *Nuovo Cimento* **18** (1960) 77.
4. M. P. Allen and D. J. Tildesley, *Computer Simulation of Liquids*, Oxford: Clarendon Press, 1987.
5. J. L. Lebowitz, J. K. Percus and L. Verlet, *Physical Review* **153** (1967) 250.

## Problems

7.1 Find the form of the full neutron scattering cross-section $d^2\sigma/d\Omega_{\vec{k}_f} d\epsilon_{\vec{k}_f}$ (without integrating on $\omega$) when the interaction is of form (7.29). Express the result in terms of functions

$$\sum_{i,j} \left\langle e^{i\vec{Q}\cdot\vec{r}_i(t)} e^{-i\vec{Q}\cdot\vec{r}_j(0)} \right\rangle \quad \text{and} \quad \sum_{i} \left\langle e^{i\vec{Q}\cdot\vec{r}_i(t)} e^{-i\vec{Q}\cdot\vec{r}_i(0)} \right\rangle$$

7.2 Solve the Yvon–Born–Green equation numerically for the case that

$$v(r) = 4\epsilon \left( \left(\frac{\sigma}{r}\right)^{12} - \left(\frac{\sigma}{r}\right)^{6} \right)$$

with $\epsilon = 125$ K and $\sigma = 3.45$ Å, suitable for argon. Take temperatures and densities in the liquid range for argon at about one atmosphere, namely $87.5\,\text{K} > T > 84\,\text{K}$ and $1.4\,\text{g/cm}^3 < \rho < 1.6\,\text{g/cm}^3$.

7.3 Write and run an MD program for a system of 100 Lennard-Jones argon atoms. The parameters for Lennard-Jones potential are $\sigma = 3.4$ Å and $\epsilon/k_B = 125$ K. The simulated system is at 85 K with a mass density of 1.4 g/cm$^3$. The time step for the simulation is 10 fs ($10^{-14}$ s) and the simulation should run for 20–30 ps. The following methods are to be implemented:
(1) periodic boundary condition,
(2) minimum image convention,
(3) Verlet algorithm for updating trajectories.
The following quantities are to be calculated:
(4) conserved total energy $E(t)$,
(5) radial distribution function $g(r)$ and the corresponding coordinate number $Z(r)$,
(6) mean square displacement $R(t)$,
(7) average pressure $P$.
Plot $g(r)$, $Z(r)$, and $E(t)$ and $R(t)$.

7.4 Prove the expression (7.68) for the fluctuations in the kinetic energy in the canonical ensemble. For a greater challenge, prove the expression (7.69) consulting, if you like, reference 5.

# 8
# Quantum liquids and solids

As discussed in Chapter 6, when the temperature is lowered in a classical liquid until the thermal wavelength becomes comparable to the interparticle spacing, then the semiclassical approximation is no longer adequate and quantum effects must be considered. In practice, most classical liquids freeze at all positive pressures before this temperature is reached. The exceptions are the helium liquids ($^3$He and $^4$He) for which the quantum effects are large enough to prevent freezing as the temperature is lowered while the liquid is kept in equilibrium with its vapor. Conveniently, $^3$He is a Fermi system and $^4$He is a Bose system. In these systems as well, phase transitions occur at low enough temperatures. But quantum effects are significant even before these transitions occur. Another system which may for some purposes be regarded as an isotropic liquid with large quantum effects is the collection of electrons in (at least some) metals. Here too, a phase transition to the superconducting state intervenes in many cases at low enough temperatures. Finally neutron stars may contain regions in which neutrons are in a liquid state with large quantum effects and white dwarf stars contain a degenerate electron gas which can be regarded as a quantum liquid. In general, the reason that quantum liquids are so hard to observe is that interactions tend to result in symmetry breaking phase transitions in high density systems at temperatures low enough to permit quantum effects to be observed. (Vapor phases at very low temperatures almost always exist. The densities are extremely low and virial expansions as discussed in Chapter 6 may be used to describe them. The metastable Bose–Einstein condensed alkali metal vapors mentioned at the end of Chapter 5 are an exception.)

Here I confine attention to some phenomenological ideas for the description of dense isotropic Fermi systems at low temperatures which are generally known as Fermi liquid theory and which are primarily due to Landau. The theory as I will describe it was designed to describe systems with short range interactions and has been extensively applied to the description of liquid $^3$He. It turns out that the theory provides a useful basis for describing the various superfluid phases of liquid $^3$He as well but I will not discuss that aspect here. Fermi liquid theory has been extended to systems with Coulomb interactions and applied to the study of electrons in metals. We will describe Fermi liquid theory in terms of a few physically reasonable assumptions essentially associated with the idea that no phase transition occurs as the temperature is lowered (or, almost equivalently, as the interactions are made stronger at fixed temperature). These assumptions have been verified (proved is probably too strong a word) by the use of more microscopically based theoretical formulations which begin with the full microscopic Hamiltonian. A theoretical formulation analogous to Fermi liquid theory has been formulated for boson fluids as well.[1] A characteristic feature of these theories is that the low energy states of the fluid have some properties like the states of the noninteracting Fermi and Bose gases respectively. This correspondence permits one to speak of these states as containing various numbers of "quasiparticles" with specified momenta. The quasiparticle concept has proved intuitively very useful (if somewhat dangerous) in thinking about these systems.

## Fundamental postulates of Fermi liquid theory

To motivate the assumptions of Fermi liquid theory, imagine a dense assembly of $^3$He atoms. The Hamiltonian is of the form

$$H = \sum_i \frac{-\hbar^2\nabla_i^2}{2m} + \sum_{i<j} v_{ij} \tag{8.1}$$

Now we multiply the interaction term by a dimensionless parameter $\lambda$ and consider the evolution of the energy eigenstates as $\lambda$ is varied between 0 and 1

$$H_\lambda = \sum_i \frac{-\hbar^2\nabla_i^2}{2m} + \lambda \sum_{i<j} v_{ij} \tag{8.2}$$

When $\lambda = 0$ there are no interactions, and the eigenstates are antisymmetrized products of spinors and plane waves (Slater determinants) of the noninteracting system. As such they can be completely specified by defining a set $\{n_{\vec{q},\sigma}\}$ of 0s and 1s which indicate which single particle states specified by $\vec{q}$, $\sigma$ are occupied. The ground state of this noninteracting system corresponds to $n_{\vec{q},\sigma} = 1$ if $|\vec{q}| < q_{\rm F}$ and $n_{\vec{q},\sigma} = 0$ otherwise where $q_{\rm F} = (3\rho\pi)^{1/3}$ and $\rho$ is the particle density. This set of $\{n_{\vec{q},\sigma}\}$ plays a special role and its members will be denoted $(n_{\vec{q},\sigma})_{\rm g}$:

$$(n_{\vec{q},\sigma})_{\rm g} = \begin{cases} 1 & |\vec{q}| < q_{\rm F} \\ 0 & |\vec{q}| > q_{\rm F} \end{cases} \tag{8.3}$$

At fixed particle number $N$, the low lying excited states of the noninteracting system consist of states in which one single particle state below but near the Fermi surface is empty while a single particle state near but above the Fermi surface is occupied, while all other single particle states retain the occupancy which they had in the ground state. These low lying excited states are called particle–hole excitations. Now let the parameter $\lambda$ increase smoothly from 0 to 1. In Fermi liquid theory, one assumes that the states of the system, which can all be completely specified by the sets $\{n_{\vec{q},\sigma}\}$ when $\lambda = 0$, evolve smoothly as $\lambda$ increases, retaining the same energy ordering that they had at $\lambda = 0$ and with no qualitatively new low lying states appearing. If the system undergoes a phase transition as a function of $\lambda$ at zero temperature then this assumption is violated. It is known for example that such phase transitions must occur in $^3$He which is in a complicated superfluid state at very low temperatures. However, these transitions occur at extremely low temperatures (of the order of millikelvins) and so when the system is at temperatures much larger than this it may be reasonable to assume that it behaves as if this assumption were true. Further, it is possible to use the postulated states which do evolve smoothly in this way as a basis for a theory describing more complicated states which occur when it is violated. If no violation occurs, the interacting ground state is still labelled by the quantum numbers $\{(n_{\vec{q},\sigma})_g\}$ of (8.3) and has evolved smoothly from the original Slater determinant of the noninteracting system as the interaction was turned on.

Now consider the energies $E_{\{n_{\vec{q},\sigma}\}}$ of all the states of the interacting system under this assumption. It is convenient to regard the energy as a function of the differences $\delta n_{\vec{q},\sigma} = n_{\vec{q},\sigma} - (n_{\vec{q},\sigma})_g$ where $(n_{\vec{q},\sigma})_g$ are given by (8.3). For the low lying states (near the ground state) it is reasonable to expand the energy in a series in these $\delta n_{\vec{q},\sigma}$:

$$E_{\{n_{\vec{q},\sigma}\}} = E_g + \sum_{\vec{q},\sigma} \epsilon_{\vec{q},\sigma} \delta n_{\vec{q},\sigma} + \frac{1}{2} \sum_{\vec{q},\sigma;\vec{q}',\sigma'} f_{\vec{q},\sigma;\vec{q}',\sigma'} \delta n_{\vec{q},\sigma} \delta n_{\vec{q}',\sigma'} + \cdots \tag{8.4}$$

where the coefficients $\epsilon_{\vec{q},\sigma}$, $f_{\vec{q},\sigma;\vec{q}',\sigma'}, \ldots$ are regarded, at the Fermi liquid theory level of the theory, as phenomenological parameters. If the assumption of smooth variation of the states with increasing interaction is true, then these parameters can, in principle, be calculated from the true wave functions using a fully microscopic theory. There have been some attempts to do this for $^3$He, but no definitive results are available.

At low enough temperatures we will argue later that only the terms shown explicitly in (8.4) need to be taken into account in calculations of the low temperature properties of the fluid. If we truncate (8.4) with only the explicitly shown linear and quadratic terms, then the statistical mechanical problem is closed, if very difficult. One must calculate the partition function associated with the spectrum (8.4).

Though it is difficult, it is quite a lot easier than the original interacting problem, because the form of the eigenvalues of the Hamiltonian is assumed known.

In fact one almost always makes a further mean field approximation in Fermi liquid theory. One may state the approximation in various ways. It amounts to assuming that only the average effects of interactions represented by the last term in (8.4) need to be taken into account. Operationally, one may obtain this approximation by replacing the last term in (8.4) by

$$\frac{1}{2}\sum_{\vec{q},\sigma;\vec{q}',\sigma'} f_{\vec{q},\sigma}(\delta n_{\vec{q},\sigma}\langle\delta n_{\vec{q}',\sigma'}\rangle + \delta n_{\vec{q}',\sigma'}\langle\delta n_{\vec{q},\sigma}\rangle) \tag{8.5}$$

and then using the resulting thermodynamic potential to find a self consistent equation for $\langle\delta n_{\vec{q},\sigma}\rangle$. Note that since the number of particles does not change as the interactions are turned on, we may assume that $\sum_{\vec{q},\sigma} n_{\vec{q},\sigma} = N$ for all the interacting states. Thus, in the approximation (8.5)

$$\Omega = -k_{\rm B}T \ln \mathrm{e}^{-\beta\delta E_{\rm g}(T)} \sum_{\{n_{\vec{q},\sigma}\}} \mathrm{e}^{-\beta(\tilde{\epsilon}_{\vec{q},\sigma}-\mu)n_{\vec{q},\sigma}} = \delta E_{\rm g}(T) - k_{\rm B}T\sum_{\vec{q},\sigma}\ln\left(1+\mathrm{e}^{(\mu-\tilde{\epsilon}_{\vec{q},\sigma})\beta}\right) \tag{8.6}$$

Here

$$\delta E_{\rm g}(T) = E_{\rm g} - \sum_{\vec{q},\sigma}\tilde{\epsilon}_{\vec{q},\sigma}(n_{\vec{q},\sigma})_{\rm g} \tag{8.7}$$

and

$$\tilde{\epsilon}_{\vec{q},\sigma} = \epsilon_{\vec{q},\sigma} + \sum_{\vec{q}',\sigma'} f_{\vec{q},\sigma;\vec{q}',\sigma'}\langle\delta n_{\vec{q}',\sigma'}\rangle \tag{8.8}$$

$E_{\rm g}(T)$ depends on temperature but not on the chemical potential. We can therefore find an equation for $\langle N\rangle$ by differentiating with respect to $\mu$:

$$\langle N\rangle = -\left(\frac{\partial\Omega}{\partial\mu}\right)_{T,V} = \sum_{\vec{q},\sigma}\frac{1}{\mathrm{e}^{(\tilde{\epsilon}_{\vec{q},\sigma}-\mu)\beta}+1} \tag{8.9}$$

and since $\langle N\rangle = \sum_{\vec{q},\sigma}\langle n_{\vec{q},\sigma}\rangle$,

$$\langle n_{\vec{q},\sigma}\rangle = \frac{1}{\mathrm{e}^{(\tilde{\epsilon}_{\vec{q},\sigma}-\mu)\beta}+1} \tag{8.10}$$

Though this looks formally like the noninteracting expression, $\tilde{\epsilon}_{\vec{q},\sigma}$ depends on $\langle n_{\vec{q}',\sigma'}\rangle$ for all $\vec{q}'$, $\sigma'$. One must regard (8.9) and (8.10) as simultaneous equations for all the $\langle n_{\vec{q},\sigma}\rangle$ and $\mu$.

From this one can see why only a few terms are adequate in the expansion. Consider the case in which the parameter $\lambda$ is small, giving weak but finite interactions which could be treated in perturbation theory. The noninteracting excited states are particle–hole excitations of the noninteracting Fermi sphere, in which the particle and the hole are both within about $k_B T$ of the Fermi surface. Now introducing the weak interaction, perturbation theory scatters these particle–hole excitations into other particle–hole excitations, conserving momentum and energy. But the conservation of total momentum and energy of the pair, together with the Pauli principle, requires that the particle–hole excitations into which the original pair is scattered also remain within $k_B T$ of the original Fermi surface. Thus $n_{\vec{q},\sigma}$ remains near its original form except when $q$ is within $k_B T$ of the original Fermi surface. The kinematic constraints imposed by the Pauli principle keep the effects of the interactions on $n_{\vec{q},\sigma}$ small so the relevant $\delta n_{\vec{q},\sigma}$ are small and the expansion is (at least qualitatively) justified. Increasing the magnitude of the coupling by increasing $\lambda$ increases the number of terms which must be considered in a perturbation expansion, but at each order, the same constraints on the final states apply and the general conclusion is sustained.

One may parametrize the constants $\epsilon_{\vec{q},\sigma}$ and $f_{\vec{q},\sigma;\vec{q}',\sigma'}$ in terms of a few numbers if one is interested in the low temperature properties of the theory. We assume that $\epsilon_{\vec{q},\sigma}$ may be expanded in a Taylor series in $|\ \vec{q}\ |$. The constant term may be absorbed in the chemical potential and the linear term must vanish if the fluid is isotropic. Thus the leading term is quadratic. It may be expressed in a suggestive way by writing

$$\epsilon_{\vec{q},\sigma} = \frac{\hbar^2 q^2}{2m^*} \tag{8.11}$$

thus introducing an "effective mass" $m^*$. This differs from the bare mass as a result of particle–particle interactions. In the application to electrons in metals and semiconductors, one can also find an effective mass (the "band mass") which differs from the bare mass as a result of interactions of the electrons with the lattice (and not with each other) and is not to be confused with this $m^*$. (We do not discuss the question of how to handle the case in which both effects are simultaneously present here.) It turns out that for the study of thermodynamic properties at low temperatures, one only needs the values of $f_{\vec{q},\sigma;\vec{q}',\sigma'}$ when both wave vectors have magnitude $q_F$. The remaining variables are the angle $\theta$ between $\vec{q}$ and $\vec{q}'$ and the relative values of the spins $\sigma$ and $\sigma'$. It is customary to define

$$f^{s}(\theta) = f_{\uparrow,\uparrow}(\theta) + f_{\uparrow,\downarrow}(\theta) \tag{8.12}$$

$$f^{a}(\theta) = f_{\uparrow,\uparrow}(\theta) - f_{\uparrow,\downarrow}(\theta) \tag{8.13}$$

One parametrizes the $\theta$ dependence in terms of the coefficients of an expansion in Legendre polynomials:

$$\mathcal{N}(0) f^{a,s}(\theta) = \sum_l F_l^{s,a} P_l(\cos\theta) \tag{8.14}$$

where $\mathcal{N}(0) = Vm^* q_F/\pi^2\hbar^3$ is the density of states of free particles (including spin) with mass $m^*$ at wave vector $q_F$. See problems 8.1 and 8.2 and reference 2 for further discussion of Fermi liquid theory.

## Models of magnets

Models for magnetic systems have played a special role in statistical mechanics for reasons which go beyond interest in magnetism itself. There are only a few models in statistical mechanics of macroscopic systems for which exact summation of the partition function has proved possible and which lead to nontrivial effects such as phase transitions. Several of these models were originally conceived as descriptions of magnetic systems, though in most cases they are too simplified to be realistic descriptions of real magnetic materials. Here we will introduce a number of the relevant models and the approaches to exact and approximate solution to some of them. We will defer discussion of the phase transitions implicit in the models to the next chapter.

### Physical basis for models of magnetic insulators: exchange

All the models to be considered here use the idea of localized spins of fixed length on a lattice interacting between nearest neighbors. In practice, even today, the parameters needed for establishing the relevance of such a model to a real material cannot be determined reliably from first quantum mechanical principles. Nevertheless, there is a general understanding of the physical origin of such a picture. At the simplest level this can be understood in terms of the Heitler–London approximation to the electronic eigenstates of a hydrogen molecule. The Hamiltonian of the hydrogen molecule is

$$H = \frac{-\hbar^2}{2m}\sum_{i,j}\nabla_i^2 + e^2\left(\frac{1}{r_{12}} - \frac{1}{r_{1\alpha}} - \frac{1}{r_{1\beta}} - \frac{1}{r_{2\alpha}} - \frac{1}{r_{2\beta}}\right) \tag{8.15}$$

where the positions of the nuclei are labelled $\alpha$, $\beta$. The Heitler–London approximation is valid when the separation between nuclei is much greater than the Bohr radius. (This is not the case for the hydrogen molecule.) Then Heitler and London approximated the wave function using a basis of products of 1s states for the ground and first excited states. Thus one formed a variational ground state of form

$$\begin{aligned}\Psi(\vec{r}_1, \sigma_2; \vec{r}_2, \sigma_2) = {}& \frac{1}{\sqrt{1 + A^2 + 2SA}}\big(\psi_{1s}(\vec{r}_1 - \vec{r}_\alpha)\psi_{1s}(\vec{r}_2 - \vec{r}_\beta) \\ &+ A\psi_{1s}(\vec{r}_2 - \vec{r}_\alpha)\psi_{1s}(\vec{r}_1 - \vec{r}_\beta)\big)\,\chi(1,2)\end{aligned} \tag{8.16}$$

where

$$S = \int d\vec{r}_1 \, \psi_{1s}(\vec{r}_1 - \vec{r}_\alpha)\psi_{1s}(\vec{r}_1 - \vec{r}_\beta) \tag{8.17}$$

$A$ is a variational parameter and $\chi$ is a normalized spin state. Calculating the expectation value of $H$ with respect to (8.17) and minimizing with respect to $A$ gives $A = \pm 1$. Then by the Pauli principle, the spin state is a singlet with $S = 0$ if $A = -1$ and a triplet with $S = 1$ if $A = +1$. The difference in energy between the singlet and the triplet is

$$\Delta E_{\mathrm{st}} = \frac{-2(VS^2 - U)}{(1 - S^4)} \tag{8.18}$$

where

$$V = \int d\vec{r}_1 \, d\vec{r}_2 \mid \psi_{1s}(\vec{r}_1 - \vec{r}_\alpha) \mid^2 \psi_{1s}(\vec{r}_2 - \vec{r}_\beta) \mid^2 \left( \frac{e^2}{r_{12}} - \frac{e^2}{r_{1\alpha}} - \frac{e^2}{r_{2\beta}} \right) \tag{8.19}$$

$$U = \int d\vec{r}_1 \, d\vec{r}_2 \, \psi_{1s}(\vec{r}_1 - \vec{r}_\alpha)\psi(\vec{r}_2 - \vec{r}_\beta)\psi_{1s}(\vec{r}_1 - \vec{r}_\beta)\psi_{1s}(\vec{r}_2 - \vec{r}_\alpha)\left( \frac{e^2}{r_{12}} - \frac{e^2}{r_{1\alpha}} - \frac{e^2}{r_{2\beta}} \right) \tag{8.20}$$

If this pair of states is the lowest lying one, then one can represent this part of the spectrum with an effective Hamiltonian $H_{\mathrm{eff}}$ of the form

$$H_{\mathrm{eff}} = -J_{\alpha\beta}\vec{S}_\alpha \cdot \vec{S}_\beta \tag{8.21}$$

where $J_{\alpha\beta} = \Delta E_{\mathrm{st}}$ and $S_\alpha$ and $S_\beta$ are spin operators identified with the *sites* $\alpha$ and $\beta$. Notice that $J_{\alpha\beta}$ takes the form identified as an exchange integral when $S = 0$. In general, however, the relationship of the constant $J_{\alpha\beta}$ to microscopic quantum mechanics is more complicated than it is in this Heitler–London example. In general all that is required for (8.21) is that the low lying states be a singlet and a triplet associated with spins localized on the two sites.

**The *n–d* models** In this context, various models have been studied for magnetic systems. The one most directly associated with the quantum mechanical picture of the last section is the Heisenberg model, obtained by summing $H_{\mathrm{eff}}$ over all pairs of localized spins on a lattice:

$$H_{\mathrm{Heis}} = -\sum_{\alpha,\beta} J_{\alpha\beta}\vec{S}_\alpha \cdot \vec{S}_\beta \tag{8.22}$$

The spin operators will obey the usual commutation relations. A further generalization consists in letting $S$, the maximum value of the projection of the spin along an arbitrary $z$ axis, be $\hbar$ times an arbitrary $1/2$ integer, instead of $\hbar/2$ as in the preceding discussion. This generalization is relevant for magnetic insulators in which the magnetic ions can be described by Russell–Saunders coupling and the orbital magnetization is quenched.

The Heisenberg model energy is invariant to a rotation of all the spins about the same angle in 3 space. However, in real magnets, physical effects are usually present which result in changes in the energy under such global rotations. Such anisotropies arise from a variety of physical causes and are parametrized by a corresponding term added to the Hamiltonian $H_{\text{Heis}}$ in the magnetic model and rendering the total model energetically favorable to alignment of the spins along one axis (in the case of uniaxial anisotropy) or possibly favorable to the confinement of the spins to a plane without prejudice to their direction within the plane. Causes of uniaxial anistropy include the effect of the classical magnetic dipole field from other spins (in noncubic materials) and the combined effects of spin orbit and electrostatic effects.

The characterization and understanding of magnetic anisotropy is more relevant to solid state physics than to statistical physics itself and we will not consider it further. However, if the (uniaxial) magnetic anisotropy is large, then the spins are essentially only observed with projections along some fixed $z$ axis in space. In that case, one can effectively ignore the transverse components of the spin and obtain in the $S = 1/2$ case, the Ising model

$$H_{\text{Ising}} = -\sum_{\alpha,\beta} J_{\alpha\beta} S^z_\alpha S^z_\beta \tag{8.23}$$

The Ising model played a huge role in the development of statistical physics. An important computational difference from the Heisenberg model is that the eigenfunctions of (8.23) are trivial to write down, whereas those of the Heisenberg model are only found after arduous numerical computation. The Ising model does not have much quantum mechanics left in it and is generally regarded as a classical model. (The statistical mechanics is not trivial though.) Similarly, if the anisotropy favors spins in the transverse $xy$ plane we obtain the $xy$ model, in which the lattice can be three dimensional but the spins are confined to a plane.

In the case of the Heisenberg and $xy$ models, we can imagine a classical limit in which $S$ gets large. Then the effects of the quantum mechanical commutation relations become negligible, all the spins commute and we have a classical model. The limit of large $S$ must be treated with some care however. One must multiply and divide the Hamiltonian by $S^2$ and consider the limit $S \to \infty$, $S^2 J \to \mathcal{J}$ finite. (In this case we usually confine attention to nearest neighbor exchange interactions so there is only one exchange coupling constant.) In this classical limit, one can imagine varying the number of components, say $n$, of the spin vectors $\vec{S}$ to any integer. If one denotes the lattice dimension by $d$ one has the set of $n$–$d$ models which have considerable historical importance in the development of the theory of critical phenomena:

$$H_{nd} = -\mathcal{J} \sum_{\alpha\beta} \vec{\mathcal{S}}_\alpha \cdot \vec{\mathcal{S}}_\beta \tag{8.24}$$

where the dot product now includes products over $n$ components of the unit length spins ($\vec{\mathcal{S}}_\alpha = \lim_{S\to\infty}(\vec{S}_\alpha/S)$).

In every case above, the addition of an external magnetic field to the model is accomplished by the simple addition of a term linear in the spins in one Euclidean direction. Schematically this means adding a term of form $-\mu \sum_i \vec{S}_i \cdot \vec{H}$. A moment's reflection reveals that while this is unambigous in the case of the Heisenberg model, in other cases more discussion is required. In the case of the Ising model, the model retains its classical character if the magnetic field is along the direction of the spin components which are present in the interacting term. If the field is normal to this ("$z$") direction then the model acquires a quantum character by virtue of the presence of the transverse spin components in the field term.

## Comparison of Ising and liquid–gas systems

It turns out that for certain purposes the Ising model can be viewed as a convenient oversimplification of the classical gas–liquid system. This is accomplished by rewriting the Ising model in terms of "lattice gas" variables $n_\alpha = S_\alpha^z + 1/2$ which take the values 0 or 1. One then interprets $n_\alpha = 1$ as the presence of an atom of the gas or liquid at the lattice site, while $n_\alpha = 0$ is interpreted as the absence of an atom at that site. With this interpretation, the magnetic field can be related to the chemical potential. The magnetic phase transition from paramagnetic to a ferromagnetic phase then maps to a gas–liquid transition to be discussed later. Even the virial expansions discussed in Chapter 6 have an analogous formal development in the magnetic case in the guise of high temperature expansions. As we will discuss later, this correspondence is believed to be quite deep and actually corresponds to an isomorphism between the two models on large length scales near the critical points of the gas–liquid and the paramagnetic–ferromagnetic phase transitions. This accounts in part for the interest in the Ising model despite the paucity of magnetic materials which can be described by it.

## Exact solution of the paramagnetic problem

The analogue to the perfect gas problem in magnetic systems is the case of zero exchange interaction $J$ in the preceding models. In each case one has

$$H_{\text{para}} = -\mu H_0 \sum_\alpha S_\alpha^1 \tag{8.25}$$

where the superscript "1" on the spin indicates a particular Euclidean direction (often called "$z$" but in the case that $n > 3$ in the $n$–$d$ models this is not very meaningful) and $H_0$ is a magnetic field. For quantum mechanical models (only defined for $n \leq 3$) the partition function is

$$Z_{\text{para}} = \sum_{\{M_s\}} \mathrm{e}^{\beta\mu H_0 \sum_\alpha M_{s,\alpha}} \tag{8.26}$$

in which $M_{s,\alpha} = -S, -S+1, \ldots, S-1, S$ at each site $\alpha$. We easily see that

$$Z_{\text{para}} = \left( \sum_{M_s=-S}^{M_s=S} \mathrm{e}^{\beta\mu H_0 M_s} \right)^N \tag{8.27}$$

and the sum can be done. Let $y = \mathrm{e}^{\beta\mu H_0}$ and $m = M_s + S$ giving

$$\sum_{M_s=-S}^{M_s=S} \mathrm{e}^{\beta\mu H_0 M_s} = \sum_{m=0}^{2S} y^{m-S} = y^{-S}\left(\frac{y^{2S+1}-1}{y-1}\right) = \frac{y^{S+1/2} - y^{-(S+1/2)}}{y^{1/2} - y^{-1/2}}$$
$$= \frac{\sinh(\beta\mu H_0(S+1/2))}{\sinh(\beta\mu H_0/2)} \tag{8.28}$$

so that

$$Z_{\text{para}} = \left( \frac{\sinh(\beta\mu H_0(S+1/2))}{\sinh(\beta\mu H_0/2)} \right)^N \tag{8.29}$$

This is the analogue of the ideal gas formula for $Z$.

## High temperature series for the Ising model

This is the magnetic analogue to the virial expansion. We give some details only for the Ising model with $S = 1/2$ in zero field and only nearest neighbor interactions but the generalizations to other cases are not very difficult. The partition function is

$$Z_{\text{Ising}} = \sum_{\{M_s\}} \mathrm{e}^{\beta \sum_{\alpha,\beta} J_{\alpha,\beta} M_{s,\alpha} M_{s,\beta}} \tag{8.30}$$

It is convenient to use the variables $\sigma_\alpha = 2M_{s,\alpha}$ and $\mathcal{J} = \beta J/4$. Then after a straightforward rearrangement:

$$Z_{\text{Ising}} = \sum_{\text{all } \sigma_\alpha = \pm 1} \prod_{\alpha,\beta} \mathrm{e}^{\mathcal{J}\sigma_\alpha\sigma_\beta} \tag{8.31}$$

Now we wish to rewrite this as a series in terms which become successively smaller at high temperatures. To do this let $\eta$ be a quantity which is $\pm 1$ and note that

$$\begin{aligned}\cosh A + \eta \sinh A = \cosh A \pm \sinh A &= (1/2)(\mathrm{e}^{A} + \mathrm{e}^{-A} \pm (\mathrm{e}^{A} - \mathrm{e}^{-A})) \\ &= \mathrm{e}^{\pm A} = \mathrm{e}^{\eta A}\end{aligned} \tag{8.32}$$

Thus

$$Z_{\text{Ising}} = \sum_{\{\sigma_\alpha\}} \prod_{\alpha,\beta} (\cosh \mathcal{J} + \sigma_\alpha \sigma_\beta \sinh \mathcal{J}) = (\cosh \mathcal{J})^{Nq/2} \sum_{\{\sigma_\alpha\}} \prod_{\alpha,\beta} (1 + \sigma_\alpha \sigma_\beta \tanh \mathcal{J}) \tag{8.33}$$

Here $q$ is the number of nearest neighbors (sometimes called the lattice coordination number). (8.33) can be rewritten

$$Z_{\text{Ising}}/(\cosh \mathcal{J})^{Nq/2} = \sum_{\{\sigma_\alpha\}} \prod_{\alpha,\beta} (1 + f_{\alpha,\beta}) \tag{8.34}$$

in which

$$f_{\alpha,\beta} = \sigma_\alpha \sigma_\beta \tanh \mathcal{J} \tag{8.35}$$

This should be compared with the corresponding expression in Chapter 6. The similarity of the resulting forms should be quite clear, but we make them more explicit in the table below:

| Gas–liquid | Nearest neighbor $S = 1/2$, $H_0 = 0$, Ising model |
|---|---|
| $Z$ | $Z_{\text{Ising}}/(\cosh J)^{\text{Nq}/2})$ |
| $f_{ij} = \mathrm{e}^{-\beta v_{ij}} - 1$ | $f_{\alpha\beta} = \sigma_\alpha \sigma_\beta \tanh \mathcal{J}$ |
| $\int \mathrm{d}^N \vec{r} \cdots$ | $\sum_{\{\sigma_\alpha\}} \cdots$ |
| $\prod_{i,j}$ | $\prod_{\alpha,\beta}$ |

$f_{\alpha,\beta}$ becomes small at large temperatures, as $f_{ij}$ does. At every stage, the detailed calculations are simpler in the Ising model because of the simpler nature of the interaction. It is clear that the terms in (8.35) may be represented by diagrams, much as they were in Chapter 6. An important simplifying feature here is that lines may only be drawn between circles representing nearest neighbor lattice sites.

Now consider the one dimensional Ising model. The diagrams are of form

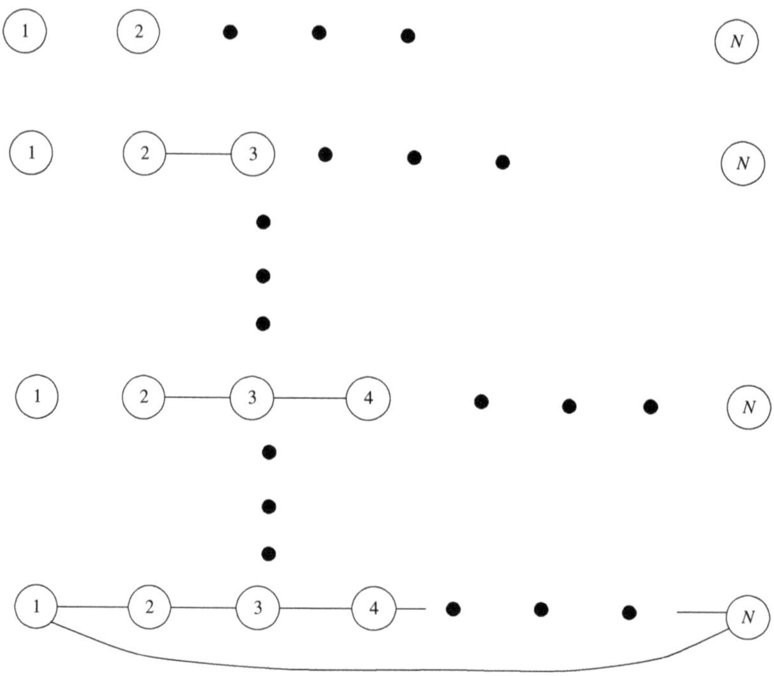


where the line connecting site 1 to site $N$ in the last diagram is present only if periodic boundary conditions are employed. The first diagram has the value $2^N$. The diagrams with untied ends are all zero because they involve a factor

$$\sum_{\sigma_\alpha=\pm 1} \sigma_\alpha \sigma_\beta \tanh \mathcal{J} = 0 \tag{8.36}$$

because $\sum_{\sigma_\alpha=\pm 1} \sigma_\alpha = 1 - 1 = 0$. The result then depends on the boundary conditions chosen for the ends of the one dimensional model. If the spins have only one neighbor ("free ends") then there are no nonzero contributions after the first and the result is

$$Z_{\text{Ising, 1}d} = 2^N \cosh^N \mathcal{J} \quad \text{"free ends"} \tag{8.37}$$

On the other hand in the case of "periodic" boundary conditions, the last spin is regarded as having the first spin as a nearest neighbor. In that case, the diagram corresponding to a fully linked loop is nonzero and has the value $2^N \tanh^N \mathcal{J}$ giving

$$Z_{\text{Ising, 1}d} = 2^N(\cosh^N \mathcal{J} + \sinh^N \mathcal{J}) \quad \text{"periodic boundary conditions"} \tag{8.38}$$

In two dimensions, it turns out that the series can also be summed exactly, but the result is much more complicated and, in particular, the model admits a phase transition. See Stanley's book for an account of this approach to the solution of the two dimensional Ising model.[3]

## Transfer matrix

The Ising model in two dimensions was originally solved by Onsager[4] using another approach, which we outline now, again applying the method only to the much simpler one dimensional Ising model. This method can be used for any problem involving discrete variables and binary interactions on a lattice. In these cases the partition function has the form

$$Z = \sum_{x_1, x_2, \ldots, x_N} \mathrm{e}^{-\beta \sum_{(\alpha,\beta)} V(x_\alpha, x_\beta)} \tag{8.39}$$

In the Ising model $V(x_\alpha, x_\beta) = -J\sigma_\alpha\sigma_\beta - (\mu H_0/2)(\sigma_\alpha + \sigma_\beta)$ and $x_\alpha = \sigma_\alpha$, $x_\beta = \sigma_\beta$. The partition function can be rewritten in terms of the two dimensional eigenvectors of the $2 \times 2$ matrix $M(x, y) = \mathrm{e}^{-\beta V(x,y)}$:

$$\sum_y M(x, y) a_\nu(y) = \lambda_\nu a_\nu(x) \tag{8.40}$$

In this case $M(x, y)$ is Hermitian and the eigenvectors can be taken to be orthogonal

$$\sum_y a_\nu^*(y) a_\mu(y) = \delta_{\nu,\mu} \tag{8.41}$$

and complete

$$\sum_\nu a_\nu^*(y) a_\nu(x) = \delta_{x,y} \tag{8.42}$$

Using these relations we can easily show that $M(x, y)$can be written as

$$M(x, y) = \sum_\nu a_\nu^*(y) \lambda_\nu a_\nu(x) \tag{8.43}$$

and the partition function is

$$Z = \sum_{x_1,\ldots,x_N} \prod_{\alpha,\beta} M(x_\alpha, x_\beta) = \sum_{x_1,\ldots,x_N} \sum_{\{\nu_{(\alpha,\beta)}\}} \prod_{\alpha,\beta} \lambda_{\nu_{(\alpha,\beta)}} a^*_{\nu_{(\alpha,\beta)}}(x_\beta) a_{\nu_{(\alpha,\beta)}}(x_\alpha) \tag{8.44}$$

There is an index $\nu_{(\alpha,\beta)}$ for each pair of nearest neighbor sites $(\alpha, \beta)$. This can be analyzed in the case of two dimensions, though it is very complicated. The result is particularly simple in the one dimensional case when each coordinate $x_\alpha$ appears only twice. Then the sums on $x_\alpha$ can all be done using the normalization condition and we obtain

$$Z = \sum_\nu \lambda_\nu^N \tag{8.45}$$

In the thermodynamic limit this will be dominated by the largest eigenvalue (which we will here suppose to be nondegenerate) and we have

$$Z \underset{N\to\infty}{\to} (\lambda_{\max})^N \tag{8.46}$$

Thus in this case the problem of finding $Z$ reduces to the problem of finding the largest eigenvalue of the matrix (including the field this time):

$$\begin{pmatrix} e^{K+C} & e^{-K} \\ e^{-K} & e^{K-C} \end{pmatrix}. \tag{8.47}$$

where $K = \beta J$ and $CV = \mu H \beta$. The eigenvalues are easily shown to be

$$\lambda_\pm = e^K \cosh C \pm (e^{2K} \sinh^2 C + e^{-2K})^{1/2} \tag{8.48}$$

The eigenvalue with the plus sign is largest for ferromagnetic interactions and the partition function becomes

$$Z = \left(e^K \cosh C + (e^{2K} \sinh^2 C + e^{-2K})^{1/2}\right)^N \tag{8.49}$$

which reduces to our earlier result (8.38) in zero field for large $N$.

## Monte Carlo methods

In cases in which exact solutions are not available, one can always find an approximate description of the equilibrium state of a system described by a classical Hamiltonian using Monte Carlo methods, which we briefly describe here for magnetic models, but the extension to other Hamiltonians is not difficult to construct. (In fact, Monte Carlo methods can also be used for some quantum systems but this requires more discussion.) The information extracted from Monte Carlo calculations is also available from molecular dynamics calculations whenever the equation of motion allows approximately ergodic motion of the system point, but the converse is not the case. Molecular dynamics provides realistic information about the dynamics of the system, whereas in most cases, Monte Carlo simulation only provides information about the distribution of states in the equilibrium state. To formulate the method generally, we consider that we have a numerical description of a state of a system. For example, for the spin $1/2$ Ising model this would consist of a specification of the values of all the spin variables ($\pm 1/2$). Call this state $\nu$. In the Monte Carlo method one selects another trial state $\nu'$ by an algorithm involving a random "flipping of a coin" and designed not to bias the trial selection. One then considers whether to add that state to an ensemble of states to be used to represent equilibrium at the end of the calculations. To determine whether the state $\nu'$ is to be included one supposes that the temporal evolution of the ensemble during the simulation is governed by a "master equation" of the form

$$dP_\nu/dt = \sum_{\nu'}(-W_{\nu\to\nu'}P_{\nu'} + -W_{\nu'\to\nu}P_\nu) \tag{8.50}$$

where the $P_\nu$ is the probability, in the ensemble, of finding the system in state $\nu$ and the $W_{\nu\to\nu'}$, $W_{\nu'\to\nu}$ are transition probabilities, to be used in generating the ensemble. When evolving the ensemble according to (8.50), one supposes that a steady state is reached such that the probabilities do not change in time. Then the time derivative is zero and one obtains from (8.50)

$$W_{\nu\to\nu'}/W_{\nu'\to\nu} = P_\nu/P_{\nu'} = e^{-\beta(E_{\nu'}-E_\nu)} \tag{8.51}$$

assuming that the final steady state describes a canonical ensemble. Notice that it is only this ratio between rates of transfer between pairs of states which needs to be preserved in order to end up with an equilibrium ensemble. Now consider the candidate state $\nu'$ and evaluate the difference $E_{\nu'} - E_\nu$. (This step is usually easy for classical systems and often very difficult for quantum ones.) Any choice of transition rates which consistently obeys (8.51) will generate an equilibrium ensemble and this leads to considerable flexibility. However, a choice which efficiently and simply generates an equilibrium ensemble (Metropoulis algorithm) is to choose $W_{\nu\to\nu'} = 1/\tau$ if $E_{\nu'} - E_\nu < 0$ and $W_{\nu\to\nu'} = e^{-\beta(E_{\nu'}-E_\nu)}/\tau$ if $E_{\nu'} - E_\nu > 0$. ($\tau$ is taken to be the time between steps so that the transition is always made in the first case and with probability $e^{-\beta(E_{\nu'}-E_\nu)}$ in the second.) To see that this will work consider a pair of states for which $E_{\nu'} - E_\nu > 0$. When the system is in state $\nu$ then the transition to state $\nu'$ takes place with probability $e^{-\beta(E_{\nu'}-E_\nu)}$ whereas when the system is in state $\nu'$ a transition to state $\nu$ takes place with probability 1 so the ratio (8.51) is preserved. (It is also possible to choose the transition rates more realistically,[5] so that the resulting time evolution during the simulation approximates the time evolution of the real system.)

## References

1. S. Kanno, *Progress in Theoretical Physics* **48** (1972) 2210.
2. For Fermi liquid theory.
3. H. E. Stanley, *Introduction to Phase Transitions and Critical Phenomena*, Oxford: Oxford University Press, 1971.
4. L. Onsager, *Physical Review* **65** (1944) 117.
5. K. Kawasaki, *Physical Review* **145** (1966) 224.

## Problems

8.1 Use the assumptions of the mean field Fermi liquid theory to derive an expression for the function $\langle \delta n_{\vec{q}} \rangle$ valid at low temperatures. In particular, show that the interactions arising from the function $f$ do not contribute to leading order.

8.2 Evaluate the low temperature susceptibility of a Fermi liquid in terms of the parametrization of the Fermi liquid model.

8.3 Find the specific heat and the susceptibility of the following models analytically and evaluate and graph them as a function of an appropriate dimensionless variable:
(a) quantum spins with $S = 1/2,\ 1,\ 3/2,\ 2$;
(b) classical spins with $d = 1,\ 2,\ 3,\ 4$;
(c) one dimensional $S = 1/2$ Ising model.

8.4 Write a Monte Carlo code to determine the magnetization of the two dimensional nearest neighbor Ising model on a square lattice as a function of temperature.

# 9

# Phase transitions: static properties

We begin with some thermodynamic considerations and then proceed to a discussion of critical phenomena. In discussing critical phenomena we first describe the phenomenology, then some general considerations concerning Landau–Ginzburg free energy functionals and mean field theory and finally an introduction to the renormalization group.

## Thermodynamic considerations

Consider a system at fixed pressure $P$ and temperature $T$. (We will not be concerned with magnetic properties yet.) We will suppose that the system contains some integer number $s$ of molecular species. We will also suppose that we have a means of distinguishing two or more phases of this system. Though this is an assumption it requires some examples and discussion. We distinguish between phases by consideration of their macroscopic properties so a spatial as well as a temporal average is involved. For example, we distinguish gas from liquid by the difference in average density and magnet from paramagnet by the existence of a finite average magnetization in the former. (We will discuss some more subtle cases later.) But if we wish to consider (as we do here) the possible coexistence of more than one phase then a problem arises concerning the length scale over which we ought to average spatially. If, in a system containing two coexisting phases, we find the average properties by averaging over the entire system, then we will always get just one number and not two and will have no means of distinguishing the phases. One must say something along the following lines: let the volume of the whole system go to infinity, as in the usual definition of the thermodynamic limit. Consider the behavior of the average properties of portions of the system as this limit is taken, in such a way that the volume fraction occupied by each portion is fixed as the total volume diverges. One must also require that the partitions into portions minimize the surface area between portions of the partition. If the average properties of some portions are different from others in the limit, then we say that the portions contain different phases. Even this is not quite enough, since by this means one could find that one of the portions contained 20% liquid and 80% gas while the other contained 20% gas and 80% liquid. So we add the following condition. Consider all such partitions of the system, taking the limit described in each case. Then that partition for which the average properties of the two portions differ by the largest amount is defined to be the partition in which each portion contains a pure phase. It should be clear that the definition could be extended to more than two coexisting phases. It should be cautioned that this approach will only work in the limit of very large volumes and when one is taking time (not ensemble) averages. One role of the large volume is to make the time infinitely long for a given phase in one region to change to the other phase in the same region. If one thinks of ensemble averages, then no such stratagem protects us from including, for example, the state with the liquid at the bottom of the container and the state with the fluid at the top in the average and the definition fails. For ensemble averages, a region of phase equilibrium is identified by the response of the system to a field or chemical potential which drives it out of phase equilibrium. For example, in the case of a magnet at low enough temperatures the magnetization of the whole system may change discontinuously from a finite positive value to a finite negative value as the field passes through zero. One infers from this nonanalyticity of the response of a calculated thermodynamic property that phase equilibrium between up and down magnetization must exist at zero field. This approach to identifying regions of phase equilibria from the nonanalyticity of thermodynamic functions calculated as ensemble averages as a function of probe fields (or chemical potentials) has been pursued with great mathematical rigor by Ruelle[1] and others.

By one or the other of these approaches we can determine that more than one phase can coexist in equilibrium at a given pressure and temperature in equilibrium. Now consider the Gibbs free energy $G$. In the case of more than one atomic or molecular spcecies in the whole system

$$G = \sum_{i=1}^{s} \mu_i N_i \tag{9.1}$$

where $\mu_i$ is the chemical potential of species $i$ and $N_i$ is the number of atoms of molecules of species $i$ in the whole system. Equation (9.1) is a simple extension of the earlier definition to the case of more than one component and is obtained by the same argument which took us from (3.43) to (3.44) starting with the expressions $(\partial G/\partial N_i)_{P,T,N_{i'\neq i}} = \mu_i$. But because we have defined each phase in a way which permits us to consider its infinite volume limit, we can imagine calculating the chemical potential $\mu_i^{(j)}$ of the $i$th species in the $j$th phase. For example, this calculation could be done by calculating the Gibbs free energy $G^{(j)}$ of each phase and taking the partial derivative $(\partial G^{(j)}/\partial N_i^{(j)})_{P,T,N_{i'\neq i}^{(j)}} = \mu_i^{(j)}$ where $N_i^{(j)}$ is the number of molecules of the $i$th species in the $j$th phase. Also, because the Gibbs free energy is extensive

$$G = \sum_{j=1}^{p} G^{(j)} = \sum_{j=1}^{p}\sum_{i=1}^{s} \mu_i^{(j)} N_i^{(j)} \tag{9.2}$$

where $p$ is the number of phases and $s$ is the number of species. In addition the numbers $N_i^{(j)}$ obey the relation $N_i = \sum_{j=1}^{p} N_i^{(j)}$ so that, picking some phase $j$:

$$N_i^{(j)} = N_i - \sum_{j'\neq j} N_i^{(j')} \tag{9.3}$$

Combining these relations

$$\mu_i = \left(\frac{\partial G}{\partial N_i}\right)_{P,T,N_{i'\neq i}} = \left(\frac{\partial G}{\partial N_i^{(j)}}\right)_{P,T,N_{i'\neq i},N_i^{j'\neq j}} = \mu_i^{(j)} \tag{9.4}$$

where we have used (9.1) in the first equality, (9.3) in the second and (9.2) in the last. The argument obviously does not depend on which phase $j$ was chosen so for each species $i$ we have a set of equations

$$\mu_i^{(1)} = \mu_i^{(2)} = \cdots = \mu_i^{(p)} \tag{9.5}$$

which must be satisfied if $p$ phases are in equilibrium. It should be emphasized that it will usually be the case that over most of the space of thermodynamic variables only one phase will be in equilibrium but we are here interested in the regions of that space in which more than one phase can coexist. To begin to analyze (9.5) consider first the case of one molecular species ($s = 1$) and suppose that the system is globally characterized by two other intensive thermodynamic parameters, which we take to be the pressure $P$ and the temperature $T$. (Extension to other systems, such as magnetic ones, is not difficult.) In that case, the chemical potentials will only depend on $P$ and $T$. To find regions of the $P$–$T$ plane in which 2 phases ($p = 2$) coexist we have from (9.5)

$$\mu_1^{(1)}(P,T) = \mu_1^{(2)}(P,T) \qquad s = 1 \quad p = 2 \tag{9.6}$$

We have one equation connecting the two variables $P, T$ so assuming that the functions $\mu_1^{(1)}(P,T)$ and $\mu_1^{(2)}(P,T)$ are well behaved we will find a one dimensional curve in the $P$–$T$ plane where the two phases 1 and 2 can coexist. Similarly in the case $s = 1$, $p = 3$ the only solutions can be points, usually called triple points. When the functions are well behaved the triple points will occur at the ends of lines of two phase coexistence and there will be no regions of four phase coexistence for

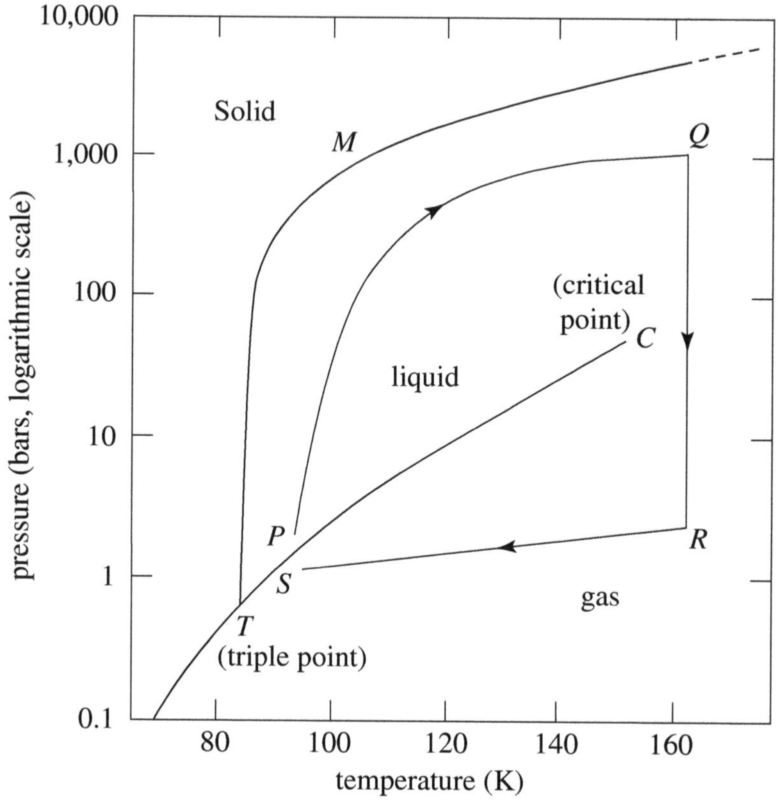


Figure 9.1 Phase diagram of argon. The path *PQRS* passes from liquid to gas continuously without crossing the coexistence line *TC*.

one component systems. All this is entirely consistent with the well known shape of the phase diagram of a simple one component system having gas, liquid and one solid phases (see Figure 9.1).

We may generalize these considerations to obtain the Gibbs phase rule, which gives the dimension $f$ of the region of phase coexistence of $p$ phases of a system of $s$ molecular species. We have to determine the dimension of the space of thermodynamic variables. For more than one component we define variables

$$x_i^{(j)} = \frac{N_i^{(j)}}{\sum_{i'=1}^{s} N_{i'}^{(j)}} \tag{9.7}$$

$x_i^{(j)}$ is the fraction of the molecules in phase $j$ which belong to species $i$. Obviously all the $x_i^{(j)}$ are 1 when there is only one species. The $x_i^{(j)}$ obey the condition

$$\sum_{i=1}^{s} x_i^{(j)} = 1 \tag{9.8}$$

so in general only $s-1$ of them in each phase are independent. The $\mu_i^{(j)}$ will depend on these $s-1$ independent $x_i^{(j)}$ in each phase. Thus the total collection of independent thermodynamic variables can be taken to be $P, T, x_1^{(1)}, \ldots, x_{s-1}^{(1)}, \ldots, x_1^{(p)}, \ldots, x_{s-1}^{(p)}$ and the dimension of the space of thermodynamic variables is $2+(s-1)p$. Now the equations (9.5) may be written more explicitly as

$$\begin{aligned}\mu_1^{(1)}(P,T,x_1^{(1)},\ldots,x_{s-1}^{(1)}) &= \mu_1^{(2)}(P,T,x_1^{(2)},\ldots,x_{s-1}^{(2)}) \\ &= \cdots = \mu_1^{(p)}(P,T,x_1^{(p)},\ldots,x_{s-1}^{(p)}) \\ &\vdots \\ \mu_s^{(1)}(P,T,x_1^{(1)},\ldots,x_{s-1}^{(1)}) &= \mu_s^{(2)}(P,T,x_1^{(2)},\ldots,x_{s-1}^{(2)}) \\ &= \cdots = \mu_s^{(p)}(P,T,x_1^{(p)},\ldots,x_{s-1}^{(p)})\end{aligned} \tag{9.9}$$

There are in general $s(p-1)$ equations here. Thus the dimension $f$ of the regions of coexistence of $p$ phases in $s$ component systems is equal to the number of variables minus the number of equations constraining them or

$$f = 2+(s-1)p - s(p-1) = 2-p+s$$

This is the Gibbs phase rule.

We can obtain further information about the nature of these regions of phase coexistence using thermodynamic relations applying to the individual phases. The simplest case is that of one component. Then the chemical potential obeys the Gibbs–Duhem relation (3.42)

$$\mathrm{d}\mu^{(j)} = v^{(j)}\mathrm{d}P - s^{(j)}\mathrm{d}T \tag{9.10}$$

where the subscripts are omitted because we are considering only one component. Then using (9.5) in the one component case one obtains

$$v^{(1)}\mathrm{d}P - s^{(1)}\mathrm{d}T = v^{(2)}\mathrm{d}P - s^{(2)}\mathrm{d}T \tag{9.11}$$

in the case of two phase equilibrium. Thus we get a relation for the slope of the two phase coexistence line in the $P$–$T$ plane in terms of differences in the properties of the two phases:

$$\frac{s^{(2)}-s^{(1)}}{v^{(2)}-v^{(1)}} = \frac{\mathrm{d}P}{\mathrm{d}T} \tag{9.12}$$

which is the Clausius–Clapeyron relation. To generalize this to the multicomponent case one needs the generalization of the Gibbs–Duhem relation. We have for the $j$th phase:

$$\mathrm{d}G^{(j)} = -S^{(j)}\mathrm{d}T + V^{(j)}\mathrm{d}P + \sum_{i=1}^{s}\mu_i^{(j)}\mathrm{d}N_i^{(j)} = \sum_{i=1}^{s}\left(N_i^{(j)}\mathrm{d}\mu_i^{(j)} + \mu_i^{(j)}\mathrm{d}N_i^{(j)}\right) \tag{9.13}$$

giving

$$\sum_{i=1}^{s} x_i^{(j)} \mathrm{d}\mu_i^{(j)} = -s^{(j)}\mathrm{d}T + v^{(j)}\mathrm{d}P \tag{9.14}$$

in which we have divided by $\sum_{i=1}^{s} N_i^{(j)}$ and used the definition of $x_i^{(j)}$. Thus in the case of two phase equilbrium the generalization of the Clausius–Clapeyron relation becomes

$$\frac{\mathrm{d}P}{\mathrm{d}T} - \sum_{i=1}^{s} \left(x_i^{(1)} - x_i^{(2)}\right) \frac{\mathrm{d}\mu_i}{\mathrm{d}T} = \frac{s^{(2)} - s^{(1)}}{v^{(2)} - v^{(1)}} \tag{9.15}$$

This provides a relation between the components of tangents to the thermodynamic coexistence hypersurface described in the space of variables $P, T, \{\mu_i^{(1)}\}$.

## Critical points

To introduce the idea of critical points, we consider the case of one component systems and the equilibrium of two phases. As discussed in the last section, the region of the thermodynamic space in which the phase equilibrium between these two phases can occur is given by solutions to the equation

$$\mu^{(1)}(P, T) = \mu^{(2)}(P, T) \tag{9.16}$$

and the solutions to this equation will, in the case of well behaved functions, describe a line in the $P$–$T$ plane. Under some circumstances, it can happen that such a line can simply end (without intersecting other lines). At this point the phases become indistinguishable. For definiteness we consider the case of a gas–liquid transition. Then the phases could be distinguished by their entropy or by their volume per particle. It is convenient to consider the volume per particle. As we pass along the coexistence line, the volume per particle, which is $(\partial\mu/\partial P)_T$ by the Gibbs–Duhem relation, is not uniquely defined but the two possible values become equal at the critical point. We can see why this corresponds to a point (and not, in the one component case, to a line) by writing the condition of indistinguishability explicitly

$$\left(\frac{\partial \mu^{(1)}}{\partial P}\right)_T = \left(\frac{\partial \mu^{(2)}}{\partial P}\right)_T \tag{9.17}$$

The last two equations then give two equations in two unknowns, resulting in general in solutions only at points in the $P$–$T$ plane. It is clear that the $\mu(P, T)$ surface will have some peculiarities near a critical point. The first derivative with respect to $P$ is undergoing a discontinuity along the line and suddenly ceases to have this discontinuity. It turns out that it solves this problem in the gas–liquid case by making the next derivative divergent in the $P$ direction:

$$\left(\frac{\partial^2 \mu}{\partial P^2}\right)_{P_c, T_c} \to \infty \tag{9.18}$$

This may be understood by noting that the derivative in question is $(\partial v/\partial P)_T$ so that the fluid is on the margin of mechanical stability when it is divergent. Generally, the thermodynamic derivatives are either zero or infinite at a critical point. Before going into this in more detail we briefly go over the same ideas in the case of magnetic phase transition.

Suppose that there are two possible phases, corresponding to magnetization "up" (denoted +) and magnetization "down" (denoted −). The free energy which has the intensive properties required for the discussion is here $-k_B T \ln \mathrm{Tr}\, e^{-\beta H}$ because the field $H_0$ enters the microscopic description naturally (unlike $P$ in the gas–liquid case). We will follow the literature and denote $F = -k_B T \ln \mathrm{Tr}\, e^{-\beta H}$. When the values of $H$ and $T$ are such that the two phases (±) are in equilibrium then

$$\mu^{(+)}(H_0, T) = \mu^{(-)}(H_0, T) \tag{9.19}$$

where $\mu_\pm = (\partial F^{(\pm)}/\partial N^{(\pm)})_{H,t}$ and $N^{(\pm)}$ are the numbers of spins in the two phases respectively. Now in lowest order in $H_0$ for small fields

$$\mu^\pm = \mu_0 \mp m(T, H_0 = 0) H_0 \tag{9.20}$$

where $m$ is the magnetization per spin so that the solution to (9.19) is $H = 0$. The coexistence line takes a very simple form. To find the critical point, we need the Gibbs–Duhem relation which in this case is

$$d\mu = -s\, dT - m\, dH_0 \tag{9.21}$$

The phases are distinguished by their magnetization so that the critical point occurs when

$$\left(\frac{\partial \mu^{(+)}}{\partial H_0}\right)_T = \left(\frac{\partial \mu^{(-)}}{\partial H_0}\right)_T \tag{9.22}$$

and $H_0 = 0$. The second derivative of $\mu$ with respect to $H_0$ also diverges, corresponding to the divergence of the susceptibility of the magnet at the critical point.

## Phenomenology of critical point singularities: scaling

As hinted, one finds that many thermodynamic derivatives are singular at critical points. This is known from experiment, numerical simulation and, as we will discuss, theory. First we summarize the known facts in the simple case of an Ising type magnet. The susceptibility along the coexistence line $\chi(T, H = 0) \propto | T - T_c |^{-\gamma}$

where $\gamma \approx 4/3$, the magnetization $m$ along the coexistence line $m \propto| T - T_c |^{\beta}$ where $\beta \approx 1/3$ (and $m = 0$ for $T > T_c$ of course). The specific heat along the coexistence line $C_H(T, H = 0) \propto| T - T_c |^{-\alpha}$ diverges weakly (though for some similar models it goes to zero so that $\alpha$ can be negative in some models). The magnetization for finite field at $T_c$ rises slowly and very nonlinearly with $H$ at $T_c$ : $m(T = T_c, H) \propto H^{1/\delta}$ where $\delta \approx 5$. Some further relations were obtained for the correlation function

$$g(\vec{r}) = \langle m(\vec{r})m(0)\rangle - \langle m\rangle^2 \tag{9.23}$$

which was found always to have the form

$$g(r) = K \frac{e^{-r/\xi}}{r^{d-2+\eta}} \tag{9.24}$$

Here $\xi$ the coherence length diverges at zero field with the temperature dependence $\xi \propto| T - T_c |^{-\nu}$. $d$ is the lattice dimension. Numerical data are available about the behavior of the Ising and other $n$–$d$ and similar models in various dimensions. Experiments can also approximate the behavior of systems of low dimensionality. Empirical values of $\nu$ for Ising like systems are about 2/3 and values of $\eta$, while finite, are small.

One notes that these numbers approximately obey the relations

$$2 - \alpha = \gamma + 2\beta = \beta(\delta + 1) \tag{9.25}$$

as well as

$$\gamma = \nu(2 - \eta) \qquad \nu d = 2 - \alpha \tag{9.26}$$

These relations were obtained for several strikingly different physical models on an empirical basis before there was much understanding of their origin. What was even more striking was that the same *values* of the exponents were obtained, both from experiment and from calculational estimates based on high temperature expansions (and exactly in the case of the two dimensional Ising model), from apparently quite diverse physical systems. An especially famous example is the case of the exponents found for the gas–liquid critical point of simple liquids and the exponents found for the three dimensional Ising model, which are exactly the same to within measuring and calculational accuracy. This phenomenon of the identity of the exponents for diverse physical systems is known as "universality" and its explanation did not come until the discovery of the renormalization group, which we describe later. One should not attach too much significance to the word "universality." Exponents are the same for some diverse sorts of physical systems but they are not the same for all systems exhibiting simple critical points. One speaks of "universality classes" of systems which have the same critical exponents, despite having superficially very different physical properties.

It was recognized that the scaling relations between the exponents represent limits of thermodynamic stability. For example, consider the first relation, which may be rewritten $\alpha + 2\beta + \gamma = 2$. Consider the specific heat at constant *magnetization* $C_M$

$$C_M = T\left(\frac{\partial S}{\partial T}\right)_M = T\left(\left(\frac{\partial S}{\partial T}\right)_H + \left(\frac{\partial S}{\partial H}\right)_T\left(\frac{\partial H}{\partial T}\right)_M\right) \tag{9.27}$$

but from the Maxwell relation

$$\left(\frac{\partial S}{\partial H}\right)_T = \left(\frac{\partial M}{\partial T}\right)_H \tag{9.28}$$

we have

$$C_M - C_H = T\frac{\left(\frac{\partial M}{\partial T}\right)_H\left(\frac{\partial H}{\partial T}\right)_M\left(\frac{\partial M}{\partial H}\right)_T}{\left(\frac{\partial M}{\partial H}\right)_T} \tag{9.29}$$

and using

$$\left(\frac{\partial H}{\partial T}\right)_M\left(\frac{\partial M}{\partial H}\right)_T\left(\frac{\partial T}{\partial M}\right)_H = -1 \tag{9.30}$$

we obtain

$$C_M = C_H - T\frac{\left(\frac{\partial M}{\partial T}\right)_H^2}{\left(\frac{\partial M}{\partial H}\right)_T} \tag{9.31}$$

Then requiring that $C_M \geq 0$ we obtain, using the singular forms for the various functions as found empirically, that (with $t = (T - T_c)/T_c$)

$$t^{-\alpha} \geq \frac{t^{2(\beta-1)}}{t^{-\gamma}} \tag{9.32}$$

or

$$2 \leq \alpha + 2\beta + \gamma \tag{9.33}$$

Another level of understanding was achieved by the following argument. Consider the free energy $f(t, h)$ per spin of the system at a temperature $T$ and field $H$ near the critical point where $t = (T - T_c)/T_c$ and $h = H - H_c$. If there are $N$ spins, this can be written

$$f(t, h) = (1/N)\left(-k_B T \ln \mathrm{Tr}_{\{S\}} e^{-\beta H}\right) \tag{9.34}$$

Now imagine that we break the trace into two parts as follows. Divide the lattice into hypercubic blocks each containing $L^d$ spins. Let the average spin of each block be

$\mathcal{S}_\alpha$ where $\alpha$ labels the block. We do the sum in two parts, first over all the variables for which the $\mathcal{S}_\alpha$ remain fixed and then over all the $\mathcal{S}_\alpha$. Then

$$f(t,h) = \frac{1}{L^d}\left(\frac{L^d}{N}\left(-k_B T \ln \mathrm{Tr}_{\mathcal{S}} e^{-\beta H_{\mathrm{eff}}}\right)\right) \tag{9.35}$$

in which

$$H_{\mathrm{eff}} = -k_B T \ln \mathrm{Tr}_{\mathrm{not}\ \mathcal{S}} e^{-\beta H} \tag{9.36}$$

and "not $\mathcal{S}$" means those other variables over which we sum before summing over $\mathcal{S}$. This is an identity. $N/L^d$ is the number of new "block spins" so

$$f' = \frac{L^d}{N}\left(-k_B T \ln \mathrm{Tr}_{\mathcal{S}} e^{-\beta H_{\mathrm{eff}}}\right) \tag{9.37}$$

can be regarded as the free energy per spin of a new "coarse grained" system in which the short wavelength degrees of freedom have been "summed out" (through the variables "not $\mathcal{S}$"). Now the physical argument is that near the critical point, the system consists of very large spatial regions of highly correlated spins. Thus the free energy per spin $f'$ of the "block system" represented by $H_{\mathrm{eff}}$ should be similar to the free energy per spin $f$ of the original system except that some of the short wavelength fluctuations have been summed out. The effect of reducing the fluctuations in this way is qualitatively like moving away from the critical point in the $(t,h)$ plane. To reproduce the scaling relations one *assumes* that this effect can be entirely taken into account by identifying $f'$ with the same free energy function $f(t,h)$ which gives the thermodynamic properties but with the variables $t,h$ at a different point $t',h'$ which is farther away from the critical point:

$$f' = f(t',h') \tag{9.38}$$

Further one assumes that $t',h'$ are related to $t,h$ through the simple relations $t' = L^x t$, $h' = L^y h$. One sees that these assumptions are consistent with the preceding qualitative remarks, though the actual form of the relation (9.38) and the scaling relations between the effective temperature and field and the real ones have not been demonstrated.

With these assumptions one has the relation (9.35, 9.37, 9.38)

$$f(t,h) = (1/L^d) f(L^y t, L^x h) \tag{9.39}$$

It is now quite straightforward to demonstrate the scaling relations cited earlier. To do so it is convenient to rewrite (9.39) for the particular choice $L = 1/t^{1/y}$

$$f(t,h) = t^{d/y} f\left(1, h/t^{x/y}\right) \tag{9.40}$$

Calling $f(1, h/t^{x/y}) \equiv \mathcal{F}(h/t^{x/y})$ we may write this

$$f(t,h) = t^{d/y}\mathcal{F}\big(h/t^{x/y}\big) \tag{9.41}$$

We will assume that the "scaling function" $\mathcal{F}(u)$ is differentiable. The specific heat for fixed field at the critical value $h = 0$ is proportional to the second derivative of $f$ with respect to $t$ so we have at once that

$$d/y - 2 = -\alpha \tag{9.42}$$

or

$$y = d/(2-\alpha) \tag{9.43}$$

The magnetization is proportional to the derivative of $f$ with respect to $h$ at fixed $t$ so

$$\beta = d/y - x/y \tag{9.44}$$

or

$$x = d(1 - \beta/(2-\alpha)) \tag{9.45}$$

To obtain the susceptibility at fixed field we take another derivative with respect to $h$, giving

$$\gamma = 2x/y - d/y \tag{9.46}$$

Combining these gives

$$\gamma + 2\beta + \alpha = 2 \tag{9.47}$$

which is consistent with the empirically determined relation between exponents. One can get a further relation by inspecting the magnetization at fixed field at $t = 0$. This involves one further idea. The magnetization is proportional to

$$t^{d/y-x/y}\mathcal{F}'\big(h/t^{x/y}\big) \tag{9.48}$$

but at $t = 0$ this must be finite at finite $h$. Therefore in this limit the function $\mathcal{F}'$ must be proportional to $(h/t^{x/y})^{(d/y-x/y)/(x/y)}$ where the exponent is selected so that the factors depending on $t$ cancel. Thus we obtain

$$1/\delta = (d-x)/x \tag{9.49}$$

giving, after inserting the result for $x$ and some algebra

$$\beta(\delta+1) = 2-\alpha \tag{9.50}$$

which is another of the empirical relations. One may make a similar argument for $g(r)$, except that the correlation function is obtained by differentiating with respect to $h(\vec{r})$ at two widely separated places, bringing down two factors of $m$. Thus in rescaling to the block spin system, one must divide by two factors of $L^d$ giving

$$g(r,t,h) = \frac{\partial^2 f(t,h(\vec{r}))}{\partial h(0)\partial h(\vec{r})} = \frac{1}{L^{2d}} \frac{\partial^2 f(t,h(\vec{r}))}{\partial L^x h(0) \partial L^x h(\vec{r}/L)}$$
$$= L^{2(x-d)} g(r/L, L^y t, L^x h) \tag{9.51}$$

from which it is quite straightforward to obtain the relations

$$\nu = 1/y = (2-\alpha)/d \tag{9.52}$$
$$\gamma = (2-\eta)\nu \tag{9.53}$$

The first of these is not obeyed by mean field theory, to be discussed next.

## Mean field theory

The first attempt to calculate the critical exponents used a set of approximations known as mean field theory. We will present this approach in a way which is intended to make clear the connection to better methods, to be discussed later. We will express the problem in terms of a magnetic problem, but it should be clear that a similar approach could be taken for other phase transitions. We consider an Ising like magnet. The free energy $F(H,T)$ is

$$F(H,T) = -k_B T \ln \mathrm{Tr}_{S_i} e^{-\beta H_{\mathrm{Ising}}} \tag{9.54}$$

where $H_{\mathrm{Ising}}$ is the Ising Hamiltonian and $H$ is the field. In general, as discussed before, we expect that the unusual properties near the critical points arise because of the behavior of the long wavelength degrees of freedom associated with the fluctuations between magnetization states. One might in this way expect that if only these long wavelength degrees of freedom matter, we would be able to ignore or eliminate the short wavelength degrees of freedom in constructing a theory. In mean field theory one makes the most extreme possible form of this assumption. We first illustrate for the case in which we are only interested in the exponents associated with thermodynamic quantities. In (9.54) we have a set of variables $\{S_i\}$. The operator associated with the magnetization of the system is $m = (1/N)\sum_i S_i$. Now suppose that we introduce new variables $m$, $\{S_i'\}$ in terms of which we take the trace in (9.54), and that we take the trace on the variable $m$ last. In a large system, $m$ will have an essentially continuous spectrum of values between $-1$ and 1. Further, near the critical point we expect that only $m$ values much smaller in magnitude than 1 will be important so we can extend the range to $-\infty$ to $\infty$ without serious error. Then the free energy becomes

$$F(H,T) = -k_B T \ln \int_{-\infty}^{\infty} dm\, \mathrm{Tr}_{\{S_i'\}} e^{-\beta H_{\mathrm{Ising}}} \tag{9.55}$$

We define the quantity

$$\mathcal{F}(m, H, T) = -k_B T \ln \mathrm{Tr}_{\{S_i'\}} \mathrm{e}^{-\beta H_{\mathrm{Ising}}} \tag{9.56}$$

This may be regarded as an effective Helmholtz free energy when both the magnetic field and the magnetization are held fixed. It is of some pedagogical importance to emphasize that no such quantity occurs in thermodynamics, because we have only two free thermodynamic fields in this system. These may be taken to be $H, T$ or $m, T$ (as they are in the magnetic analogue to the Gibbs free energy) but never all three variables $m, H, T$ at once. We can write $F(H, T)$ using the last two equations as

$$F(H, T) = -k_B T \ln \int_{-\infty}^{\infty} \mathrm{d}m\, \mathrm{e}^{-\beta \mathcal{F}(m, H, T)} \tag{9.57}$$

So far we have introduced no approximations except the inessential one of extending the range of the integral on $m$. The essential approximation consists in assuming that $\mathcal{F}(m, H, T)$ is an analytical function of $m$ for fixed $H, T$. (This turns out not to be correct at low spatial dimension.) With this mean field assumption we obtain

$$\mathcal{F} = \mathcal{F}_0 - Hm(H, T) + A(T, H)m^2(H, T) + C(H, T)m^3(H, T) + B(H, T)m^4(H, T) \tag{9.58}$$

We have anticipated that at small fields, the linear term will be proportional to the field. To lowest order in $H$ we may fix the other coefficients at their $H = 0$ values. Because the Hamiltonian is invariant to a change of sign of all the spins in the absence of magnetic field, we may assume that $C(H = 0, T) = 0$. Then we have, to lowest order in the field $H$

$$\beta\mathcal{F} = \beta\mathcal{F}_0 - Hm(H, T) + A(T, H = 0)m^2(H, T) + B(T, H = 0)m^4(H, T) \tag{9.59}$$

In order that the integral on $m$ in (9.57) be finite, we require that $B(H = 0, T) > 0$. Then the possible forms of $\mathcal{F}$ as a function of $m$ for zero and for small fields are sketched in Figures 9.2 and 9.3.

The presence of two minima when $A < 0$ can be associated with the presence of a critical point. Recall that in the canonical ensemble, the presence of a phase transition was signaled by a discontinuity in the derivative of $F$ with respect to $H$ as a function of the field $H$. Note also that, because $\mathcal{F}$ is proportional to the number of spins, the lowest lying minimum will be overwhelmingly favored in the integral as soon as the free energy difference between the minima, which is proportional to $NH$, is less than $k_B T$. Thus for any $A < 0$ and in the $N \to \infty$ limit we will have a discontinuity in $m$ as a function of $H$ whereas no such discontinuity will occur for

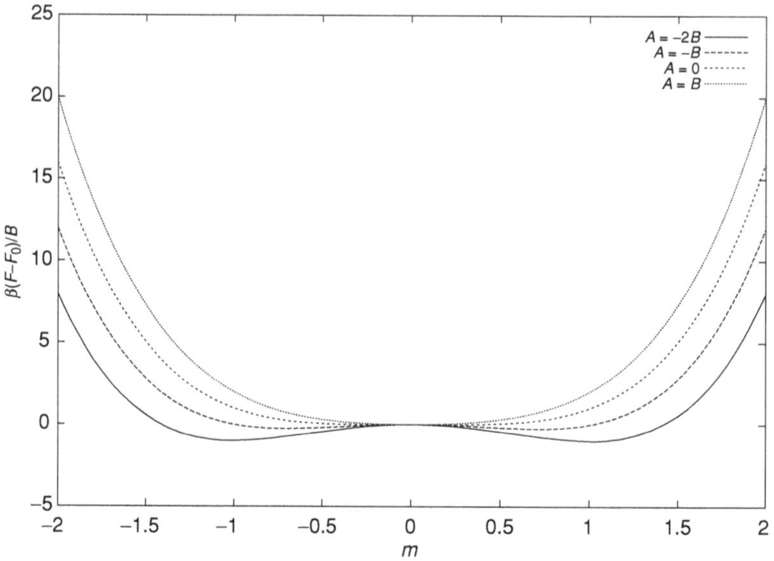


Figure 9.2 Landau–Ginzburg functional with $H = 0$.

$A > 0$. It follows that the critical point corresponds to $A = 0$. Finally we suppose that the $A$ is analytic in $T$ around $T_c$. The correctness of this assumption (which is still made in a better theory) will be discussed in more detail later. Then we can write $A(T) = a(T - T_c)$ where $a$ is a constant. Finally we take $B$ to be constant with respect to $T$ around $T_c$, expanding $B(T)$ around $T_c$ and noting that there must be a positive constant term. Thus finally we have

$$F = -k_B T \ln \int_{-\infty}^{\infty} dm \, e^{-(a(T-T_c)m^2 + Hm - bm^4 + \beta\mathcal{F}_0)} \tag{9.60}$$

When, as here, the minima are very deep (because of the proportionality of the exponent to the number of spins $N$), we can get an excellent approximation to the integral by expanding the integrand around its minima. For $T > T_c$ the minimum for $H = 0$ is at $m = 0$ and we obtain

$$\langle m \rangle = \left( \frac{\partial F}{\partial H} \right)_T = \frac{\int_{-\infty}^{\infty} m \, e^{-a(T-T_c)m^2} dm}{\int_{-\infty}^{\infty} e^{-a(T-T_c)m^2} dm} = 0 \tag{9.61}$$

whereas for $T < T_c$ the minima are at $m = \pm\sqrt{a(T_c - T)/2b}$ and we obtain, if $H \to 0^+$:

$$\langle m \rangle = +\sqrt{\frac{a(T_c - T)}{2b}} + \frac{\int_{-\infty}^{\infty} d\delta m \, e^{+a^2(T_c-T)^2/4b - 2a(T_c-T)\delta m^2} \delta m}{\int_{-\infty}^{\infty} d\delta m \, e^{+a^2(T_c-T)^2/4b - 2a(T_c-T)\delta m^2}} = \sqrt{\frac{a(T_c - T)}{2b}} \tag{9.62}$$

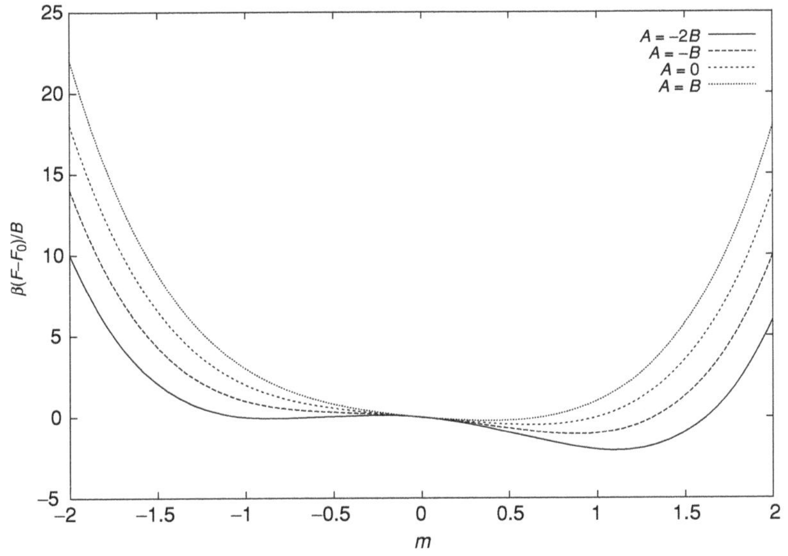


Figure 9.3 Landau–Ginzburg functional with $H/B = 1$.

whereas the same result with the opposite sign is obtained by taking the limit $H \to 0^-$. Thus $\beta = 1/2$ in this mean field theory.

The susceptibility, generally expressed as

$$\chi = \left(\frac{\partial^2 F}{\partial H^2}\right)_T = \frac{N}{k_B T}(\langle m^2 \rangle - \langle m \rangle^2) \tag{9.63}$$

is given above $T_c$ by

$$\chi = \frac{N}{a(T - T_c)} \frac{\int_{-\infty}^{\infty} u^2 e^{-u^2} du}{\int_{-\infty}^{\infty} e^{-u^2} du} \tag{9.64}$$

and for $T$ below $T_c$ by

$$\chi = \frac{N}{2a(T_c - T)} \frac{\int_{-\infty}^{\infty} u^2 e^{-u^2} du}{\int_{-\infty}^{\infty} e^{-u^2} du} \tag{9.65}$$

so that $\gamma = 1$. To compute $\delta$ take $T = T_c$ and find by a similar computation that

$$\langle m \rangle(T_c) = \text{sgn}\, h \left(\frac{|h|}{3b}\right)^{1/3} \tag{9.66}$$

so that $\delta = 3$. Note that these satisfy the scaling relations but are not the same as the experimental values.

Now we consider the calculation of $g(r)$ by mean field theory. For this it is not adequate to integrate out everything but the magnetization. Instead we employ a coarse graining like that used for the discussion of scaling, resulting in a free energy expressed in terms of block spins $\mathcal{S}_\alpha$. Then we make a continuum field $m(\vec{r})$ defined at every point in space such that $\mathcal{S}_\alpha = m(\vec{r} = \vec{r}_\alpha)$ and varying smoothly in between the block lattice points. If the blocks are big enough they will behave like the bulk system so the resulting effective free energy functional will have terms like those of the bulk:

$$\mathcal{F}(m(\vec{r})) = \int \mathrm{d}\vec{r}\,\left(a(T - T_\mathrm{c})m(\vec{r})^2 + bm(\vec{r})^4 - h(\vec{r})m(\vec{r})\right)\cdots \tag{9.67}$$

with an additional term arising from the interaction between blocks. To lowest order in the differences in the magnetization of adjoining blocks, the latter can be seen to have to be of the form $(\mathcal{S}_\alpha - \mathcal{S}_{\alpha+\delta})^2$ or in terms of the continous function $m$, of the form $c(\nabla m(\vec{r}))^2$. Thus finally we have

$$F = -k_\mathrm{B}T \ln \int \prod_{\text{all } \vec{r}} \mathrm{d}m(\vec{r})\,\mathrm{e}^{-\beta\mathcal{F}(\{m(\vec{r})\}} \tag{9.68}$$

in which

$$\beta\mathcal{F}(m(\vec{r})) = \int \mathrm{d}\vec{r}\,\left(a(T - T_\mathrm{c})m(\vec{r})^2 + bm(\vec{r})^4 - h(\vec{r})m(\vec{r}) + c(\nabla m(\vec{r}))^2\right) \tag{9.69}$$

The formally infinite factors in the integral in $F$ are not a problem when physically observable quantities are computed. Such integrals are called functional integrals and are sometimes denoted

$$\int \prod_{\text{all } \vec{r}} \mathrm{d}m(\vec{r})\cdots = \int \mathcal{D}(\{m(\vec{r})\})\cdots \tag{9.70}$$

There is a more detailed discussion in the book by Feynman and Hibbs.[2] To evaluate the free energy we can again take the minimum of $\mathcal{F}$ as a first approximation and then evaluate the effects of fluctuations away from it. This gives

$$h(\vec{r}) = 2a(T - T_\mathrm{c})m(\vec{r}) + 4bm^3(\vec{r}) - 2c\nabla^2 m(\vec{r}) \tag{9.71}$$

for the function which minimizes $\mathcal{F}$ and this equation also is easily shown to give the expected value $\langle m(\vec{r})\rangle$ of the magnetization when the fluctuations are neglected. Now to obtain $g(\vec{r} - \vec{r}\,')$ we consider the variation $h(\vec{r}) \to h(\vec{r}) + \delta h(\vec{r})$ and first write an exact equation to first order for the corresponding change in $\langle m(\vec{r})\rangle$:

$$\delta m(\vec{r}) = \beta \int \mathrm{d}\vec{r}\,'\delta h(\vec{r}\,')(\langle m(\vec{r})m(\vec{r}\,')\rangle - \langle m(\vec{r})\rangle\langle m(\vec{r}\,')\rangle) = \beta \int \mathrm{d}\vec{r}\,'\delta h(\vec{r}\,')g(\vec{r}, \vec{r}\,') \tag{9.72}$$

On the other hand in mean field theory by use of (9.71) we find

$$\delta h(\vec{r}) = 2a(T - T_\mathrm{c})\delta m(\vec{r}) + 12b\langle m(\vec{r})\rangle^2\delta m(\vec{r}) - 2c\nabla^2\delta m(\vec{r}) \tag{9.73}$$

Using the above exact expression for $\delta m(\vec{r})$ in this we obtain

$$k_B T \int d\vec{r}' \delta(\vec{r}' - \vec{r}) \delta h(\vec{r}') = \int \left(2a(T - T_c) + 12b \langle m(\vec{r}) \rangle^2 - 2c\nabla_{\vec{r}}^2\right) g(\vec{r}, \vec{r}') \delta h(\vec{r}')\, d\vec{r}' \tag{9.74}$$

Because this must be true for any $\delta h(\vec{r})$ we obtain an equation for $g(\vec{r}, \vec{r}')$:

$$\left(2a(T - T_c) + 12b\langle m(\vec{r}) \rangle^2 - 2c\nabla_{\vec{r}}^2\right) g(\vec{r}, \vec{r}') = k_B T \delta(\vec{r} - \vec{r}') \tag{9.75}$$

Solution of this by Fourier transform gives

$$g(r) = \frac{k_B T}{4\pi c r} e^{-r/\xi} \tag{9.76}$$

in which

$$\xi^2 = c/a(T - T_c) \qquad T > T_c \tag{9.77}$$

and

$$\xi^2 = c/2a(T_c - T) \quad T < T_c \tag{9.78}$$

Thus $\nu = 1/2$ and $\eta = 0$ in mean field theory.

## Renormalization group: general scheme

We have found that while mean field theory gives results which are approximately correct and which satisfy the empirically established scaling relations, it does not give the right numbers for observed critical exponents. This is now understood to arise because the fluctuations which occur around the minima in the free energy functional as found in the last section are too large to ignore. Indeed one can see that in this respect the mean field theory is not self consistent for all lattice dimensions, providing a hint concerning the approach to its correction. For self consistency the fluctuations given by $g(r)$, defined by

$$g(r) \equiv \langle m(r)m(0) \rangle - \langle m^2 \rangle = \frac{K e^{-r/\xi}}{r^{d-2+\eta}} \tag{9.79}$$

should be smaller in mean field theory than the square of the magnetization $\langle m \rangle^2$ below $T_c$. But, by evaluating (9.79) at $r = \xi$ we find this requires that

$$2\beta < \nu(d - 2 + \eta) \tag{9.80}$$

In the scaling theory this is an equality, but in mean field theory we find using the values $\beta = 1/2$, $\nu = 1/2$ and $\eta = 0$ just obtained by neglecting the fluctuations that

$$d > 4 \tag{9.81}$$

is implied by the preceding equation. Thus, the fluctuations cannot be neglected for three dimensions. Equation (9.80) is sometimes called the Landau–Ginzburg criterion for validity of mean field theory and (9.81), which gives the lowest dimension for which mean field theory is valid, establishes what is called an upper critical dimensionality for the model.

For the same reason, the problem of neglected fluctuations cannot be handled below the upper critical dimension by doing some form of perturbation theory starting at mean field theory and adding fluctuation corrections. This is because the fluctuations are more divergent than the mean values and the resulting series are not convergent. Instead, one pursues the idea by which the scaling forms were motivated. Begin with a microscopic Hamiltonian such as the Ising model or with a coarse grained free energy such as the Landau–Ginzburg free energy. In each case the effective "Hamiltonian" which occurs in the trace defining the free energy can be characterized as belonging to a space of free energy functionals. For example in the case of the Ising model, we can envision a space of all free energy functionals of the form $\beta H = \sum K(n)\sigma_{i_1}\cdots\sigma_{i_n}$ in which the sum extends over all clusters of $n$ spins where $n = 1, 2, \ldots$. Then the starting point is the Ising model with $K(1) = -\beta H$ and $K(2) = -\beta J$. Similarly in the case of the Landau–Ginzburg model one can characterize the Landau–Ginzburg free energy functional as a special case of the set of all free energy functionals which are the spatial integral of arbitrary local polynomials in the magnetization and its derivatives. Let us denote an arbitrary point in the parameter space which characterizes such a space of effective free energy functions by $\mathcal{K}$ and the starting point (which depends on the temperature and the field) by $\mathcal{K}_0(t, h)$. Now we select a systematic way to sum out the short wavelength degrees of freedom for the free energy in such a model. For example we can fix the average spins in each block of a coarse grained lattice and sum out the remaining variables as discussed in motivating the scaling theory. Alternatively, in the case of the Landau–Ginzburg formulation it is convenient to sum out the short wavelength components of the free energy functional. In this or other cases, we will at this stage have an expression for the free energy per spin of the form

$$f(t, h)/k_{\mathrm{B}}T = -\frac{1}{N}\ln \mathrm{Tr}_{\mathcal{S}}\mathrm{e}^{-\beta F(\mathcal{K}_0)} = -\frac{1}{N}\ln \mathrm{Tr}_{\mathcal{S}'}\mathrm{e}^{-\beta F(\mathcal{K})} \tag{9.82}$$

in which

$$\beta F(\mathcal{K}) = -\ln \mathrm{Tr}_{\mathcal{S}\text{ not }\mathcal{S}'}\mathrm{e}^{-\beta F(\mathcal{K}_0)} \tag{9.83}$$

defines the new free energy functional. Equation (9.83) defines a transformation in the space of free energy functionals (sometimes called "Hamiltonians" in the literature, but this is misleading)

$$\mathcal{R}(\mathcal{K}_0) = \mathcal{K} \tag{9.84}$$

It should be clear that the transformation can be repeated, though in general the resulting walk through the space of models will at first get more complicated as one goes to larger and larger length scales. This set of transformations generally obeys a multiplicative law. That is, the combination of two transformations leads to another of the same sort, but unique inverses do not exist so the resulting mathematical structure is a semigroup. This set of rescaling operations is nevertheless often called the renormalization group in this context.

Now near critical points, though the model will become at first more complicated, it is eventually expected to simplify as the systematic coarse graining gets to a scale at which the large fluctuations characteristic of criticality are described. At this point, one expects that the model may again depend on only a few parameters. Indeed, if one goes all the way to an infinite volume then only the thermodynamic variables should remain in any case. Let us consider what will happen to the coherence length, $\hat{\xi}(\mathcal{K})$, measured in units of the lattice spacing at the current level of renormalization, in such a set of renormalizations. If the length scale changes by a (dimensionless) factor $L$ in one renormalization then the relation

$$\hat{\xi}(\mathcal{K}_{i+1}) = (1/L)\hat{\xi}(\mathcal{K}_i) \tag{9.85}$$

will obtain between successive renormalizations and the relationship between the original $\hat{\xi}$ and the renormalized one will be

$$\hat{\xi}(\mathcal{K}_0(t,h)) = \lim_{n\to\infty} L^n \hat{\xi}(\mathcal{K}_n) \tag{9.86}$$

Now one can see how to characterize a critical point in terms of this renormalization process. At a critical point (only) the coherence length $\hat{\xi}(\mathcal{K}_0(t,h))$ diverges and, according to (9.86), this can only happen if $\xi(\mathcal{K}_n)$ approaches a finite value not equal to zero as the renormalization proceeds to infinity. This will happen if the starting thermodynamic variables $t$, $h$ are at a critical point. The resulting final set of parameters $\mathcal{K}^* = \lim_{n\to\infty} \mathcal{K}_n$ is called an unstable fixed point. The reason for the nomenclature is understood by considering the case in which starting thermodynamic variables are not at a critical point. Then the coherence length is finite and this can only be achieved if the renormalized coherence length goes to zero. This will generally happen as the parameter space variables $\mathcal{K}$ go to limits in which the renormalized free energies describe totally ordered or totally disordered phases which are characteristic of the behavior on either side of the fixed point. These fixed points characterizing phases on either side of the critical point are said to be stable because, under renormalization, the system is driven to them over a range of starting thermodynamic variables.

In summary we suppose that we have a set of operations $\mathcal{R}$ which take one effective free energy into another while increasing the length scale according to the relation

$$\mathcal{K}_i = \mathcal{R}^i(\mathcal{K}_0) \tag{9.87}$$

If the starting thermodynamic point is a critical point then this process will lead to an unstable fixed point at which the infinitely renormalized coherence length is finite. Otherwise it will lead to a stable fixed point (of which there are two) at which the infinitely renormalized coherence length is 0. The fixed points may be found by solving the equation

$$\mathcal{K}^* = \mathcal{R}(\mathcal{K}^*) \tag{9.88}$$

Now to use this structure to find critical exponents, we consider a "trajectory" in the $\mathcal{K}$ space which starts at a thermodynamic point near, but not at, a critical point. We expect it eventually to go to a stable fixed point, but by continuity, we expect that it will first pass close to the unstable fixed point. We consider a region near the fixed point and *linearize* the renormalization group transformation $\mathcal{R}$ in the region near the fixed point as the trajectory passes. We write

$$\mathcal{K}_i = \mathcal{K}^* + k_i \tag{9.89}$$

and

$$k_{i+1} = \mathcal{L}k_i \tag{9.90}$$

For a linear operator we can consider the eigenfunctions

$$\mathcal{L}\phi_\nu = \Lambda_\nu\phi_\nu \tag{9.91}$$

We assume that these are complete so that the $k_i$ of (9.89) and (9.90) can be written

$$k_i = \sum_\nu u_{i\nu}\phi_\nu \tag{9.92}$$

Now consider repeating the operation (9.90) $m$ times:

$$k_{i+m} = \sum_\nu u_{i\nu}\Lambda_\nu^m\phi_\nu \tag{9.93}$$

It now becomes evident how the variables get "thinned out" at large scales in this scheme. If $|\Lambda_\nu| < 1$ then if $m$ is large, the corresponding variable $\phi_\nu$ will have disappeared from the sum in (9.93) whereas if $|\Lambda_\nu| > 1$ the corresponding variable must be kept. The variables that survive are termed "relevant" to the determination of the critical behavior. The others, with $|\Lambda_\nu| < 1$ are called irrelevant and those with eigenvalues with modulus 1 are called marginal. The existence of a few "relevant" operators which survive renormalization provides the qualitative explanation for the "universality" of exponents mentioned earlier. After renormalization, the short wavelength degrees of freedom have all been integrated out and the remaining description, which as we show below allows the calculation of the critical exponents, involves only the relevant operators. No matter how diverse the physical properties of two systems on short and intermediate wavelengths may be, as long as the surviving relevant operators after renormalization behave in the same mathematical way, the calculated critical exponents will be the same. This is a mathematical way of describing the idea that systems in the same universality class "look" the same near critical points, if they are observed on very long length scales. (The word "look" is actually quite appropriate: optical wavelengths correspond to thousands of atomic or spin spacings, approaching the appropriate large scale limit. The first observations of critical fluctuations were the observation of "critical opalescence" visually in fluids, and many subsequent studies of critical properties have been carried out with light scattering.)

Now to illustrate the procedure by which critical exponents are calculated using this formalism, consider again the case of the coherence length, supposing that we start the trajectory with $h = 0$ and with $t$ small but finite. We suppose that the iteration has proceeded long enough at the $i$th iteration so that only a variable proportional to $t$, which will certainly be relevant, is left. We designate the label $\nu = 1$ for this variable. Then we imagine iterating (9.93) $n$ more times using (9.86) with the result

$$\xi(\mathcal{K}_0(h = 0, t)) = L^i \xi(u_{i1}t, \ldots) = L^{i+n} \xi\left(\Lambda_1^n u_{i1}t, \ldots\right) \tag{9.94}$$

But if only the temperature variable is left in the second and third expressions, then all three expressions for $\xi$ must vary as the $-\nu$ power of their arguments so that

$$L^i (u_{1i}t)^{-\nu} = L^{i+n}\left(u_{1i}\Lambda_1^n t\right)^{-\nu} \tag{9.95}$$

Thus

$$L^n \Lambda_1^{-n\nu} = 1 \tag{9.96}$$

so that taking the power $1/n$ and the ln we have

$$\nu = \frac{\ln L}{\ln \Lambda_1} \tag{9.97}$$

In this way, we can determine the critical exponent $\nu$ from one of the eigenvalues (Problem (9.91)).

## Renormalization group: the Landau–Ginzburg model

As an illustration of the method we consider Wilson's original treatment of the Landau–Ginzburg model in which the free energy functional of the starting model is taken to be (9.69). The renormalization $\mathcal{R}$ operation is taken to be the integration of the Fourier components $m_{\vec{q}}$ corresponding to the largest 1/2 of the wave vectors. Already we run into a problem here because in the continuum model there is an infinite set of wave vectors. In nonrelativistic condensed matter physics the problem is only apparent (though it is quite serious in some corresponding relativistic field theory problems which we will not discuss). In fact, in the original Ising model on a lattice, the set of wave vectors was confined to the first Brillouin zone. Therefore it is completely reasonable in the corresponding continuum model to introduce an upper limit on the momentum components which are nonzero. Before renormalization starts, this upper limit should be taken to be of the order of the reciprocal of the lattice constant. As long as we are interested in critical properties, which are a manifestation of the long wavelength properties of the model, it should not matter very much exactly how this short wavelength cutoff is implemented. It is first helpful to rewrite (9.69) in terms of dimensionless variables. If, as we suppose, $m(\vec{r})$ is dimensionless then we can rewrite

$$\beta\mathcal{F} = \int \mathrm{d}\vec{r}\,\left(c \mid \nabla m(\vec{r}) \mid^2 + a' t m(\vec{r})^2 + b m^4\right) \tag{9.98}$$

in which $a' = T_c a$. $a'$ has dimension $1/\text{length}^d$ and $c$ has dimension $1/\text{length}^{d-2}$ so it is evident that the ratio $c/a'$ has the dimensions of a length squared. We refer to this length as the zero temperature or "bare" coherence length and denote it $\xi_0$. We multiply and divide (9.98) by $\xi_0^d$ and obtain

$$\beta\mathcal{F} = \int \mathrm{d}\vec{r}'\,\left(\mid \nabla_{\vec{r}'} \tilde{m}(\vec{r}') \mid^2 + t\tilde{m}(\vec{r})^2 + \tilde{b}\tilde{m}^4\right) \tag{9.99}$$

Here we have defined a dimensionless length scale $\vec{r}' = \vec{r}/\xi_0$, and written $\tilde{m}(\vec{r}') = \xi_0^{d/2} a'^{1/2} m(\vec{r})$ (which remains dimensionless and still has the range $-\infty$ to $\infty$) and $\tilde{b} = b/\xi_0^d a'^2$. Now Fourier transforms are defined as

$$m_{\vec{q}} = \int d^d\vec{r}' \mathrm{e}^{-\mathrm{i}\vec{q}\cdot\vec{r}'} \tilde{m}(\vec{r}') \tag{9.100}$$

The inverse of (9.100) is

$$\tilde{m}(\vec{r}') = \int \frac{\mathrm{d}^d\vec{q}}{(2\pi)^d} \mathrm{e}^{\mathrm{i}\vec{q}\cdot\vec{r}'} m_{\vec{q}} \tag{9.101}$$

in which, as noted, we wish to cut off the integrals at large $q$ at a value $\Lambda_a\xi_0$ where $\Lambda_a$ is of the order of the reciprocal of the lattice constant. However, because we do not expect the exact value of this cutoff to be significant and $\xi_0$ is of the same order as the lattice constant, it is very convenient, and should make no essential difference, if we take this cutoff to be 1. That is, the wave vector integrals are over the range $0 <| \vec{q} |< 1$. Using (9.101), 9.69 is easily rewritten as

$$\beta\mathcal{F} = \int \frac{\mathrm{d}^d\vec{q}}{(2\pi)^d}\left((q^2+t)m_{\vec{q}}m_{-\vec{q}} + h_{-\vec{q}}m_{\vec{q}}\right) + \tilde{b}\int \frac{\mathrm{d}^d\vec{q}}{(2\pi)^d}\int \frac{\mathrm{d}^d\vec{q}_2}{(2\pi)^d}\int \frac{\mathrm{d}^d\vec{q}_3}{(2\pi)^d} m_{\vec{q}}m_{\vec{q}_2}m_{\vec{q}_3}m_{-\vec{q}-\vec{q}_2-\vec{q}_3} \tag{9.102}$$

For simplicity, we will here only consider the case $T > T_{\mathrm{c}}$ and unless otherwise stated we will take $h_{-\vec{q}} = 0$.

To assist in following the argument, let us first anticipate its outlines. It will turn out that, after repeated renormalizations, the free energy functional will again be of form (9.102) but the quadratic term only survives (is relevant) when the dimension $d$ of the system is less than 4. Thus for $d$ less than 4, the relevant parameters in the set $\mathcal{K}$ turn out to be $t$ and $\tilde{b}$ and the goal of the calculation is to obtain a linearized recursion relation for them near the unstable fixed point. Because $\tilde{b}$ only becomes relevant when $d < 4$ and the only way to deal analytically with the $\tilde{b}$ term is by a perturbation expansion, it is useful to consider the artificial situation in which $d$ is *infinitesimally* less than 4 so that the perturbations will be in some sense small. One defines the variable $\epsilon = 4 - d$ and seeks expansions for the needed quantities in $\epsilon$. One thus has two expansions: one in the $\tilde{b}$ term and one in $\epsilon$, each of which is needed in order to obtain the linearized recursion relations. The first expansion is carried out around the model which results from setting $\tilde{b} = 0$ in (9.99). This is called the Gaussian model. It is closely related to but not quite the same as the mean field theory discussed earlier. We first discuss a few features of the Gaussian model, including how to get its recursion relations, and then proceed to an expansion in $\tilde{b}$ around the Gaussian model and the extraction of recursion relations to leading order in $\epsilon$ from it.

From the preceding discussion, the free energy of the Gaussian model is

$$f/k_{\mathrm{B}}T = -\ln\int \prod_{0<|\vec{q}|<1} \mathrm{d}m_{\vec{q}}\, \mathrm{e}^{-\int \frac{\mathrm{d}^d\vec{q}}{(2\pi)^d}(q^2+t)m_{\vec{q}}m_{-\vec{q}}} \tag{9.103}$$

where the $\vec{q}$ integrals are over the range $0 <| \vec{q} |< 1$. Renormalization consists in separating the integral into two ranges $0 <| \vec{q} |< 1/2$ and $1/2 <| \vec{q} |< 1$. Then the free energy separates into a nonsingular part arising from the short wavelengths and the long wavelength part:

$$f/k_{\mathrm{B}}T = -\ln\int \prod_{0<|\vec{q}|<1/2} \mathrm{d}m_{\vec{q}}\, \mathrm{e}^{-\int_{0<|\vec{q}|<1/2} \frac{\mathrm{d}^d\vec{q}}{(2\pi)^d}(q^2+t)m_{\vec{q}}m_{-\vec{q}}} + \text{nonsingular part} \tag{9.104}$$

To see how the parameters rescale, one must cast the factor in the exponential into the original form. For this, one defines $\vec{q}\,' = 2\vec{q}$ so that $0 <| \vec{q}\,' |< 1$. We define a rescaled spin variable $m'_{\vec{q}'} = m_{\vec{q}}/\zeta$ giving for the factor in the exponent:

$$\frac{\zeta^2}{2^d}\int_{0<|\vec{q}'|<1}\frac{\mathrm{d}^d\vec{q}'}{(2\pi)^d}\left(\frac{q'^2}{4}+t\right)m'_{\vec{q}'}m'_{-\vec{q}'} \tag{9.105}$$

which is put in precisely the form before normalization by writing $t' = 4t$, $\zeta^2/2^{d+2} = 1$ giving

$$\int_{0<|\vec{q}'|<1}\frac{\mathrm{d}^d\vec{q}'}{(2\pi)^d}(q'^2+t')m'_{\vec{q}\,'}m'_{-\vec{q}'} \tag{9.106}$$

Thus in this case the renormalization operation involves only $t$ and we have the relation

$$\mathcal{L}(t) = 4t(= 2^2 t) \tag{9.107}$$

where the second form reminds us that $L = 2$ here in the previous language. In terms of the scaling formulation, this establishes that the exponent $y = 2$. Thus the eigenvalue $\Lambda_1$ of $t$ is $L^2$ and the exponent $\nu$ is ((9.91) and (9.97))

$$\nu = \frac{\ln L}{2\ln L} = 1/2 \tag{9.108}$$

We may also obtain the scaling with respect to the field by writing the field term as

$$-\int \mathrm{d}\vec{r}\,'\tilde{m}(\vec{r}\,')h = -m_{\vec{q}=0}h \tag{9.109}$$

before renormalization and using $m'_{\vec{q}'} = \zeta m_{\vec{q}}$ and $\zeta^2/2^{d+2} = 1$:

$$-m'_{\vec{q}'=0}h' \tag{9.110}$$

after renormalization where $h' = 2^{d/2+1}h$. Thus

$$\mathcal{L}(h) = 2^{d/2+1}h \tag{9.111}$$

and the eigenvalue $\Lambda_2$ associated with the field is $2^{d/2+1}$. This gives the value $x = d/2 + 1$ for the scaling theory for the Gaussian model. Here we see a difference between the Landau mean field theory and the Gaussian model. In the Landau theory, the value of $\delta$ was 3, independent of lattice dimension, whereas here using $\delta = x/(d - x)$ from the scaling theory we obtain $\delta = (d + 2)/d - 2)$ which is 5 when $d = 3$ and 3 when $d = 4$. The results are the same at the upper critical dimension after which we expect mean field theory to work.

Now we turn to the consideration of the $\tilde{b}$ term in (9.99). We write (9.99) as

$$\beta\mathcal{F} = \beta(\mathcal{F}_{\mathrm{G}} + \mathcal{F}_{\mathrm{I}}) \tag{9.112}$$

where $\mathcal{F}_G$ is the Gaussian part just considered:

$$\beta\mathcal{F}_G = \int \frac{d^d\vec{q}}{(2\pi)^d}(q^2 + t)m_{\vec{q}}m_{-\vec{q}} \tag{9.113}$$

and $\mathcal{F}_I$ is the interaction term

$$\mathcal{F}_I = \tilde{b} \int \frac{d^d\vec{q}}{(2\pi)^d} \int \frac{d^d\vec{q}_2}{(2\pi)^d} \int \frac{d^d\vec{q}_3}{(2\pi)^d} m_{\vec{q}}m_{\vec{q}_2}m_{\vec{q}_3}m_{-\vec{q}-\vec{q}_2-\vec{q}_3} \tag{9.114}$$

The total free energy is

$$f/k_BT = -\ln \int \prod_{0<|\vec{q}|<1} dm_{\vec{q}}\, e^{-\beta(\mathcal{F}_G+\mathcal{F}_I)} \tag{9.115}$$

As before we renormalize by summing out the short wavelengths. This time it is harder. We denote the set of $m_{\vec{q}}$ with $0 <| \vec{q} |< 1/2$ by $\{m_0\}$ and the set with $1/2 <| \vec{q} |< 1$ by $\{m_1\}$. We have

$$\mathcal{F}_G = \mathcal{F}_G(\{m_0\}) + \mathcal{F}_G(\{m_1\}) \tag{9.116}$$

as in the Gaussian model but the interacting part does not separate. To handle this part we expand the exponent, hoping as discussed above, to control the convergence of the resulting series by keeping the dimension near 4. The free energy is then of form

$$f/k_BT = -\ln \int \prod_{0<|\vec{q}|<1/2} dm_{\vec{q}}\, e^{-\beta\mathcal{F}_G(\{m_0\})} \times \int \prod_{1/2<|\vec{q}|<1} dm_{\vec{q}}\, e^{-\beta\mathcal{F}_G(\{m_1\})}(1 - \beta\mathcal{F}_I + (1/2)(\beta\mathcal{F}_I)^2 + \cdots) \tag{9.117}$$

The first factor is exactly the same as in the Gaussian model. In the second factor one must carry out averages of the various interacting terms with respect to the Gaussian weight $e^{\beta\mathcal{F}_G(\{m_1\})}$. The integrals can all be done term by term. The rules for carrying out an integral of the form

$$\frac{\int \prod_{1/2<|\vec{q}|<1} dm_{\vec{q}}\, e^{-\beta\mathcal{F}_G(\{m_1\})} m_{\vec{q}_1}\cdots m_{\vec{q}_k}}{\int \prod_{1/2<|\vec{q}|<1} dm_{\vec{q}}\, e^{-\beta\mathcal{F}_G(\{m_1\})}} \equiv \overline{m_{\vec{q}_1}\cdots m_{\vec{q}_k}} \tag{9.118}$$

where all the $\vec{q}$ satisfy $1/2 <| \vec{q} |< 1$ are as follows.

(1) If $k$ is odd the answer is zero.
(2) If $k$ is even form all possible pairings of the factors $m_{\vec{q}}$ on the line.
(3) For each pair within each pairing, there is a factor $(2\pi)^d\delta^{(d)}(\vec{q}_1 + \vec{q}_2)/2(q_1^2 + t)$.

Details are to be found in the paper by Wilson and Kogut.[3] In order to obtain the renormalized free energy, we need the logarithm of the series and recognize that we are dealing with another version of the cumulant series discussed in Chapter 6. Here we will only consider the first two terms which give

$$\beta\mathcal{F}'(\{m'_{\vec{q}'}\}) = \beta\mathcal{F}_{\mathrm{G}}(\{m_0\}) + \overline{\beta\mathcal{F}_{\mathrm{I}}} - (1/2)\left(\overline{\beta\mathcal{F}_{\mathrm{I}}^2} - \overline{\beta\mathcal{F}_{\mathrm{I}}}^2\right) + \cdots \tag{9.119}$$

A diagrammatic notation can be introduced for the resulting terms. The diagrams represent terms in the series with the following rules.

Each power of $\tilde{b}$ is represented by a dot $\cdot$.
Each line with a free end represents a power of $m_{\vec{q}}$ with $0 <\mid \vec{q} \mid< 1/2$.
Each line with ends connected to dots represents a factor

$$\frac{\int \prod_{1/2<|q|<1} \mathrm{d}m_{\vec{q}}\, m_{\vec{q}_1} m_{\vec{q}_2} \mathrm{e}^{-\sum_{1/2<|q'|<1} m_{\vec{q}'} m_{-\vec{q}'}(q'^2+t)}}{\int \prod_{1/2<|q|<1} \mathrm{d}m_{\vec{q}}\, \mathrm{e}^{-\sum_{1/2<|q'|<1} m_{\vec{q}'} m_{-\vec{q}'}(q'^2+t)}} = (2\pi)^d \delta^{(d)}(\vec{q}_1 + \vec{q}_2)/2(q_1^2 + t) \tag{9.120}$$

The trickiest part of writing the terms in the series correctly is to get the combinatorial factors associated with counting the numbers of each type of term right. I refer to Wilson's article for some details. To second order in $\tilde{b}$, the series looks like this diagrammatically:

$$\beta\mathcal{F}' = \beta\mathcal{F}_{\mathrm{G}}(\{m_0\}) +$$

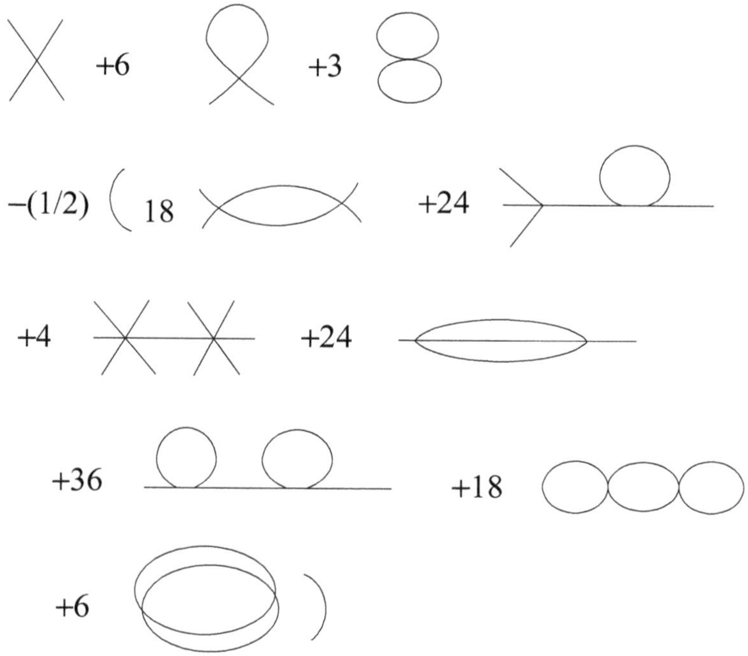


(9.121)

It turns out that the leading terms when $d$ is near 4 are the "tadpole-like" diagram which contributes to the renormalized $t$ and the term

in the expression for the renormalized $\tilde{b}$. The calculation of the quadratic terms proceeds much as in the case of the Gaussian model. We have, using the "tadpole" term

$$\int \frac{\mathrm{d}^d q'}{(2\pi)^d}(q'^2 + t')m'_{\vec{q}'}m'_{-\vec{q}'}$$
$$= \int \frac{\mathrm{d}^d q'}{(2\pi)^d}\left(q'^2/4 + t + 6\tilde{b}\int \frac{\mathrm{d}^d q_3}{(2\pi)^d}\frac{1}{2(q_3^2+t)}\right)\left(\frac{\zeta^2}{2^d}\right)m'_{\vec{q}'}m'_{-\vec{q}'} \quad (9.122)$$

in which the integral over $q_3$ is over the interval $1/2 <| q_3 |< 1$. Thus the renormalization of $t$ gives

$$t' = \mathcal{R}(t) = 4t + 12\tilde{b}\int \frac{\mathrm{d}^d q_3}{(2\pi)^d}\frac{1}{(q_3^2+t)} \quad (9.123)$$

(Our $\tilde{b}$ is related to the $u$ of Wilson and Kogut by $\tilde{b} = 4u$.) The first term is the same as for the Gaussian model and one sees that to this order $\zeta = 2^{d/2+1}$ as it was in the Gaussian model. However, $t$ will transform differently than it did in the Gaussian model if $\tilde{b}$ is relevant. To find out if and when it is relevant we write the terms that give the renormalization of $\tilde{b}$. We have

$$\tilde{b}'\int \frac{\mathrm{d}^d q_1'}{(2\pi)^d}\int \frac{\mathrm{d}^d q_2'}{(2\pi)^d}\int \frac{\mathrm{d}^d q_3'}{(2\pi)^d}m'_{\vec{q}_1'}m'_{\vec{q}_2'}m'_{\vec{q}_3'}m'_{-\vec{q}_1'-\vec{q}_2'-\vec{q}_3'}$$
$$= \frac{\tilde{b}\zeta^4}{2^{3d}}\int \frac{\mathrm{d}^d q_1'}{(2\pi)^d}\int \frac{\mathrm{d}^d q_2'}{(2\pi)^d}\int \frac{\mathrm{d}^d q_3'}{(2\pi)^d}m'_{\vec{q}_1'}m'_{\vec{q}_2'}m'_{\vec{q}_3'}m'_{-\vec{q}_1'-\vec{q}_2'-\vec{q}_3'}$$
$$\times\left(1 - 9\tilde{b}\int \frac{\mathrm{d}^d q_4}{(2\pi)^d}\frac{1}{(q_4^2+t)((\vec{q}_1+\vec{q}_2+\vec{q}_4)^2+t)}\right) \quad (9.124)$$

where again the integral over $q_4$ is over $1/2 <| q_4 |< 1$. Thus

$$\tilde{b}' = \mathcal{R}(\tilde{b}) = \frac{\zeta^4}{2^{3d}}\tilde{b}\left(1 - 9\tilde{b}\int \frac{\mathrm{d}^d q_4}{(2\pi)^d}\frac{1}{(q_4^2+t)((\vec{q}_1+\vec{q}_2+\vec{q}_4)^2+t)}\right) \quad (9.125)$$

The prefactor is of particular interest. Since $\zeta = 2^{d/2+1}$, it is $2^{-d+4}$. Thus $\tilde{b}$ will disappear under repeated iterations if $d > 4$ but not if $d < 4$, consistent with the

Landau–Ginzburg criterion (9.81). In terms of $\epsilon = 4 - d$ the last relation becomes

$$\tilde{b}' = \mathcal{R}(\tilde{b}) = 2^{\epsilon}\tilde{b}\left(1 - 9\tilde{b}\int \frac{\mathrm{d}^d q_4}{(2\pi)^d}\frac{1}{(q_4^2 + t)((\vec{q}_1 + \vec{q}_2 + \vec{q}_4)^2 + t)}\right) \tag{9.126}$$

Equations (9.123) and (9.126) constitute a renormalization group transformation for the Landau–Ginzburg model. We will analyze these equations following the procedure outlined in the last section. For demonstration that the remaining terms in the perturbation series give irrelevant contributions when $d$ is near 4 we refer to the paper by Wilson and Kogut. The fixed point equations are

$$t^* = 4\left(t^* + 3\tilde{b}^*\int \frac{\mathrm{d}^d q_3}{(2\pi)^d}\frac{1}{(q_3^2 + t^*)}\right)$$

$$\tilde{b}^* = 2^{\epsilon}\tilde{b}^*\left(1 - 9\tilde{b}^*\int \frac{\mathrm{d}^d q_4}{(2\pi)^d}\frac{1}{(q_4^2 + t^*)((\vec{q}_1 + \vec{q}_2 + \vec{q}_4)^2 + t^*)}\right) \tag{9.127}$$

We will suppose that the dependence on the small wave vectors $q_1$ and $q_2$ in the last equation can be ignored so that it becomes

$$\tilde{b}^* = 2^{\epsilon}\tilde{b}^*\left(1 - 9\tilde{b}^*\int \frac{\mathrm{d}^d q_4}{(2\pi)^d}\frac{1}{(q_4^2 + t^*)^2}\right) \tag{9.128}$$

It will turn out that for small $\epsilon = 4 - d$ the fixed point of interest has $t^*$ and $\tilde{b}^*$ both of $\mathcal{O}(\epsilon)$. Working to this order we can then expand the right hand sides of the equations for the fixed point in powers of $\epsilon$ and obtain:

$$\tilde{b}^* = \frac{\epsilon \ln 2}{9\int \frac{\mathrm{d}^d q_4}{(2\pi)^d}\frac{1}{q_4^4}} \tag{9.129}$$

$$t^* = -\frac{4}{9}\epsilon \ln 2 \frac{\int \frac{\mathrm{d}^d q_4}{(2\pi)^d}\frac{1}{q_4^2}}{\int \frac{\mathrm{d}^d q_4}{(2\pi)^d}\frac{1}{q_4^4}} \tag{9.130}$$

In the next step we must linearize the renormalization transformation (9.123) and (9.126) about the fixed point. To lowest order in $\epsilon$ we denote $\int \frac{\mathrm{d}^d q_4}{(2\pi)^d}\frac{1}{q_4^2} = \int \frac{1}{p^2}$ and $\int \frac{\mathrm{d}^d q_4}{(2\pi)^d}\frac{1}{q_4^4} = \int \frac{1}{p^4}$ and find

$$\begin{pmatrix} t' - t^* \\ \tilde{b}' - \tilde{b}^* \end{pmatrix} = \begin{pmatrix} 4 - 12b^*\int \frac{1}{p^4} & 12\int \frac{1}{p^2} \\ 18b^{*2}\int \frac{1}{p^2} & 2^{\epsilon}(1 - 18b^*)\int \frac{1}{p^4} \end{pmatrix}\begin{pmatrix} t - t^* \\ \tilde{b} - b^* \end{pmatrix} \tag{9.131}$$

Note that the matrix is not symmetric and that one must distinguish between right and left eigenvectors. The right eigenvectors (meaning those that give back the eigenvalue times themselves when multiplied from the left by the matrix) are the ones of interest here and one sees that to lowest order in $\epsilon$ (using the fact that $b^*$ is of order $\epsilon$) the eigenvalue problem has the structure

$$\begin{pmatrix} A & B \\ 0 & D \end{pmatrix} \begin{pmatrix} u \\ v \end{pmatrix} = \Lambda \begin{pmatrix} u \\ v \end{pmatrix} \tag{9.132}$$

with a solution

$$\begin{pmatrix} u_1 \\ v_1 \end{pmatrix} = \begin{pmatrix} 1 \\ 0 \end{pmatrix} \tag{9.133}$$

with $\Lambda_1 = A = 4 - (4/3)\epsilon \ln 2$. The other eigenvalue is easily found to be $\Lambda_2 = D = 1 - \epsilon \ln 2$ (because the determinant of the matrix minus $\Lambda$ times the identity is zero). Thus the second eigenvector is irrelevant and the first is simply $t$ as anticipated in the general discussion given earlier. But the mixing of $t$ and $\tilde{b}$ has affected the eigenvalue associated with the temperature. As a result, following precisely the analysis in the section on the general formulation of the renormalization group analysis, the lowest order estimate of the exponent $\nu$ is, using (9.97),

$$\nu = \frac{\ln 2}{\ln \Lambda_1} = \frac{1}{2} + \frac{\epsilon}{12} + \mathcal{O}(\epsilon^2). \tag{9.134}$$

## References


1. D. Ruelle, *Statistical Mechanics*, New York: W. A. Benjamin, 1969.
2. R. P. Feynman and A. R. Hibbs, *Quantum Mechanics and Path Integrals*, New York: McGraw-Hill, 1965.
3. K. Wilson and J. Kogut, *Physics Reports* **12C** (1975) 75.


## Problems

9.1 Consider a monatomic substance for which the chemical potential as a function of the volume per particle and the temperature is

$$\mu = -k_B T \ln \left( \frac{(v - v_c)}{\lambda^3} \right) - \frac{a}{v}$$

Here $v_c$ is a constant positive parameter with the dimensions of volume and $a$ is a constant positive parameter with the dimensions of energy times volume. $\lambda$ is the thermal wavelength, $\lambda = \sqrt{2\pi\hbar^2/mk_B T}$.

(a) Discuss the physical significance of this model for $\mu$ with particular attention to the parameters $v_c$ and $a$.

(b) Find the condition or conditions under which this model for $\mu$ leads to phase separation.

(c) Find the critical point for this model (values of $T_c$, $v_{\text{critical}}$ (not to be confused with the parameter $v_c$ of the model) and $P_c$) in terms of the given parameters.

(d) Show explicitly that at the critical point, $\left(\frac{\partial P}{\partial v}\right)_{T,N} = 0$ and $\left(\frac{\partial^2 P}{\partial v^2}\right)_{T,N} = 0$.

(e) Evaluate $v$ as a function of $T$ near the critical point and thus find the exponent $\beta$ for this model.

9.2 Demonstrate that the scaling relations ((9.52) and (9.53))

$$\nu = 1/y = (2-\alpha)/d$$
$$\gamma = \nu(2-\eta)$$

follow from the scaling form for $g(r,t,h)$ given in equation (9.51)

$$g(r,t,h) = L^{2(x-d)} g(r/L, L^y t, L^x h)$$

9.3 In magnetic models, one can sometimes use a renormalization operation $\mathcal{R}$ consisting of summing out a fraction of the spins, leaving the remainder to take account of the longer wavelength degrees of freedom. The simplest version of such a transformation is called "decimation." In the one dimensional Ising model, it consists of summing out every other spin and, unlike most such transformations, the resulting transformation can be written out exactly. Carry out this program for the one dimensional Ising model and show that applying the analysis described above leads to correct conclusions.

9.4 Consider a model for phase transitions in which the Landau–Ginzburg free energy functional is

$$\beta\mathcal{F} = \int \frac{d^d\vec{q}}{(2\pi)^d}\left((q^\mu + t)m_{\vec{q}}m_{-\vec{q}} + h_{-\vec{q}}m_{\vec{q}}\right)$$
$$+\tilde{b}\int \frac{d^d\vec{q}}{(2\pi)^d}\int \frac{d^d\vec{q}_2}{(2\pi)^d}\int \frac{d^d\vec{q}_3}{(2\pi)^d}(m_{\vec{q}}m_{\vec{q}_2})(m_{\vec{q}_3}m_{-\vec{q}-\vec{q}_2-\vec{q}_3})$$

Here $\mu$ is a positive number (not 2 and not necessarily an integer).

(a) Consider the "Gaussian" approximation in which the term involving $\tilde{b}$ is dropped. Work out the Wilson renormalization group transformation for this functional in that case. Find the upper critical dimension and the exponent $\nu$.

(b) Calculate the correlation function $\langle m_{\vec{q}}m_{-\vec{q}}\rangle$ exactly within this approximation for $t > 0$. Use result in the limits $t \to 0$ at finite $q$ and $q \to 0$ at finite $t$ to find the values of the exponents $\gamma$ and $\eta$.

(c) For the case that $\tilde{b}$ is not zero, work out the renormalization group transformation, find the fixed points and the value of $\nu$ to lowest order in $\epsilon$ (appropriately defined!).

9.5 Consider a model for a magnetic system which is described by the Landau–Ginzburg free energy functional, but with $n$ components of the magnetization instead of 1:

$$\beta\mathcal{F} = \int \frac{d^d\vec{q}}{(2\pi)^d}\left((q^2 + t)\mathbf{m}_{\vec{q}}\cdot\mathbf{m}_{-\vec{q}} + h_{-\vec{q}}m_{\vec{q},1}\right)$$
$$+\tilde{b}\int \frac{d^d\vec{q}}{(2\pi)^d}\int \frac{d^d\vec{q}_2}{(2\pi)^d}\int \frac{d^d\vec{q}_3}{(2\pi)^d}(\mathbf{m}_{\vec{q}}\cdot\mathbf{m}_{\vec{q}_2})(\mathbf{m}_{\vec{q}_3}\cdot\mathbf{m}_{-\vec{q}-\vec{q}_2-\vec{q}_3})$$

Both for $t > 0$ and for $t < 0$ you are to treat this free energy in Gaussian approximation, in which only terms to second order in the magnetization fluctuations are kept.

(a) Calculate the correlation function $\langle m_{\vec{q},\nu} m_{-\vec{q},\nu} \rangle$ exactly within this Gaussian approximation for $t > 0$.
(b) Use the result from (a) in the limits $t \to 0$ at finite $q$ and $q \to 0$ at finite $t$ to find the values of the exponents $\gamma$ and $\eta$.
(c) Consider the same questions for $t < 0$. Here you must keep $h$ small but finite to specify about which of the free energy minima you are expanding. Take care to note that the answer is different depending on whether you are considering $\langle m_{\vec{q},\nu} m_{-\vec{q},\nu} \rangle$ for $\nu \neq 1$ or $\langle m_{\vec{q},1} m_{-\vec{q},1} \rangle - \langle m_{\vec{q},1} \rangle \langle m_{-\vec{q},1} \rangle$ where $\nu = 1$ denotes the direction of the magnetic field.

# Part III

Dynamics

# 10

# Hydrodynamics and definition of transport coefficients

## General discussion

In general, by a hydrodynamic description of a many body fluid we mean a description valid at long wavelengths and low frequencies and which is based on closure of the local conservation laws of the fluid by use of a linear relation between fluxes and the gradients of densities. The coefficients of the linear relation are transport coefficients and they are phenomenological parameters of the hydrodynamic theory, calculable in principle from a theory describing the system at shorter length and time scales. The resulting hydrodynamic theory is generally a set of nonlinear partial differential equations of which the Navier–Stokes equations for the hydrodynamics of a simple fluid are a familiar example.

The reason that hydrodynamic theories accurately describe slow motions on large length scales is that global conservation laws link long distances to long times. Physically, for example, conservation of mass results in a diffusion equation in which the distance which particles diffuse increases with the square root of the time (see Problem 10.1). Although this link guarantees that *some* of the slow variables of the system are described by the hydrodynamic equations, it does not ensure that *all* of the slow variables can be so described. Near critical points associated with second order phase transitions, there are very slow changes in the fluid which are not described by the conservation laws of hydrodynamics, but which arise because of the very slow development and decay of large, almost stable regions looking like one of the (two or more) phases between which the system is slowly fluctuating. (For example, at the gas–liquid critical point, these are gas and liquid like regions.) This slow motion is not a consequence of the conservation laws on which hydrodynamics is based but arises from other features of the dynamics of the fluid.

## Hydrodynamic equations for a classical fluid

The basic variables of ordinary fluid mechanics are the following conserved quantities: the number density $\varrho(\vec{r}, t)$, the momentum density $\vec{J}(\vec{r}, t)$, and the energy density $e(\vec{r}, t)$. The hydrodynamic equations arise from the fundamental microscopic equations of motion through the conservation laws for the densities of particles, momentum and energy. Though these are usually introduced at the classical level, it is easy to show that they also are exactly correct in a quantum fluid. In particular we define the operator

$$\rho(\vec{r}) = \sum_i \delta(\vec{r} - \vec{r}_i) \tag{10.1}$$

Then a direct application of the equation of motion

$$\frac{\partial \rho}{\partial t} = \frac{\mathrm{i}}{\hbar}[H, \rho] \tag{10.2}$$

with the Hamiltionian

$$H = \sum_i \frac{-\hbar^2 \nabla_i^2}{2m} + (1/2) \sum_{i \neq j} \phi(\vec{r}_i - \vec{r}_j) \tag{10.3}$$

gives the continuity equation for particle conservation in operator form

$$\frac{\partial \rho}{\partial t} = -\nabla \cdot \vec{J}/m \tag{10.4}$$

in which

$$\vec{J}/m = (1/2) \sum_i \left( \frac{\vec{p}_i}{m} \delta(\vec{r} - \vec{r}_i) + \delta(\vec{r} - \vec{r}_i) \frac{\vec{p}_i}{m} \right) \tag{10.5}$$

is the quantum mechanical operator describing the particle current density. Similarly

$$\frac{\partial \vec{J}}{\partial t} = -\nabla \cdot \mathbf{\Pi} \tag{10.6}$$

In which the operator $\mathbf{\Pi}$ is a tensor describing the flux of momentum:

$$\mathbf{\Pi} = (1/4) \sum_i \left( \frac{\vec{p}_i \vec{p}_i}{m} \delta(\vec{r} - \vec{r}_i) + 2 \frac{\vec{p}_i}{m} \delta(\vec{r} - \vec{r}_i) \vec{p}_i + \delta(\vec{r} - \vec{r}_i) \frac{\vec{p}_i}{m} \vec{p}_i \right)$$
$$+ (1/2) \sum_{i \neq j} \vec{r}_{ji} \nabla_i \phi(\vec{r}_i - \vec{r}_j) \delta(\vec{r} - \vec{r}_i) \tag{10.7}$$

Finally the conservation of energy density can also be written in operator form. One defines the energy density $e$ as

$$e(\vec{r}) = \sum_i \left\{ (1/2)[T_i \delta(\vec{r} - \vec{r}_i) + \delta(\vec{r} - \vec{r}_i) T_i] + (1/2) \sum_{i \neq j} \phi(\vec{r}_i - \vec{r}_j) \delta(\vec{r} - \vec{r}_i) \right\} \tag{10.8}$$

where $T_i = -\hbar^2\nabla_i^2/2m$. Then by direct calculation one gets the energy conservation equation as

$$\frac{\partial e}{\partial t} = -\nabla \cdot \vec{S} \tag{10.9}$$

in which

$$\vec{S} = \vec{S}_K + \vec{S}_\phi + \vec{S}_v \tag{10.10}$$

and

$$\begin{aligned}\vec{S}_K = (1/4)\sum_i \Bigg[ T_i \left( \frac{\vec{p}_i}{m}\delta(\vec{r}-\vec{r}_i) + \delta(\vec{r}-\vec{r}_i)\frac{\vec{p}_i}{m} \right) \\ + \left( \frac{\vec{p}_i}{m}\delta(\vec{r}-\vec{r}_i) + \delta(\vec{r}-\vec{r}_i)\frac{\vec{p}_i}{m} \right) T_i \Bigg]\end{aligned} \tag{10.11}$$

$$\vec{S}_\phi = (1/4)\sum_i \left( \frac{\vec{p}_i}{m}\phi(\vec{r}_i - \vec{r}_j)\delta(\vec{r}-\vec{r}_i) + \phi(\vec{r}_i - \vec{r}_j)\delta(\vec{r}-\vec{r}_i)\frac{\vec{p}_i}{m} \right) \tag{10.12}$$

$$\vec{S}_v = (1/4)\sum_i \sum_{j\neq i} \left( \frac{\vec{p}_i}{m} \cdot \nabla_{ij}\phi(\vec{r}_i - \vec{r}_j)\vec{r}_{ji}\delta(\vec{r}-\vec{r}_i) + \vec{r}_{ji}\delta(\vec{r}-\vec{r}_i)\nabla_{ij}\phi(\vec{r}_i - \vec{r}_j) \cdot \frac{\vec{p}_i}{m} \right) \tag{10.13}$$

The equations (10.4), (10.6) and (10.9) are formally correct but the quantities in them are extremely singular because of the delta functions $\delta(\vec{r} - \vec{r}_i)$ in the definitions of the densities and fluxes. (Mathematically these are really relations between "distributions," not functions.)

To obtain hydrodynamic equations from these conservation laws, one must average them over spatial regions which are large compared to the microscopic distances between atoms and small compared to the wavelengths of interest. Here we simply assume that the average is a linear process, so that the average $\langle \ldots \rangle(\vec{R}, t)$ of a quantity is defined in a coarse grained region labelled $\vec{R}$ and that $\langle \nabla_{\vec{r}} A(\vec{r}) \rangle(\vec{R}, t) = \nabla_{\vec{R}} \langle A \rangle(\vec{R}, t)$ where $\nabla_{\vec{R}}$ is a coarse grained gradient. We then define the fluid velocity field $\vec{v}(\vec{R}, t)$ as $\vec{v}(\vec{R}, t) = \frac{\langle \vec{J} \rangle(\vec{R},t)/m}{\langle \rho \rangle(\vec{R},t)}$ and rewrite the conservation laws by writing $\vec{p}_i/m = \vec{v} + \delta v_i$. This gives

$$\begin{aligned}\frac{\partial \rho}{\partial t} + \nabla \cdot (\varrho \vec{v}) = 0 \\ \frac{\partial \vec{J}}{\partial t} + \nabla \cdot (m\rho\vec{v}\vec{v} - \tilde{\sigma}) = 0 \\ \frac{\partial e}{\partial t} + \nabla \cdot (\vec{Q} + e\vec{v} - \vec{v} \cdot \tilde{\sigma}) = 0\end{aligned} \tag{10.14}$$

Here we have written

$$\langle \mathbf{\Pi} \rangle = m\langle \rho \rangle \vec{v}\vec{v} - \tilde{\sigma} \tag{10.15}$$

and

$$\langle \vec{S} \rangle = \vec{Q} + \langle e \rangle \vec{v} - \vec{v} \cdot \tilde{\sigma} \tag{10.16}$$

$\vec{Q}$ and $\tilde{\sigma}$ depend only on the differences $\vec{p}_i/m - \vec{v}$ between the particle velocities and the local hydrodynamic velocity. The tensor $\tilde{\sigma}$ is called the stress tensor and $\vec{Q}$ is the heat current. Because the heat current and stress tensor $\vec{Q}$ and $\tilde{\sigma}$ appear on the right hand side of the conservation laws, the last three equations are not closed and merely represent the first in a hierarchy of equations describing the microscopic dynamics of the fluid. Notice, however, that because the three conservation laws are all of the form $\partial A/\partial t = -\nabla \cdot \vec{B}$, then unless the fluxes (generically $\vec{B}$) have a very peculiar dependence on position, low frequency disturbances of the variables in these equations will lead to long wavelength excitation and vice versa (long wavelength disturbances will lead to low frequency response). When the equations are linearized this can be shown quite rigorously (and is easy to see by taking Fourier transforms in space and time). Thus the confinement of the description of the dynamics to low frequencies and long wavelengths can be self consistent. This is the merit of basing the hydrodynamic theory on conservation laws. Actually, however, the nonlinear terms in the hydrodynamic equations can couple long to short wavelengths, and this can lead to conditions in which this self consistency breaks down when the amplitude of the disturbance is large. This breakdown leads to the phenomenon of turbulence, which is extremely important technically and of great scientific interest and which is only partly understood at the fundamental level. We will not discuss turbulence further here. In the hydrodynamic limit of low frequencies and long wavelengths, these equations can be closed to obtain a hydrodynamic theory containing only three equations for the densities $\varrho$, $\vec{J}$ and $e$ by relating $\tilde{\sigma}$ and $\vec{Q}$ to the gradients of these densities through phenomenological constants which are usually called transport coefficients. To do this, the stress tensor $\sigma_{ij}$ is written as

$$\sigma_{ij} = -p\delta_{ij} + \sigma'_{ij} \tag{10.17}$$

where $p$ is the local pressure (defined by the same averaging procedure which defines the densities) and (10.17) defines $\tilde{\sigma}'$, which is known as the viscous part of the stress tensor. Closure is obtained by assuming that $\sigma'_{ij}$ is proportional to the gradients of the velocity $\vec{v}$ as one expects from the elementary intuitive description of viscosity. By use of the isotropy of the fluid on large length and long time scales, one can show[1] that the only possible form of $\sigma'$ is

$$\sigma'_{ij} = \eta_{\mathrm{s}}\left[\nabla_i v_j + \nabla_j v_i - \frac{2}{3}\nabla \cdot \vec{v}\delta_{ij}\right] + \eta_{\mathrm{v}}\nabla \cdot \vec{v}\delta_{ij} \tag{10.18}$$

The two coefficients $\eta_s$ and $\eta_v$ are called the shear and bulk viscosities, respectively. (Various other symbols are used for $\eta_v$ including $\zeta$ and $\mu_B$). The bulk viscosity describes the viscous or dissipative part of the response to a compression while $\eta_s$ describes the response to shear. The use of the isotropy of the fluid and the requirement that $\sigma'$ be traceless reduces the number of constants in this expression from $3^4 = 81$ to 2. (In the Navier–Stokes form of the hydrodynamic equations, the term in $\nabla \cdot \vec{v}$ is dropped.)

The diffusive heat flux $\vec{Q}$ is proportional to the gradient of the local temperature $\nabla T$ and obeys Fourier's law

$$\vec{Q} = -\Lambda \vec{\nabla} T \tag{10.19}$$

The quantity $\Lambda$ is called the thermal conductivity. According to this law, heat flows from regions of high temperature to regions of low temperature, that is, in a direction opposite to the temperature gradient. The temperature $T$ can be related to the energy density $e$ through the application of thermodynamic relations. The applicability of thermodynamics here follows because the quantities $\rho$, $\vec{J}$ , $e$, $p$ are all defined as averages of microscopic expressions over spatial regions which are large compared to interatomic distances but small compared to the relevant hydrodynamic wavelengths and over times large compared to the time for local thermodynamic equilibrium to be established in these regions but small compared to the inverse of the relevant hydrodynamic frequencies. Analogous use of local thermodynamic equilibrium occurs in the formulation of other hydrodynamic theories.

### Fluctuation–dissipation relations for hydrodynamic transport coefficients

The general idea associated with fluctuation–dissipation theorems is very simply illustrated by the case of a simple magnet for which the intensive thermodynamic variables are $H, T$ and the static susceptibility at zero field is

$$\left(\frac{\partial M}{\partial H}\right)_T = (\langle M^2\rangle - \langle M\rangle^2)/k_B T \tag{10.20}$$

(This is easily obtained by use of $\langle M\rangle = -(\partial F/\partial H)_T$ and $F = -k_B T \ln \sum\{S\}\exp(-(H_0 - H\sum_i S_i^z))$ for example. The magnetic moment per unit spin is absorbed in the definition of the field $H$ so that $H$ has the dimensions of energy.) In (10.20), the left hand side is representative of the *response* of the system to an external magnetic field, whereas the right hand side, evaluated at zero field, is representative of the *fluctuations* of the magnetization in the absence of that external field. Generically, this type of relation occurs repeatedly in statistical mechanics. The response to a small external field is proportional to the magnitude of the equilibrium fluctuations (in the absence of the field) of the quantity which the external field probes. (The relations are called fluctuation–dissipation theorems rather than fluctuation response theorems because at finite frequency they are often expressed in terms of the imaginary, out of phase response of the system, corresponding to the energetically dissipative part of the response.) A somewhat more involved example is the relationship between the response of a many body system to a beam of neutrons (expressed as a differential cross-section) to fluctuations in the density of the system as described for liquids in Chapter 7.

Because the transport coefficients $\eta_v$, $\eta_s$, $\Lambda$ of the hydrodynamic equations are associated with energy dissipation in the system, one might expect that a type of fluctuation–dissipation theorem might be obtained for them, and we will pursue that line of investigation here. One can see in equations (10.18) and (10.19) that the transport coefficients relate momentum and energy currents to gradients of the velocity and temperature respectively so, in a sense, they are response functions. However, the fields (gradients of velocity and temperature) to which the response corresponds are more difficult to characterize theoretically (and experimentally) than an external magnetic field or a neutron beam, each of which can be characterized by parameters in the underlying Hamiltonian without any kind of self consistent calculation. In the hydrodynamic case, the gradients of velocity and temperature which are the driving fields are quantities averaged over short length and time scales. That it is nevertheless possible to derive fluctuation–dissipation theorems which give the Navier–Stokes transport coefficients in terms of correlation functions describing the fluctuations of the fluid in the absence of gradients in temperature and velocity was first shown by Kadanoff and Martin.[2] We will review that work here. The idea is to think of the fluid in a state of hydrodynamic flow from the present into the future ($t > 0$) as being produced by external forces which were slowly turned on in the past ($t < 0$) in such a way that there is a small perturbation $\delta p(\vec{r})$ in the pressure, but otherwise each part of the fluid is in local equilibrium at $t = 0$, when the external forces are turned off. Now the behavior of the fluid for $t > 0$ can be calculated in two ways. (1) As a response of the system to the external perturbation (now shut off, but still influencing the time dependent behavior). This calculation goes much like the calculations of the susceptibility and the neutron scattering cross-section and results in a response which is proportional to correlation functions characterizing fluctuations of the equilibrium fluid and (2) as a solution of the hydrodynamic equations with appropriate initial conditions. By equating the result of (1) to the result of (2) one obtains expressions involving correlation functions of the equilibrium fluid (from (1)) to hydrodynamic transport coefficients (from (2)). The advantage of this procedure is that the response of the fluid after $t = 0$ can be described by the hydrodynamic equations (because the disturbance was produced by a slow, long wavelength perturbation) but because the state was produced by a set of weak external forces, one can also calculate the response to the external fields directly, getting results in terms of correlation functions of microscopic operators. We will now illustrate this in some detail.

Because the external fields we apply will be small, a linearized form of the hydrodynamic equations should be adequate. To obtain the linearized form of the equations, one substitutes $\varrho = \varrho_0 + \varrho_1$, $e = e_0 + e_1$, $T = T_0 + T_1$, $\vec{g} = \vec{g}_0 + \vec{g}_1$, $\vec{v} = \vec{v}_0 + \vec{v}_1$ into the equations and retains all terms that are no higher than first order in the fluctuations. Note that $\vec{v}_0 = 0$ because the hydrodynamic velocity is zero in equilibrium and that $e = (e_{\text{int}} + (1/2)m\rho v^2)$ so that the first order fluctuation $e_1$ in $e$ refers only to the internal energy $e_{\text{int}}$ and not to the kinetic energy associated with the flow. Applying these considerations we find the linearized equations:

$$\begin{aligned}
&\frac{\partial \rho_1}{\partial t} + \rho_0 \nabla \cdot \vec{v}_1 = 0 \\
&m\rho_0 \frac{\partial \vec{v}_1}{\partial t} = -\nabla p_1 + \eta_s \nabla^2 \vec{v}_1 + \left(\eta_v + \frac{1}{3}\eta_s\right) \vec{\nabla}(\vec{\nabla} \cdot \vec{v}_1) \\
&\frac{\partial e_1}{\partial t} + (e_0 + p_0)\vec{\nabla} \cdot \vec{v}_1 = \Lambda \nabla^2 T_1
\end{aligned} \tag{10.21}$$

Now take the divergence of the second (momentum conservation) equation and use the first (particle number) equation in the second two equations to eliminate $\vec{v}_1$. Thus one finds

$$m\left(\frac{\partial^2}{\partial t^2} - D_l \nabla^2 \frac{\partial}{\partial t}\right) \rho_1 = \nabla^2 p_1 \tag{10.22}$$

$$\frac{\partial}{\partial t}\left(e_1 - \left(\frac{e_0 + p_0}{\rho_0}\right) \rho_1\right) = \Lambda \nabla^2 T_1 \tag{10.23}$$

Here $D_l = (\eta_v + (4/3)\eta_s)/\rho_0 m$ is called the longitudinal diffusion constant. The first of these equations will become a wave equation describing sound waves and the second will become a diffusion equation describing heat diffusion in appropriate cases. To close the equations one must invoke the assumption of local thermodynamic equilibrium to relate the four quantities $\rho_1$, $T_1$, $P_1$ and $e_1$ to one another in terms of two independent thermodynamic variables. A convenient choice is to express $e_1$ and $p_1$ and $T_1$ in terms of $\rho_1$ and $q_1$, a heat fluctuation defined as ($V$ is the system volume)

$$q_1 \equiv e_1 - \left(\frac{e_0 + p_0}{\rho_0}\right) \rho_1 = \frac{T_0}{V} S_1 \tag{10.24}$$

The second equality follows from elementary thermodynamics at constant particle number $N$. $S_1$ is the fluctuation in the total entropy. Thus choosing the entropy and the number density as independent variables we have from the chain rule that

$$T_1 = \left(\frac{\partial T}{\partial \rho_1}\right)_S \rho_1 + \frac{V}{T}\left(\frac{\partial T}{\partial S}\right)_\rho q_1 \tag{10.25}$$

$$p_1 = \left(\frac{\partial p}{\partial \rho_1}\right)_S \rho_1 + \frac{V}{T}\left(\frac{\partial p}{\partial S}\right)_\rho q_1 \tag{10.26}$$

Inserting these equations into the two previous ones gives

$$\left(\frac{\partial^2}{\partial t^2} - D_l \nabla^2 \frac{\partial}{\partial t} - c_{\mathrm{s}}^2 \nabla^2\right)\rho_1 - \frac{\alpha \rho c_{\mathrm{s}}^2}{C_p}\nabla^2 q_1 = 0 \tag{10.27}$$

$$\frac{\partial}{\partial t} q_1 - \gamma D_T \nabla^2 q_1 - \frac{\Lambda}{\alpha \rho_0}(\gamma - 1)\nabla^2 \rho_1 = 0 \tag{10.28}$$

Here $c_{\mathrm{s}}^2 = (1/m)(\partial p/\partial \rho)_S$ is the adiabatic sound velocity, $\alpha = -(1/\rho_0)(\partial \rho/\partial T)_P$ is the thermal expansion coefficient, $C_p = (1/V)(T\partial S/\partial T)_P$ is the isobaric specific heat per unit volume, $\gamma = C_P/C_V$ is the specific heat ratio and $D_T = \Lambda/C_P$. Notice that at low temperatures, the thermal expansion coefficient becomes small and the term in the first equation which couples the density fluctuations to the heat fluctuations becomes small. Then the first equation just describes adiabatic sound propagation.

Further, the damping of the sound is of order $q^2 D_l$ while the frequency is $c_{\mathrm{s}} q$ so the damping becomes weaker at long wavelengths. At the same time the coupling term in the second equation also becomes small at low temperatures, so that the second equation reduces to the diffusive heat equation (because $C_p \approx C_V$.)

It remains to treat the initial conditions in these equations in an appropriate way so that they can be used to calculate the density correlation function . As mentioned above, the general idea is to connect the correlation function to the hydrodynamic response through a fluctuation–dissipation theorem. We impose a perturbation of the form

$$H'(t) = -\int \mathrm{d}\vec{r}\; \delta p(\vec{r})\rho(\vec{r}, t)\mathrm{e}^{\epsilon t}/\langle \rho \rangle \tag{10.29}$$

for $t < 0$ (and $H'(t) = 0$ for $t > 0$). Here $\delta p(\vec{r})$ is a small $c$-number function which may be regarded as a slowly imposed fluctuation in the pressure. $\epsilon$ is an infinitesimally small positive number describing the very slow rate at which the perturbation is turned on. $\rho(\vec{r}, t)$ is the Heisenberg representation of the number density operator of the fluid. By quite standard manipulations of time dependent perturbation theory one can show that the change in the density $\delta \rho(\vec{r}, t)$ at time $t > 0$, resulting from this perturbation is

$$\delta\rho(\vec{r},t) = \frac{\mathrm{i}}{\hbar}\int_{-\infty}^{0} \mathrm{d}t'\mathrm{e}^{\epsilon t'}\int \mathrm{d}\vec{r}\,\frac{\delta p(\vec{r}\,')}{\langle\rho\rangle}\langle[\rho(\vec{r}\,',t'),\rho(\vec{r},t)]\rangle_T \tag{10.30}$$

Here $\langle\ldots\rangle_T$ means a thermal average in the unperturbed system, which we assume is described by the canonical ensemble as $t \to -\infty$. This is a fluctuation–dissipation theorem in the sense discussed earlier because it relates the response $\delta\rho(\vec{r},t)$ to an externally imposed disturbance (characterized by $\delta p(\vec{r}\,')$ and the slow introduction of the perturbation in time up to $t = 0$) through the equilibrium correlation functions $\langle[\rho(\vec{r}\,',t'),\rho(\vec{r},t)]\rangle$ describing the fluctuations of the undisturbed system. It is not hard, for example by introducing a set of exact energy eigenstates, to show that the time Fourier transform of the commutator on the right hand side of this expression is related to the density density correlation function by

$$\begin{aligned}&\int_{-\infty}^{\infty}\mathrm{d}\tau\,\mathrm{e}^{-\mathrm{i}\omega\tau}\langle[\rho(\vec{r}\,',0),\rho(\vec{r},\tau)]\rangle_T \equiv \langle[\rho(\vec{r}\,'),\rho(\vec{r})]\rangle_T(\omega)\\ &= (1-\mathrm{e}^{-\hbar\omega\beta})\int_{-\infty}^{\infty}\mathrm{e}^{-\mathrm{i}\omega\tau}\langle\rho(\vec{r}\,',0)\rho(\vec{r},\tau)\rangle_T\mathrm{d}\tau \equiv (1-\mathrm{e}^{-\hbar\beta\omega})\langle\rho(\vec{r}\,')\rho(\vec{r})\rangle_T(\omega)\end{aligned} \tag{10.31}$$

If we assume that the perturbation $H'(t)$ left the system in a state of local equilibrium at time $t = 0$ with pressure $p + \delta p(\vec{r})$ in place of the equilibrium pressure, then we can use the hydrodynamic equations to calculate the response $\delta\rho(\vec{r},t)$ for $t > 0$. This assumption can be established[2] for small enough $\epsilon$. To use the result to calculate the density–density correlation function, we take the complex Laplace transform of (10.30) for a complex variable $z$ in the upper half of the complex plane:

$$\begin{aligned}\delta\rho(\vec{r},z) &\equiv \int_0^{\infty}\mathrm{d}t\,\mathrm{e}^{\mathrm{i}zt}\delta\rho(\vec{r},t)\\ &= \int\frac{\delta p(\vec{r}\,')}{\langle\rho\rangle}(-1/\hbar)\int_{-\infty}^{\infty}\frac{\mathrm{d}\omega}{2\pi}\frac{1}{(\omega-\mathrm{i}\epsilon)}\frac{1}{(\mathrm{i}(z-\omega))}\langle[\rho(\vec{r}\,'),\rho(\vec{r})]\rangle_T\mathrm{d}\vec{r}\,'(\omega)\end{aligned} \tag{10.32}$$

Then by taking $z \to \omega + \mathrm{i}\epsilon$, using the identity $1/(x+\mathrm{i}\epsilon) \to P(1/x) - \mathrm{i}\pi\delta(x)$ and defining spatial Fourier transforms as $f(\vec{q}) = \int \mathrm{d}\vec{r}\mathrm{e}^{-\mathrm{i}\vec{q}\cdot\vec{r}} f(\vec{r})$ one finds

$$\frac{2\hbar\omega V}{(1-\mathrm{e}^{-\beta\hbar\omega)})}\mathrm{Re}(\delta\rho(\vec{q},z\to\omega+\mathrm{i}\epsilon)\langle\rho\rangle/\delta p(\vec{q})) = \langle\rho\rho\rangle_T(\vec{q},\omega) \tag{10.33}$$

in which the Fourier transform of the equilibrium density–density correlation function is defined by

$$\langle\rho\rho\rangle_T(\vec{q},\omega) = \int \mathrm{d}(t-t') \int \mathrm{d}(\vec{r}-\vec{r}\,')\,\mathrm{e}^{\mathrm{i}\omega(t-t')-\mathrm{i}\vec{q}\cdot(\vec{r}-\vec{r}\,')}\langle\rho(\vec{r},t)\rho(\vec{r}\,',t')\rangle_T \quad (10.34)$$

Thus the program is simply to solve the hydrodyamic equations for the Laplace transform with initial condition $p_1(\vec{r}, t=0) = \delta p(\vec{r})$ and then use (10.33) to compute the correlation function. The equations for the Laplace transform take the form:

$$\left(-z^2 + \mathrm{i}zD_l\nabla^2 - c_\mathrm{s}^2\nabla^2\right)\rho_1(\vec{r},z) - \frac{\alpha\rho_0 c_\mathrm{s}^2}{C_P}\nabla^2 q_1(\vec{r},z) = (-D_l\nabla^2 - \mathrm{i}z)\rho(\vec{r},t=0) \quad (10.35)$$

$$(-\mathrm{i}z - \gamma D_T\nabla^2)q_1(\vec{r},z) - \frac{\Lambda}{\alpha\rho_0}(\gamma-1)\nabla^2\rho_1(\vec{r},z) = q_1(\vec{r},t=0) \quad (10.36)$$

Now we take the spatial Fourier transform and write

$$\rho_1(\vec{q},t=0) = \left(\frac{\partial\rho}{\partial p}\right)_T \delta p(\vec{q}) \quad (10.37)$$

$$q_1(\vec{q},t=0) = \left(\frac{T}{V}\right)\left(\frac{\partial S}{\partial p}\right)_T \delta p(\vec{q}) \quad (10.38)$$

Here we are using the previously discussed assumption concerning the preparation of the initial state with the perturbation $H'(t)$. Then the two equations become:

$$(-z^2 - \mathrm{i}zq^2D_l + (c_\mathrm{s}q)^2)\rho_1(\vec{q},z) + \frac{\alpha\rho_0 c_\mathrm{s}^2}{C_P}q^2 q_1(\vec{q},z) = (q^2D_l - \mathrm{i}z)\left(\frac{\partial\rho}{\partial p}\right)_T \delta p(\vec{q}) \quad (10.39)$$

$$\frac{\Lambda}{\alpha\rho_0}(\gamma-1)q^2\rho_1(\vec{q},z) + (-\mathrm{i}z + \gamma D_T q^2)q_1(\vec{q},z) = \left(\frac{T}{V}\right)\left(\frac{\partial S}{\partial p}\right)_T \delta p(\vec{q}) \quad (10.40)$$

These are two linear equations for $\rho_1(\vec{q},z)$ and $q_1(\vec{q},z)$. The solution for $\rho_1(\vec{q},z)$ is easily written out as a ratio of $2\times 2$ determinants:

$$\rho_1(\vec{q},z) = \frac{(q^2D_l - \mathrm{i}z)(q^2\gamma q^2 - \mathrm{i}z)\left(\frac{\partial\rho}{\partial p}\right)_T - \frac{\alpha\rho_0 c_\mathrm{s}^2}{C_P}q^2\left(\frac{T}{V}\right)\left(\frac{\partial S}{\partial p}\right)_T}{(-z^2 - \mathrm{i}zq^2D_l - (c_\mathrm{s}q)^2)(-\mathrm{i}z + \gamma D_T q^2) - D_T c_\mathrm{s}^2 q^4(\gamma-1)}\delta p(\vec{q}) \quad (10.41)$$

For given values of the hydrodynamic and thermodynamic parameters one can now substitute $z \to \omega + \mathrm{i}\epsilon$ in this and take the real part. At the cost of some more work, one gets considerably greater insight by rearranging this expression and making some approximations. First, one may make an expansion of the roots of the denominator in powers of $q$. To second order in $q$ one then finds poles at

$$z = \pm c_s q - \mathrm{i}(\Gamma/2)q^2 \qquad - \mathrm{i}D_T q^2 \tag{10.42}$$

These poles correspond to the propagation of sound in the fluid. $\Gamma = (\gamma - 1)D_T + D_l$ is a measure of decay rate of the sound mode. It turns out that the correlation function $\langle\rho\rho\rangle_T(\vec{q}, \omega)$ is directly proportion to the rate of inelastic light scattering from a fluid and, at the long wavelengths and low frequencies for which the hydrodyamic theory is valid, this is a particularly useful way to measure $\langle\rho\rho\rangle_T(\vec{q}, \omega)$. In such an experiment, the first two poles described in (10.42) correspond to the Stokes and anti-Stokes peaks in the light scattering amplitude, corresponding to scattering angles giving momentum transfers $\vec{q}$ to or from the fluid respectively and corresponding to the absorption or emission of energy by or from the liquid from or by the incident light beam respectively. The third pole corresponds to the central "Rayleigh peak" associated with the diffusion of heat in the fluid. (Notice, that by careful treatment of the factors $\gamma$ here one finds that the central peak is proportional to $1/C_V$ and not to $1/C_P$.) To obtain a physically transparent form for $\langle\rho\rho\rangle_T(\vec{q}, \omega)$ it is useful to rewrite the factor $\frac{\alpha\rho_0}{C_P}\left(\frac{T}{V}\right)\left(\frac{\partial S}{\partial p}\right)_T$ which appears in the rightmost term of the numerator as

$$\frac{\alpha\rho_0 c_s^2}{C_P}\left(\frac{T}{V}\right)\left(\frac{\partial S}{\partial p}\right)_T = C_P\left(-1 + \frac{1}{\gamma}\right)\left(\frac{\partial\rho}{\partial p}\right)_T \tag{10.43}$$

by use of several thermodynamic identities. This rightmost term in the denominator then becomes simply $(c_s q)^2(1 - 1/\gamma)$. Now if we suppose that $c_s q \gg D_T q^2, \Gamma q^2$ then the peaks as a function of $\omega$ will be well separated and we can approximate the weight of each pole by simply evaluating all the finite factors at the position of the pole. In this way we find using (10.33) and (10.41),

$$\begin{aligned}\langle\rho\rho\rangle_T(\vec{q}, \omega) &\\ = \frac{k_B T \rho_0 \chi_T}{\pi}&\left[(1/\gamma)\left(\frac{\Gamma q^2}{(\omega - c_s q)^2 + (\Gamma q^2/2)^2} + \frac{\Gamma q^2}{(\omega + c_s q)^2 + (\Gamma q^2/2)^2}\right)\right.\\ &\left. + (1 - 1/\gamma)\frac{D_T q^2}{(\omega^2 + (D_T q^2)^2)}\right]\end{aligned} \tag{10.44}$$

which has a simple interpretation.

From this one can extract various relations between transport coefficients and the equilibrium correlation function. For example:

$$\frac{1}{\Gamma} = \lim_{q\to 0} \frac{q^2\gamma\pi\langle\rho\rho\rangle_T(\vec{q}, \omega = c_s q)}{k_B T\rho\chi_T} \tag{10.45}$$

$$\frac{1}{D_T} = \lim_{q\to 0}\lim_{\omega\to 0} \frac{q^2\pi\langle\rho\rho\rangle_T(q, \omega)}{(1 - 1/\gamma)k_B T\rho\chi_T} \tag{10.46}$$

$$\Gamma/\gamma + (1 - 1/\gamma)D_T = \lim_{\omega\to 0}\lim_{q\to 0} \frac{\gamma\pi\langle\rho\rho\rangle_T(q, \omega)\omega^2}{q^2 k_B T\rho\chi_T} \tag{10.47}$$

Note that the order of taking limits is very important. Some work has been done to use these relations to make computations of transport coefficients.[3]

**Superfluid hydrodynamics** The hydrodyamics of superfluid $^4$He is described by a two fluid model.[4] The basic idea for the derivation of the two fluid hydrodynamics is the same as the one used in the derivation of the classical hydrodynamic equations. One writes down local conservation laws and uses general symmetry arguments to close the equations by expressing the resulting currents in terms of gradients of the densities of the conserved quantities and phenomenological transport coefficients. The difference between the classical hydrodynamics and the two fluid theory is that there is an additional slow variable in the two fluid theory. The additional slow variable arises because the lambda transition leading to superfluidity resulted in the breaking of a symmetry in the fluid.

Unfortunately, this symmetry, which we will discuss shortly, is somewhat difficult to visualize, so we will begin with a brief discussion of the partly analogous situation in a Heisenberg ferromagnet. In a Heisenberg ferromagnet, at any temperature below the Curie temperature, the magnetization of the magnet can point in any of an infinite number of directions on the unit sphere. In practice, the magnetization is found in one of these infinitely many possible states, at least over macroscopic distances, though a large sample may contain many of these macroscopic domains. Thus the thermodynamic state of the system is described by a density matrix which does not include all the states of the system, but only those consistent with this particular magnetization direction. Such a density matrix will not be invariant under the simultaneous rotations of all the spin directions although the Hamiltonian of the system is invariant under such rotations. Once the magnet is in such a state, the total magnetization does not change in time, because the Hamiltonian commutes with the total magnetization. This is a global conservation law (conservation of magnetization) which is a result of the rotational invariance of the Hamiltonian. (The magnetization could change if the system is not isolated, as a result of interactions with the thermal environment if the magnet is of finite size. But the time for such changes is extremely long for macroscopic samples well below the Curie point.)

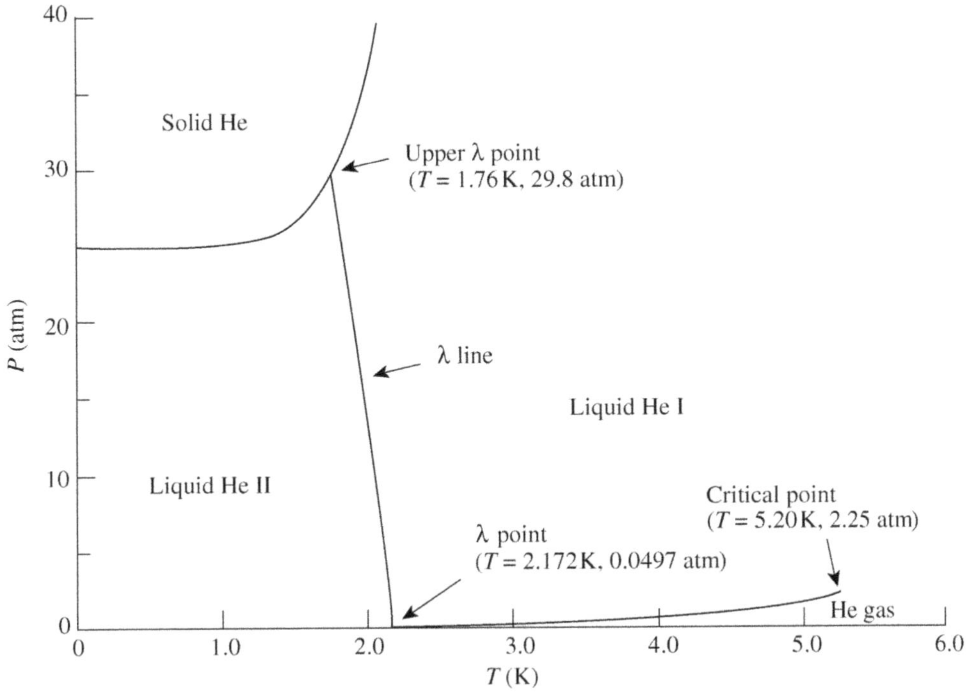


Figure 10.1 Phase diagram of $^4$He.

Now consider making a hydrodynamic theory of the ferromagnet.[5] One considers a region of the magnet which is large compared with interatomic distances but small compared with the long wavelengths of interest and supposes that such a region is in local thermodynamic equilibrium in the constrained sense that the density matrix describes the system with a fixed local magnetization direction. However, we suppose that the whole system is slightly out of global equilibrium, so neighboring regions of similar size in the magnet are also in local equilibrium but have a different local magnetization. Terms of the Hamiltonian coupling spins within one of these regions will conserve the local magnetization, but terms coupling spins in regions of different local magnetization will change the local magnetization directions, consistent with global magnetization conservation. From this picture one can derive a new conservation law, describing the flow of magnetization. This new conservation equation arises as a result of the broken rotational symmetry of the equilibrium density matrix for the system below the Curie point. Much as the conservation of momentum leads to sound waves in the Navier–Stokes equations, this conservation law leads to spin waves in the hydrodynamics of a Heisenberg ferromagnet.

Now consider the case of liquid $^4$He. We show an experimental phase diagram in Figure 10.1. In addition to the solid, "normal" He I liquid and vapor phases, there is another phase, termed superfluid or He II phase, which exists at pressures below about 25 atmospheres and temperatures below about 2 K. The experimental discovery of this phenomenon began in 1911 when a density anomaly was discovered by the group of Kammerlingh Onnes. The low temperature phase was found to have anomalous flow properties in the 1930s and, in 1938, London suggested that Bose condensation was associated with onset of the He II phase. However, liquid $^4$He in this regime is much too dense for use of the low density Gross–Pitaevskii–Bogoliubov theory described in Chapter 5. A full microscopic theoretical description of the superfluid phase is to this day somewhat incomplete. However, by understanding some basic symmetry properties one can construct a correct hydrodynamic description. The symmetry breaking analogous to the rotational symmetry breaking occuring in a ferromagnet is the selection by the low temperature liquid of a quantum mechanical phase which is defined as follows. We begin with a zero temperature description. In a basis of energy eigenstates the density matrix may be described as

$$\rho = |\Psi_{N,0}\rangle\langle\Psi_{N,0}| \tag{10.48}$$

where $|\Psi_{N,0}\rangle$ is the ground state of the system. This may be projected onto eigenstates $|\vec{r}_1,\ldots,\vec{r}_N\rangle$ to give

$$\begin{aligned}\rho(\vec{r}_1{}',\ldots,\vec{r}_N{}';\vec{r}_1,\ldots,\vec{r}_N) &= \langle\vec{r}_1,\ldots,\vec{r}_N||\Psi_{N,0}\rangle\langle\Psi_{N,0}|\vec{r}_1{}',\ldots,\vec{r}_N{}'\rangle \\ &= \Psi^*_{N,0}(\vec{r}_1,\ldots,\vec{r}_N)\Psi_{N,0}(\vec{r}_1{}',\ldots,\vec{r}_N{}')\end{aligned} \tag{10.49}$$

To display the feature which distinguishes the superfluid from an ordinary fluid, one sets $\vec{r}_2,\ldots,\vec{r}_N = \vec{r}_2{}',\ldots,\vec{r}_N{}'$ here and integrates over $\vec{r}_2,\ldots,\vec{r}_N$ to obtain the one particle density matrix $\rho_1(\vec{r}_1{}',\vec{r}_1)$:

$$\rho_1(\vec{r}_1{}',\vec{r}_1) = N\int \mathrm{d}\vec{r}_2\cdots\mathrm{d}\vec{r}_N\,\Psi^*_{N,0}(\vec{r}_1,\vec{r}_2,\ldots,\vec{r}_N)\Psi_{N,0}(\vec{r}_1,\vec{r}_2,\ldots,\vec{r}_N) \tag{10.50}$$

(The factor $N$ is included so that $\rho_1(\vec{r}_1{}',\vec{r}_1)$ reduces to the particle density when $\vec{r}_1 = \vec{r}_1{}'$.) Now take the limit in which $|\vec{r}_1 - \vec{r}_1{}'|$ becomes large (macroscopic). In a superfluid, $\rho_1(\vec{r}_1,\vec{r}_1)$ approaches a finite limit of factors independent of the difference $\vec{r}_1 - \vec{r}_1{}'$

$$\rho_1(\vec{r}_1,\vec{r}_1{}') \to N\psi_0^*(\vec{r}_1{}')\psi_0(\vec{r}_1) \tag{10.51}$$

In a noninteracting Bose gas, $\psi_0$ is simply the eigenfunction of the lowest eigenstate of the single particle Hamiltonian. More generally, it has an amplitude which is denoted $\sqrt{n_0(\vec{r})/V}$ and a phase $\phi(\vec{r})$ so that

$$\rho_1(\vec{r}_1,\vec{r}_1{}') \to (N/V)\sqrt{n_0(\vec{r}_1)n_0(\vec{r}_1{}')}\exp(+\mathrm{i}(\phi(\vec{r}_1) - \phi(\vec{r}_1{}')) \tag{10.52}$$

In the ground state of a homogeneous bulk superfluid, $\phi$ and $n_0$ are constants and $\rho_1(\vec{r}_1, \vec{r}_1') \to (Nn_0/V)$ which is a fraction of the total density of the fluid, termed the "condensate fraction." If the fluid is not interacting then $n_0 = 1$ whereas in the presence of interactions $n_0 < 1$. In this bulk limit and the ground state, the quantum mechanical phases $\phi$ (not to be confused with the thermodynamic phases) do not enter the limiting form of $\rho_1$. In particular, a change $\delta\phi$ in the phase everywhere would not change $\rho_1$ or anything else observable about the system. Now consider a low lying time dependent wave function, associated with a hydrodyamic flow state of the fluid, which is inhomogeneous on scales much larger than interatomic distances. Now the phases can be defined as above, but they will be position and time dependent:

$$\rho_1(\vec{r}_1, \vec{r}_1', t) = N \int \mathrm{d}\vec{r}_2 \cdots \mathrm{d}\vec{r}_N \Psi_N^*(\vec{r}_1', \ldots, \vec{r}_N', t)\Psi_N(\vec{r}_1, \ldots, \vec{r}_N, t)$$
$$\to (N/V)\sqrt{n_0(\vec{r}_1)n_0(\vec{r}_1')}\exp(+\mathrm{i}(\phi(\vec{r}_1, t) - \phi(\vec{r}_1', t)) \qquad (10.53)$$

The new hydrodynamic equation involves the derivatives of the phase. One defines

$$\vec{v}_s(\vec{r}, t) \equiv (\hbar/m)\nabla_r\phi \qquad (10.54)$$

which has the dimensions of a velocity. It is called the "superfluid velocity." However, it is important to note that because the phase difference between two points in the fluid has only been defined when the two points are far apart (in practice much farther than interatomic distances), the derivative in the last equation can only make sense as a finite difference. To get an equation for $\partial\phi/\partial t$ we write $\rho_1(\vec{r}_1, \vec{r}_1')$ a different way

$$\rho_1(\vec{r}_1, \vec{r}_1') = N\langle\Psi_N|\psi^\dagger(\vec{r}_1')\psi(\vec{r}_1)|N\rangle \qquad (10.55)$$

for the equilibrium state where $\psi(\vec{r}) = \sum_\nu \phi_n(\vec{r})a_n$ is the wave operator discussed in connection with the Gross–Pitaevskii–Bogoliubov theory and $\phi_n(\vec{r})$ is any complete one particle basis. The time dependent $\rho_1$ for a hydrodynamic state is, in this language

$$\rho_1(\vec{r}_1, \vec{r}_1', t) = N\langle\Psi_N|\psi^\dagger(\vec{r}_1, t)\psi(\vec{r}_1', t)|N\rangle \qquad (10.56)$$

where $\psi(\vec{r}_1', t)$ is the Heisenberg operator

$$\psi(\vec{r}_1', t) = \mathrm{e}^{\mathrm{i}Ht/\hbar}\psi(\vec{r}_1')\,\mathrm{e}^{-\mathrm{i}Ht/\hbar} \qquad (10.57)$$

Insert a complete set of $N-1$ particle states in this form of $\rho_1(\vec{r}_1, \vec{r}_1', t)$ and take the limit of large separations of $\vec{r}_1$ and $\vec{r}_1'$. We assume that, with an appropriate choice of basis for the complete set, only one intermediate state $|\Psi_{N-1}\rangle$ survives the limit giving

$$\rho_1(\vec{r}_1, \vec{r}_1', t) \to N\langle\Psi_N|\psi^\dagger(\vec{r}_1', t)|\Psi_{N-1}\rangle\langle\Psi_{N-1}|\psi(\vec{r}_1, t)|\Psi_N\rangle \tag{10.58}$$

allowing the identification

$$\sqrt{n_0(\vec{r})}\,\mathrm{e}^{\mathrm{i}\phi(\vec{r},t)} = \langle\Psi_{N-1}|\psi(\vec{r}_1, t)|\Psi_N\rangle \tag{10.59}$$

We are supposing that the time dependence on the left hand side is all in the phase. Taking the time derivative of this equation gives

$$\frac{\partial\phi(\vec{r}, t)}{\partial t} = -(E_N - E_{N-1})/\hbar \tag{10.60}$$

if $|\Psi_N\rangle$ and $\langle\Psi_{N-1}|$ are energy eigenstates. More generally the right hand side is

$$(\langle\Psi_{N-1}|H\psi(\vec{r}_1, t)|\Psi_N\rangle - \langle\Psi_{N-1}|\psi(\vec{r}_1, t)H|\Psi_N\rangle)/\langle\Psi_{N-1}|\psi(\vec{r}_1, t)|\Psi_N\rangle\hbar \tag{10.61}$$

It is reasonable to identify this with the negative of a local chemical potential, since it represents the negative of the energy cost of adding a particle to the system at $\vec{r}_1$:

$$\frac{\partial\phi(\vec{r}, t)}{\partial t} = \frac{-\mu(\vec{r}, t)}{\hbar} \tag{10.62}$$

and taking the gradient and using the definition of $\vec{v}_{\mathrm{s}}$

$$m\frac{\partial\vec{v}_{\mathrm{s}}}{\partial t} = -\nabla\mu(\vec{r}, t) \tag{10.63}$$

which is the desired new equation. These relations may be interpreted to mean that an increase in time of the phase of the condensate wave function is associated with the local addition of a particle to the system and the associated energy is $\mu$ so the time scale is $\hbar/\mu$. A gradient in the phase means that particles are being added or subtracted at different rates in neighboring parts of the fluid, resulting in fluid flow. Thus the superfluid flow is driven by the gradient of the chemical potential. (This relation is attributed to Josephson.)

All this discussion may be extended to finite temperatures. In the language used here the equilibrium one particle density matrix is

$$\rho_1(\vec{r}_1, \vec{r}_1') = N\int \mathrm{d}\vec{r}_2\cdots\mathrm{d}\vec{r}_N\sum_\nu \mathrm{e}^{(-E_{N,\nu}\beta)}\Psi^*_{N,\nu}(\vec{r}_1', \ldots, \vec{r}_N')\Psi_{N,\nu}(\vec{r}_1, \ldots, \vec{r}_N) \tag{10.64}$$

where $\Psi_{N,\nu}$ is a complete set of many body eigenstates and the condensate wave function is defined in exactly the same way

$$\rho_1(\vec{r}_1, \vec{r}_1') \to N\psi_0^*(\vec{r}_1')\psi_0(\vec{r}_1) \tag{10.65}$$

which is temperature dependent so obviously the "condensate wave function" is not a wave function in the usual sense. There is a possibility that there will be no finite term of order $N$ when $\vec{r}_1 - \vec{r}_1'$ is large. This will occur when the temperature is large enough and the temperature at which it first occurs is the lambda transition temperature.

It is sometimes convenient to think of the definition of phase in another, equivalent, way due to Onsager and Penrose.[6] Regard the one particle density matrix as an integral operator acting on functions $\psi(\vec{r})$ and consider the eigenvalue equation:

$$\int d\vec{r}'\rho_1(\vec{r}, \vec{r}')\psi_\lambda(\vec{r}') = n_\lambda\psi_\lambda(\vec{r}) \tag{10.66}$$

It is easy to establish that the operator is Hermitian so the eigenvalues $n_\lambda$ are real. For a Hermitian operator the operator can be rewritten as

$$\rho_1(\vec{r}, \vec{r}') = \sum_\lambda \psi_\lambda(\vec{r})n_\lambda\psi_\lambda^*(\vec{r}') \tag{10.67}$$

Then the lambda transition occurs when one of the terms in this sum has a macroscopically large eigenvalue $n_\lambda$. Call this eigenvalue $n_{\lambda_0}$. Then the relationship to the previous formulation is $\psi_{\lambda_0} = \psi_0(\vec{r})$ and $n_{\lambda_0} = n_0 N$. The phase $\phi$ of $\psi_{\lambda_0}(\vec{r})$ is the phase used to define the superfluid velocity as before. This formulation in terms of "Penrose orbitals" is useful for study of transitions to the strong coupling analogue of Bose condensation in finite and heterogeneous systems. Unless the many body problem has been completely solved, however, the Penrose orbitals are not easy to calculate.

The quantum mechanical phase will only survive at large separations if only one term dominates the sum on $\nu$ in $\rho_1$. Thus in the limit, although $n_0$ is temperature dependent, the phase $\phi$ is associated with just one many body wave function, whose weight in the total density matrix gets smaller as the temperature rises. Thus the essential elements of the derivation of the equation for $\vec{v}_s$ are unchanged except that there are additional, uncontrolled terms which cause the phase to degrade in time. One takes account of these with a dissipative term:

$$m\frac{\partial \vec{v}_s}{\partial t} = -\nabla(\mu(\vec{r}, t) + h) \tag{10.68}$$

Now we combine (10.68) with the conservation laws which led to the Navier–Stokes equations in order to obtain the superfluid hydrodynamics.

$$\frac{\partial \rho}{\partial t} = -\nabla \cdot \vec{J}/m \tag{10.69}$$

$$\frac{\partial \vec{J}}{\partial t} = -\nabla \cdot \mathbf{\Pi} \tag{10.70}$$

$$\frac{\partial e}{\partial t} = -\nabla \cdot \vec{S} \tag{10.71}$$

in which

$$\vec{J}/m = (1/2)\sum_i \left(\frac{\vec{p}_i}{m}\delta(\vec{r}-\vec{r}_i) + \delta(\vec{r}-\vec{r}_i)\frac{\vec{p}_i}{m}\right) \tag{10.72}$$

$$\begin{aligned}\mathbf{\Pi} = (1/4)\sum_i &\left(\frac{\vec{p}_i}{m}\vec{p}_i\delta(\vec{r}-\vec{r}_i) + 2\frac{\vec{p}_i}{m}\delta(\vec{r}-\vec{r}_i)\vec{p}_i + \delta(\vec{r}-\vec{r}_i)\frac{\vec{p}_i}{m}\vec{p}_i\right)\\ &+ (1/2)\sum_{i\neq j}\vec{r}_{ji}\nabla_i\phi(\vec{r}_i-\vec{r}_j)\delta(\vec{r}-\vec{r}_i)\end{aligned} \tag{10.73}$$

$$\vec{S} = \vec{S}_K + \vec{S}_\phi + \vec{S}_v \tag{10.74}$$

and

$$\vec{S}_K = (1/4)\sum_i\left[T_i\left(\frac{\vec{p}_i}{m}\delta(\vec{r}-\vec{r}_i) + \delta(\vec{r}-\vec{r}_i)\frac{\vec{p}_i}{m}\right) + \left(\frac{\vec{p}_i}{m}\delta(\vec{r}-\vec{r}_i) + \delta(\vec{r}-\vec{r}_i)\frac{\vec{p}_i}{m}\right)T_i\right] \tag{10.75}$$

$$\vec{S}_\phi = (1/4)\sum_i\left(\frac{\vec{p}_i}{m}\phi(\vec{r}_i-\vec{r}_j)\delta(\vec{r}-\vec{r}_i) + \phi(\vec{r}_i-\vec{r}_j)\delta(\vec{r}-\vec{r}_i)\frac{\vec{p}_i}{m}\right) \tag{10.76}$$

$$\vec{S}_v = (1/4)\sum_i\sum_{j\neq i}\left(\frac{\vec{p}_i}{m}\cdot\nabla_{ij}\phi(\vec{r}_i-\vec{r}_j)\vec{r}_{ji}\delta(\vec{r}-\vec{r}_i) + \vec{r}_{ji}\delta(\vec{r}-\vec{r}_i)\nabla_{ij}\phi(\vec{r}_i-\vec{r}_j)\cdot\frac{\vec{p}_i}{m}\right) \tag{10.77}$$

as before.

In experimental practice, the part of the flow of the fluid which arises from $\vec{v}_s$ can be distinguished from any other flow in the system by the fact that, after any external forces are removed, it decays much more slowly in time than any other flow. This existence of superfluid currents is the central phenomenon of superfluidity and arises because a change in the coherent gradient $\nabla\phi$ of the phase of a macroscopically large number of the particles would require a simultaneous coherent change in the phases of all of them at once and this is unlikely and energetically expensive. (One can get more insight into this by study of the Gross–Pitaevskii–Bogoliubov model discussed in Chapter 5.) Thus both computationally and experimentally one can distinguish a component $\vec{J}_{\rm s}$ of the momentum current which survives after removal of any external forces and depends linearly on $\vec{v}_{\rm s}$:

$$\vec{J}_{\rm s} = \lim_{t\to\infty} \vec{J}(t) = m\rho_{\rm s}\vec{v}_{\rm s} \tag{10.78}$$

where the coefficient $\rho_{\rm s}$ is defined by this equation. (Here we have defined $\rho_{\rm s}$ as a number density though in much, but not all, of the literature on superfluidity $\rho_{\rm s}$ is this quantity multiplied by $m$.) At shorter times, or in the presence of continual external driving forces, the current has additional terms which we write in the form:

$$\vec{J} = m(\rho\vec{v}_{\rm s} + (\rho - \rho_{\rm s})\vec{v}_{\rm n}) \tag{10.79}$$

This equation is to be regarded as defining $\vec{v}_{\rm n}$. It is convenient to write the coefficient of $\vec{v}_{\rm n}$ as $\rho_{\rm n} = \rho - \rho_{\rm s}$. One refers to this formulation as two fluid hydrodyamics, with the densities and velocities of the "superfluid" and "normal" components of the fluid being respectively $\rho_{\rm s}$, $\vec{v}_{\rm s}$, $\rho_{\rm n}$, $\vec{v}_{\rm n}$. This picture is intuitively appealing but it must be treated with considerable caution, returning when in doubt to the original definitions. Averaging the three conservation laws of ordinary hydrodynamics and writing $\langle\tilde{\Pi}\rangle = \mathbf{1}p - \sigma'$ as before gives

$$\frac{\partial\rho}{\partial t} = -\nabla\cdot\vec{J}/m \tag{10.80}$$

$$\frac{\partial\vec{J}}{\partial t} = -\nabla p + \nabla\cdot\tilde{\sigma}' \tag{10.81}$$

$$\frac{\partial e}{\partial t} = -\nabla\vec{j}^{\epsilon} \tag{10.82}$$

and the new equation

$$m\frac{\partial\vec{v}_{\rm s}}{\partial t} = -\nabla(\mu(\vec{r},t) + h) \tag{10.83}$$

The correct forms for $\vec{j}^{\epsilon}$, $\tilde{\sigma}'$ and $h$ are given by Khalatnikov. They are strongly constrained by the requirements of Galilean invariance, local thermodynamic equilibrium and the increase of entropy with time:

$$\tilde{\sigma}'_{ij} = \eta((\nabla_i v_{{\rm n},j} + \nabla_j v_{{\rm n},i}) - (2/3)\nabla\cdot\vec{v}_{\rm n}\delta_{ij} + \delta_{ij}(\zeta_2 - m\rho\zeta_1)\nabla\cdot\vec{v}_{\rm n}) \tag{10.84}$$

$$\vec{j}^{\epsilon} = \mu\vec{J}/m + Ts\vec{v}_{\rm n} - \Lambda\nabla T \tag{10.85}$$

$$h = \zeta_3\nabla\cdot(\vec{J} - \rho\vec{v}_{\rm n}) + \zeta_1\nabla\cdot\vec{v}_{\rm n} \tag{10.86}$$

The equations are closed by the relations

$$\vec{J} = m(\rho_{\rm s}\vec{v}_{\rm s} + \rho_{\rm n}\vec{v}_{\rm n}) \tag{10.87}$$

$$\rho = \rho_{\rm s} + \rho_{\rm n} \tag{10.88}$$

which were already introduced. New transport coefficients $\zeta_1, \zeta_2, \zeta_3$ have been introduced. $s$ is the entropy per unit *volume.* (Elsewhere we have used the same symbol to denote the entropy *per particle.*) We imagine inserting the definitions of $\tilde{\sigma}'_{ij}$ (10.84), $\vec{j}^{\epsilon}$ and $h$ (10.86) into equations (10.81), (10.82) and (10.83). Then the remaining equations are the continuity equation (10.80) (1), the momentum equation (10.81) (3), the superfluid velocity equation (10.83) (3), the energy conservation equation (10.82) (1) and the relations (10.87) (3) and (10.88) (1) which relate the current and density to their superfluid and normal compenents for a total of 12 equations relating the variables $e, s, T, \rho, p, \mu$ (six thermodynamic variables) and $\vec{J}, \vec{v}_s, \vec{v}_n, \rho_s$ and $\rho_n$ (11 other variables) for a total of 17 variables. Thermodynamic relations between the six thermodynamic variables reduce the number of independent thermodynamic variables to two independent variables and the constraint $\nabla \times \vec{v}_s = 0$ reduces the number of independent components of $\vec{v}_s$ to two so the equations can be closed ($17 - 4 - 1 = 12$ independent variables and 12 equations).

Relations for the transport coefficients in terms of equilibrium correlation functions of the currents, analogous to equations (10.46) and (10.47) were worked out by Hohenberg and Martin.[7]

## References

1. H. Jeffreys, *Cartesian Tensors*, Cambridge: Cambridge University Press, 1957, Chapters VII and IX, particularly pp. 70 and 84.
2. L. Kadanoff and P. Martin, *Annals of Physics* **24** (1963) 419. Reprinted in *Annals of Physics* **281** (2000) 800 and available from Science Direct.
3. R. Stadler, D. Alfe, G. Kresse, G. A. de Wijs and M. J. Gillan, *Journal of Non-Crystalline Solids* **250–252** (1999) 82.
4. I. M. Khalatnikov, *Introduction to the Theory of Superfluidity*, New York: W. A. Benjamin, 1965.
5. D. Forster, *Hydrodynamic Fluctuations, Broken Symmetry, and Correlation Functions*, Reading, MA: Benjamin, 1975.
6. L. Onsager and O. Penrose, *Physical Review* **104** (1956) 576.
7. P. C. Hohenberg and P. C. Martin, *Annals of Physics* **281** (2000) 638. Reprinted from *Annals of Physics* **34** (1965) 291.

## Problems

10.1 Consider a monatomic fluid moving in a porous material in such a way that the momentum is not conserved.

(a) Write the conservation law for mass in this system in terms of $\rho = \sum_i \delta(\vec{r} - \vec{r}_i)$. Define the quantities which appear carefully.

(b) Imagine averaging this equation over short time scales as discussed to get a hydrodynamic equation. Here there is only one hydrodynamic quantity, the average density $\langle\rho\rangle(\vec{r},t)$. What is the appropriate way to close the equation, analogous to what was done for a fluid conserving momentum? (Hint: you must expand the current to lowest order in gradients of the hydrodynamic variables and introduce a transport coefficient.) What is the name of the equation you get this way, and what is the name of the transport coefficient?

(c) Now deduce some Kubo relations for this system. Introduce a perturbation

$$H' = -\int \frac{\delta p(\vec{r})}{\rho_0}\rho(\vec{r},t)$$

exactly as for the hydrodyamic case. You may use the relation

$$\frac{2\hbar\omega}{\left(1-\mathrm{e}^{-\beta\hbar\omega)}\right)}\mathrm{Re}(\delta\rho(\vec{q},z\to\omega+\mathrm{i}\epsilon)\langle\rho\rangle/p(\vec{q})) = \langle\rho\rho\rangle_T(\vec{q},\omega)$$

where

$$\langle\rho\rho\rangle_T(\vec{q},\omega) = \int \mathrm{d}(t-t')\int \mathrm{d}(\vec{r}-\vec{r}\,')\mathrm{e}^{\mathrm{i}\omega(t-t')-\mathrm{i}\vec{q}\cdot(\vec{r}-\vec{r}\,')}\langle\rho(\vec{r},t)\rho(\vec{r}\,',t')\rangle_T$$

as in (10.34). Use this and your hydrodynamic equation to find an expression for $\langle\rho\rho\rangle_T(\vec{q},\omega)$ in terms of $\omega$, $\beta$, the equilibrium density, $\vec{q}$, an equilibrium thermodynamic derivative and your transport coefficient. (There should be no reference to $\delta p$ in the answer.)

(d) Use the answer to (c) to find two different expressions for the transport coefficient in terms of $\langle\rho\rho\rangle_T(\vec{q},\omega)$ and the other quantities using the limits $\lim_{\omega\to 0}\lim_{q\to 0}$.

10.2 Linearize the equations of superfluid hydrodynamics and solve them in the case of zero viscosities. Show that there are two propagating, sound like modes. These are called first and second sound.

10.3 Use the Gross–Pitaevskii–Bogoliubov theory for a weakly interacting Bose gas, as described in Chapter 5, to calculate the one particle density matrix, thus identifying the condensate wave function in this case (at zero temperature).

# 11

# Stochastic models and dynamical critical phenomena

## General discussion of stochastic models

In the hydrodynamic theories, one writes equations which describe the system at hand on long time and length scales, using local conservation laws and the presence of broken symmetries to determine the identity of these slowly varying quantities. However, one would like a way to take account of the effects of the more rapidly varying degrees of freedom in such theories. Implicitly, these more rapid degrees of freedom are present in the integrals which determine the transport coefficients in the hydrodynamic equations by way of Kubo relations, but no method is presented to calculate the required integrands. One way to take account of the other degrees of freedom is to take on the entire many body problem all the way down to the atomic or electronic level, as one does in molecular dynamics simulations of various sorts. However, it is useful to have some approximate analytical ways to attack this problem as well. Here we review the basis for the most common such approach, the Langevin equation. For most of this discussion, we will assume that the identity of the slow variable or variables is known and that the separation of time scales is extreme so that the faster variables are essentially instantaneous in a sense we will discuss. These assumptions are not often particularly well justified. However, the resulting formulation has yielded very useful insights and is an important part of the subject. Even after these assumptions, the Langevin equation (and its relatives) is not particularly simple to solve because it contains a stochastic noise term.

## Generalized Langevin equation

We suppose that we know that a set of selected variables, generically labelled $\psi_i(t)$, is known to be "slow." We suppose that these can be expressed in terms of the coordinates and momenta which are used to write down the Hamiltonian or Lagrangian of the system. For example, these variables could be the hydrodynamic quantities $\rho(\vec{r},t)$, $\vec{v}(\vec{r},t)$ and $e(\vec{r},t)$ which enter the Navier–Stokes equations. However, here we ignore the spatial dependence which can be taken into account later by use of the index $i$. In a system in which quantum mechanical effects are important at low frequencies, the $\psi_i(t)$ could be treated as operators without changing very much of what follows. However, we will assume that they are classical variables. In general, if one knows a correct model for the system, expressed as a Hamiltonian or Lagrangian, one can write down equations for the time derivatives of these selected variables. However, the expressions obtained for the time derivatives will be very complicated functions of the original coordinates and momenta and will not involve only the $\psi_i$ variables originally selected. (Unless, of course, one has selected the entire collection of coordinates and momenta to be the $\psi$. But then their dynamics will not be "slow" in any sense and one might as well do a molecular dynamics simulation.) In fact we carried out the process of finding the time derivatives of the hydrodynamic variables in the last chapter and found expressions which did not "close." That is, the time derivatives involved quantities which were not expressible in terms of the hydrodynamic variables themselves. In deriving the Navier–Stokes equations we closed the equations by assuming that unknown currents could be expressed in terms of gradients of densities in a way which introduced phenomenological transport coefficients. To go beyond that assumption here, we suppose that the time derivatives can be written in the form

$$\frac{\partial \psi_i}{\partial t} = -\sum_j \int_0^t \mathrm{d}\bar{t}\; K_{ij}(t-\bar{t})\psi_j(\bar{t}) + f_i(t) \tag{11.1}$$

This is equivalent to separating off a term of the form shown in the first part of the right hand side from the time derivative. This is supposed to be the "slow" part. It is assumed to be linear in the $\psi_j$ but the possibility of a time delay is taken into account. The factor $K_{ij}(t-\bar{t})$ is called a "memory function" since it is interpreted to mean that $\partial\psi_i/\partial t$ is affected by values of $\psi_j$ in the "past" ($\bar{t} < t$). The second term on the right, $f_i(t)$, is called the "noise." It is assumed (and to some extent can be shown) to be "fast." That is, its dynamics will involve frequencies (much) higher than those resulting from the first term alone.

To proceed further we suppose that our system is in local equilibrium, so that one can define an average, termed a "thermal average" and denoted $\langle \ldots \rangle$, in which one averages the time dependent quantities over times which are short compared to the long periods of all the motions of interest. So far, the functions $K_{ij}(t-\bar{t})$ and $f_i(t)$ are quite arbitrary but one finds some useful constraints on them if one supposes that

$$\langle \psi_i(t=0) f_j(t) \rangle = 0 \tag{11.2}$$

for all positive $t$. One says that the noise is not correlated with the (initial) values of the $\psi_i$. With these constraints one can show quite generally that the correlation function of the "noise" is related to the "memory function":

$$\langle f_i(t) f_j(t')\rangle = \sum_l K_{il}(t-t')\langle \psi_l \psi_j\rangle \tag{11.3}$$

Furthermore if one separates off a kind of "mean field" term from $K_{il}(t-t')$

$$K_{il}(t-t') = \delta(t-t')\sum_j \left\langle \psi_j \frac{\partial \psi_i}{\partial t}\right\rangle (\langle \psi_j \psi_l\rangle)^{-1} + K_{il}^{(\mathrm{d})}(t-t') \equiv K_{il}^{(\mathrm{s})} + K_{il}^{(\mathrm{d})}(t-t') \tag{11.4}$$

then only $K_{il}^{(\mathrm{d})}(t-t')$ contributes to (11.3):

$$\langle f_i(t) f_j(t')\rangle = \sum_l K_{il}^{(\mathrm{d})}(t-t')\langle \psi_l \psi_j\rangle \tag{11.5}$$

We provide a sketch of the proof of these results (more details appear in references 1–3). One defines a Laplace transform $F(z)$ of any function $F(t)$ as

$$F(z) = -\mathrm{i}\int_0^\infty \mathrm{e}^{\mathrm{i}zt} F(t)\,\mathrm{d}t \tag{11.6}$$

(where $z$ has a small positive imaginary part). Then the Laplace transform of (11.1) is

$$z\psi_i(z) - \sum_j K_{ij}(z)\psi_j(z) = \psi_i(t=0) + \mathrm{i} f_i(z) \tag{11.7}$$

(We have made use of the assumption that $\psi_j(t)$ varies slowly to omit a small part of the second term.) Multiply this by $\psi_l(t=0) \equiv \psi_l$ and average using (11.2):

$$\sum_j (z\delta_{i,j} - K_{ij}(z))\langle \psi_l \psi_j(z)\rangle = \langle \psi_l \psi_i\rangle \tag{11.8}$$

It is useful to define

$$C_{ij}(t-t') = \Theta(t-t')\langle \psi_j(t)\psi_i(t')\rangle \tag{11.9}$$

where $\Theta(t-t')$ is the Heaviside function (equal to 1 when $t > t'$ and 0 otherwise). The Fourier transform of $C_{ij}(t-t')$ defined as

$$C_{ij}(\omega) = \int_\infty^\infty \mathrm{e}^{\mathrm{i}\omega(t-t')} C_{ij}(t-t')\,\mathrm{d}(t-t') \tag{11.10}$$

(note that only half the interval contributes) is related to the quantity $\langle \psi_i(z_1)\psi_i(z_2)\rangle$ by the relation

$$\langle \psi_i(z_1)\psi_j(z_2)\rangle = \int_{-\infty}^{+\infty} \frac{\mathrm{d}\omega}{2\pi} \frac{C_{ij}(\omega)}{(z_1-\omega)(z_2+\omega)} \tag{11.11}$$

which is in turn given in terms of the Laplace transforms of $C_{ij}(t-t')$ as

$$\langle\psi_i(z_1)\psi_j(z_2)\rangle = \frac{1}{z_1+z_2}(C_{ij}(z_1)+C_{ji}(z_2)) \tag{11.12}$$

Now consider the correlation function of the Laplace transform of the noise. Using (11.7) to eliminate the $f$ and (11.8) several times one gets

$$\langle f_i(z_1)f_j(z_2)\rangle = -\sum_{l,k}(z_1\delta_{il}-K_{il}(z_1))(z_2\delta_{jk}-K_{jk}(z_2))\langle\psi_l(z_1)\psi_k(z_2)\rangle + \langle\psi_i\psi_j\rangle \tag{11.13}$$

or making use again of (11.12) and (11.8) and using $C_{ij}(z)=\langle\psi_j\psi_i(z)\rangle$

$$\langle f_i(z_1)f_j(z_2)\rangle = -\frac{1}{z_1+z_2}\sum_k(K_{jk}(z_1)\langle\psi_k\psi_i\rangle + K_{ik}(z_2)\langle\psi_k\psi_j\rangle) \tag{11.14}$$

Inverting the Laplace transform gives (11.3). To show that only $K_{il}^{(\mathrm{d})}(t-t')$, and not $K_{il}^{(\mathrm{s})}(t-t')$, contributes to (11.3), one can use (11.14) to write the contribution of $K_{il}^{(\mathrm{s})}(t-t')$ by inserting the Laplace transform of $K_{il}^{(\mathrm{s})}(t-t')$, which is just a constant equal to $\sum_j\langle\psi_j\partial\psi_i/\partial t\rangle\langle\psi_j\psi_k\rangle^{-1}$ giving for the contribution of $K_{il}^{(\mathrm{s})}(t-t')$ to the sum in (11.14)

$$\langle\psi_i\partial\psi_j/\partial t\rangle + \langle(\partial\psi_i/\partial t)\psi_j\rangle \tag{11.15}$$

But it is quite easy to show that this is zero using the time reversal properties of $\langle\psi_i(t)\psi_j(t')\rangle$ (Problem 11.1).

Now consider the limiting case, which is the most common one studied, in which the time scale of the noise is much shorter than the time scales of interest. Then one often approximates

$$\langle f_i(t)f_j(t')\rangle = \delta(t-t')\Gamma\delta_{ij} \tag{11.16}$$

(More generally this does not have to be diagonal in $i$, $j$.) We invert (11.5) to give

$$K_{ij}^{(\mathrm{d})}(t-\bar{t}) = \delta(t-\bar{t})\Gamma\langle\psi_i\psi_j\rangle^{-1} \tag{11.17}$$

in which $\langle\psi_i\psi_j\rangle^{-1}$ is the $ij$ element of the inverse of the matrix $\langle\psi_i\psi_j\rangle$. Then the Langevin equation in this "Markov" approximation becomes using (11.4)

$$\frac{\partial\psi_i}{\partial t} = -\sum_j K_{ij}^{(\mathrm{s})}\psi_j(t) - \sum_j\Gamma\langle\psi_i\psi_j\rangle^{-1}\psi_j(t) + f_i(t) \tag{11.18}$$

in which the weight of the noise is related to the second term on the right by (11.16). In practice one can arrange, for example, that the terms involving $K_{ij}^{(\mathrm{s})}$ give, in the hydrodynamic case, the (linearized) Navier–Stokes equations. Then this prescription permits one to extend the Navier–Stokes equations to take account of high frequency noise, providing that one remembers that to do this consistently; one must also include a damping term (the second one on the right in the last equation) which is determined by the weight of the noise and the static correlation functions $\langle\psi_i\psi_j\rangle$ of the fluid.

## General discussion of dynamical critical phenomena

At critical points, the fluctuations of the properties of the system become large at long wavelengths, as discussed in Chapter 9. However, the scale of the fluctuations becomes large not only in space but also in time, that is they "slow down." To understand this critical "slowing down" we start with a hydrodyamic theory describing conservation laws plus equations for the dynamics of the fluctuations of the quantity, such as magnetization or phase, which characterizes the broken symmetry of the ordered phase below the critical point. However, it turns out to be essential also to take into account the effects of the dynamics of the other degrees of freedom in the system, which have characteristically faster dynamics. From the preceding discussion, we saw that including the "noise" associated with these other degrees of freedom also necessarily, and for consistency, leads to additional dissipative terms which can affect the slow dynamics and whose magnitude is related to the magnitude of the noise. Because the separation of time scales just discussed is very large near critical points, the Langevin theory, which assumed such a time separation, is a particularly good place to start to understand dynamical critical phenomena. To be a little more specific, one can refer to equation (11.19), specialized to just one variable $\psi$:

$$\frac{\partial\psi}{\partial t} = -K^{(s)}\psi(t) - \Gamma\langle\psi\psi\rangle^{-1}\psi(t) + f(t) \tag{11.19}$$

where $\Gamma$ is the weight of the noise. Now suppose for definiteness that $\psi$ is the deviation of the magnetization of a magnet with no local conservation laws from its equilibrium value $\psi = M - \langle M\rangle \equiv \delta M$. (If there are no local conservation laws, then at low frequencies, the "mean field" term (also known as the Poisson bracket term) gives no contribution.) Consider the susceptibility $\chi$ of such a magnet:

$$\chi = \frac{\partial\langle M\rangle}{\partial H}_T = \frac{\partial^2 F}{\partial H^2} = \frac{\langle(\delta M)^2\rangle}{k_B T} \tag{11.20}$$

(This is a very simple form of the fluctuation–dissipation theorem.) But near the critical point $\chi = C/|T - T_c|^\gamma$ so the second term on the right hand side of (11.19) is $(-\Gamma|T - T_c|^\gamma/C)\delta M$ which gets very small (that is slow) as $T \to T_c$. Thus we have critical slowing down from this term. Basically, this little example has the elements of a theory due to Van Hove of critical slowing down. The theory does account for the qualitative fact that slowing down occurs but quantitatively it does not give the right exponents to describe the reduction in time scales as a function of the distance in thermodynamic parameter space (e.g. temperature or field) from the critical point. In fact it turns out that the critical fluctuations can cause $\Gamma$ to be significantly temperature dependent and to diverge at $T_c$ in some models. To go further, one needs a more detailed theory which incorporates hydrodynamic degrees of freedom, as well as the order parameter dynamics. However, before embarking on that discussion we need to discuss the phenomenological description of dynamical critical phenomena which, like the static critical phenomena, can be based on a scaling hypothesis.

**Scaling description of dynamical critical phenomena** To characterize dynamical behavior associated with the large, long wavelength fluctuations which occur near critical points, it is convenient to think about dynamical correlation functions. These can be defined for many variables, though we will be primarily interested in the variable which acquires a value which breaks a global symmetry of the Hamiltonian by acquiring a finite value and thus drastically reducing phase space over which the density matrix is nonzero below $T_c$. This variable is often called the "order parameter." Examples are the magnetization in magnetic models and the density in the gas–liquid transition. (More specifically for the gas–liquid transition it is the difference between the density of the gas phase and the liquid phase when the pressure is fixed along the coexistence line. The subtleties associated with identifying separate phases in a system with which we are not already quite familiar were briefly discussed in Chapter 9.) In the last chapter, we described the calculation of such a dynamical correlation function ($\langle\rho\rho\rangle(\vec{q},\omega)$) for a fluid in the limit of long wavelengths and low frequencies, and using the hydrodynamic Navier–Stokes equations. We argued there that at long wavelengths the assumptions that led to the hydrodynamic theory were justified. The essential assumption is that by averaging properties over regions of size which are small compared to the wavelength considered, one can describe the average local properties with the assumption of local thermodynamic equilibrium. This assumption was used to justify the use of local thermodynamic relations in the derivation of the Navier–Stokes equations. Near critical points, these assumptions remain valid only if the observations are confined to larger and larger length scales, because equilibrium fluctuations associated with longer and longer wavelengths become important. Thus it is because the closure of the hydrodynamic equations required the repeated use of the assumption of local thermodynamic equilibrium (though the fundamental conservation laws which were the starting point of the hydrodynamic theory were obtained from exact operator identities), that the region of validity of hydrodynamics narrows near critical points. Near critical points, we have seen that there are large amplitude fluctuations (in the order parameter but sometimes also in other variables coupled

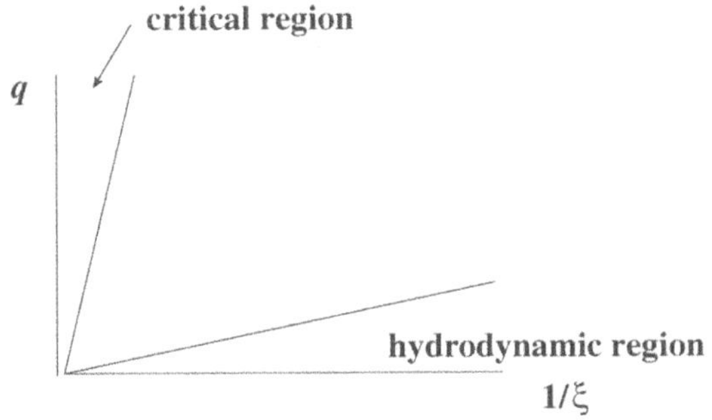


Figure 11.1 Illustration of the hydrodynamic and critical regions.

to it) over very large length scales. In order to describe the thermodynamic state near a critical point, we must therefore choose a region of the system which is large compared to these fluctuations, which means that the volume of the regions over which we average the conservation equations to obtain the hydrodynamic theory must be larger than the coherence length which characterizes the fluctuations. If we do not choose it that large, we still get the conservation equations but we cannot use thermodynamic relations to close them because the regions are not in local thermal equilibrium. Thus the hydrodynamic equations only make sense if the wave vector $q$ characterizing the scale of the observation is much smaller than the inverse $1/\xi$ of the coherence length $\xi$ (Figure 11.1). At the critical point itself, $\xi \to \infty$ and the region of hydrodynamically allowed $q$ shrinks to zero: hydrodynamics becomes useless.

To obtain a phenomenological characterization of how the dynamics evolves as one leaves the hydrodynamic regime, one needs a characterization of the dominant frequency scale of the fluctuations. This may be done in a variety of essentially equivalent ways. Here we follow Halperin and Hohenberg[4] and rewrite the correlation functions of form $\langle\psi\psi\rangle(\vec{q},\omega)$ as

$$\langle\psi\psi\rangle(\vec{q},\omega) = \frac{2\pi}{\omega_\psi(\vec{q})}\langle\psi\psi\rangle(\vec{q})f(\omega/\omega_\psi(\vec{q})) \tag{11.21}$$

where

$$\langle\psi\psi\rangle(\vec{q}) = \int \mathrm{d}\vec{r}\,\exp(-\mathrm{i}\vec{q}\cdot\vec{r})\langle\psi(\vec{r},t=0)\psi(\vec{r}=0,t=0)\rangle \tag{11.22}$$

is related to

$$\langle\psi\psi\rangle(\vec{q},\omega) = \int \mathrm{d}\vec{r}\,\mathrm{d}t\,\mathrm{e}^{-\mathrm{i}\vec{q}\cdot\vec{r}+\mathrm{i}\omega t}\langle\psi(\vec{r},t)\psi(\vec{r}=0,t=0)\rangle \tag{11.23}$$

by the relation

$$\int \frac{\mathrm{d}\omega}{2\pi} \langle \psi\psi \rangle(\vec{q}, \omega) = \langle \psi\psi \rangle(\vec{q}) \tag{11.24}$$

The integral on $\omega$ is over the interval $-\infty < \omega < \infty$. Then with the form (11.21) one has for the function $f$ the equivalent relation

$$\int_{-\infty}^{\infty} \mathrm{d}x \; f(x) = 1 \tag{11.25}$$

Now the frequency $\omega_\psi(\vec{q})$ in (11.21) is fixed by requiring that

$$\int_{-1}^{1} f(x) \, \mathrm{d}x = 1/2 \tag{11.26}$$

This is easily shown to be a sensible (though not unique) way of characterizing the frequency of the fluctuations near a critical point. It is better than using a frequency moment of the correlation function for this purpose, because for some models, the most natural moments to choose turn out to be divergent.

Now if $\psi$ represents the fluctuations of the order parameter (magnetization for a ferromagnet above the Curie point, magnetization minus the average magnetization below the Curie point, for example) then the scaling form for $\langle \psi(\vec{r}, t = 0)\psi(\vec{r} = 0, t = 0) \rangle = g(\vec{r})$ was discussed in Chapter 9 and can be written in the form

$$g(r) = r^{-(d-2)+\eta} F(r/\xi) \tag{11.27}$$

or for the Fourier transform

$$S(q) = \int \mathrm{d}\vec{r} \exp(-\mathrm{i}\vec{q} \cdot \vec{r}) g(r) \tag{11.28}$$

the scaling form is

$$S(q) = q^{\eta - 2} \mathcal{F}(q\xi) \tag{11.29}$$

We have argued that the dynamics of the critical fluctuations will also depend on $q\xi$. Accordingly, Halperin and Hohenberg postulated the scaling relation, analogous to (11.29)

$$\omega_\psi(q) = q^z \Omega(q\xi) \tag{11.30}$$

The function $\Omega(q\xi)$ can be expected to take different forms in the hydrodynamic and critical regions shown in Figure 11.1, without affecting the validity of (11.30). One new exponent, the dynamical critical exponent $z$, has been introduced.

For example, we can use this hypothesis to draw conclusions about the critical dynamics at the gas–liquid critical point. From the density–density correlation function $S(q, \omega)$ for a classical hydrodynamic fluid which was worked out in Chapter 9 we found two modes, a sound mode and a zero frequency overdamped mode corresponding to entropy fluctuations in the fluid. Near the critical point, the entropy fluctuation mode associated with heat diffusion dominates the critical behavior. Its damping rate is

$$\omega_{q_1}(q) = D_T q^2 \tag{11.31}$$

and represents the rate of decay of fluctuations in the heat, defined (equation (10.23)) as $q_1 = e_1 - (e_0 + p_0)(\rho_1/\rho_0)$. Here $D_T = \Lambda/C_P$ has the generic form of a transport coefficient divided by a static susceptibility as mentioned earlier. $C_P \propto \epsilon^{-\gamma}$ near the critical point. We must, as discussed earlier, consider the possibility that the transport coefficient $\Lambda$ also may diverge at the critical point. Assuming $\Lambda \propto \epsilon^{-l}$ one has

$$\omega_\rho(q) \propto \epsilon^{-l+\gamma} q^2 \propto \xi^{(l-\gamma)/\nu} q^2 \propto q^{2-(l-\gamma)/\nu} (q\xi)^{(l-\gamma)/\nu} \tag{11.32}$$

This means that the dynamical exponent $z$ is

$$z = 2 - (l - \gamma)/\nu \tag{11.33}$$

In the critical dynamics region $q\xi \gg 1$ the function $\Omega(q\xi)$ must approach a finite limit: otherwise, the dynamical frequency will become either zero or infinity independent of $q$ in this region. Thus the scaling hypothesis predicts

$$\omega_\rho(q) = \Omega(\infty) q^{2-(l-\gamma)/\nu} \tag{11.34}$$

when $q\xi \gg 1$ near the critical point. This $q$ dependence can be checked experimentally if $l$ is known. However, these relations do not give the value of $l$ or, equivalently within the dynamical scaling hypothesis, of $z$. For that, an extension of the renormalization group approach to the case of dynamical response functions is required. For applications of the dynamical scaling hypothesis to ferromagnets and antiferromagnets near their critical points, see reference 4.

Note that, in the Van Hove theory, the dynamical exponent $z$ for the gas–liquid case would be $z = 2 + \gamma/\nu = 4 - \eta$, that is if $l$, characterizing the hypothesized singularity in the transport coefficient (the thermal conductivity in the gas–liquid case), were zero. This is characteristic of systems in which the order parameter is conserved but the hydrodynamic mode is overdamped, so that there is a leading $q^2$ in the characteristic frequency (arising basically from the diffusive form which the equation of motion of the order parameter takes in those cases). In the case that the order parameter is not conserved, there is no leading 2 and the Van Hove prediction for the dynamical exponent is $z = \gamma/\nu = 2 - \eta$. A third possibility is that the lowest frequency hydrodyamic mode is an underdamped sound like mode. This occurs in an isotropic antiferromagnet. Then the dynamical exponent takes the general form $z = 1 - l/\nu + \gamma/\nu$. In the antiferromagnet the sound velocity goes to zero as $\xi^{-1/2}$ and the transverse susceptibility has no singularity so that $l/\nu = -1/2$, $\gamma = 0$ and $z = 3/2$. The Van Hove theory would give $z = 1$ for the antiferromagnet, not close to the known answer. However in the latter case, $z$ is fixed by the form of the hydrodynamic theory and elaborate calculations (briefly discussed below) are not required to obtain it.

Experimentally one finds that the Van Hove predictions are not always well satisfied. Van Hove theory works well for many magnetic models (though not for antiferromagnets as mentioned above). However, for the gas–liquid transition, $l \approx \nu$ and $z \approx 3 - \eta$. The thermal conductivity diverges approximately as $\xi$ near the critical point in three dimensions. To account for such differences one must consider the calculation of the transport coefficient near the critical point more carefully. Though the general notion that the dynamical critical exponent does not depend on the microscopic details of the system still holds ("universality") one can see that, for a given set of static critical exponents characterizing a "universality class" with respect to these static exponents, there can be more than one result for the dynamical exponent $z$. Hence the number of "universality classes" proliferates for dynamical properties and they are not particularly "universal" in the ordinary sense of the word. (A review of a large set of models for dynamical critical properties is given by Hohenberg and Halperin, where some commonly used names are assigned to them.[5])

Here we will confine attention to the simplest of these models, which has the same static critical behavior as the Landau–Ginzburg model for the Ising model which was discussed in Chapter 9. (This is "Model A" with $n = 1$ in reference 5.) The models are formulated as Langevin equations for the relevant variables. Though, in principle, the Langevin equations could be derived from microscopic Hamiltonians, along the lines described at the beginning of this chapter, in practice they have been postulated to be correct. Generally one requires that (1) the static critical behavior be consistent with the known critical behavior of the system of interest and that (2) the model reduces in the hydrodynamic regime to the known hydrodynamic behavior. The model we consider is described by the Langevin equation

$$\frac{\partial m(\vec{r}, t)}{\partial t} = -\Gamma_0 \frac{\delta \mathcal{F}}{\delta m} + f(\vec{x}, t) \tag{11.35}$$

in which

$$\beta \mathcal{F} = \int d\vec{r} (| \nabla_{\vec{r}} \bar{m}(\vec{r}\,') |^2 + r\bar{m}(\vec{r})^2 + b\bar{m}^4) \tag{11.36}$$

is exactly the same as the free energy functional defined in equation (9.99) (and with the same dimensionless choice of units) except that to avoid confusion between the temperature variable and the time we have used the variable $r$ in place of the variable $t$ in (9.99). Also the coupling constant $\tilde{b}$ of (9.99) is denoted $b$ here and the variable $\tilde{m}$ is denoted $\bar{m}$. One can see that (11.35) is nearly of the form (11.18) with the discrete index $i$ in (11.18) replaced by the continuous position label $\vec{r}$. However, because of the term proportional to $\tilde{b}$ in $\beta\mathcal{F}$, (11.35) is *nonlinear* unlike (11.18) and this makes it significantly more difficult to solve, as well as to justify. Mazenko[1] has shown that one can obtain a derivation of such equations by use of a generalized Fokker–Planck equation. Fokker–Planck equations have content closely related to that described by Langevin equations but are expressed in terms of the probability distribution $P(\{m\})$ of the "slow" or macroscopic variables $\{m\}$. To be specific, in the case of the Ising like magnet, we identify

$$P(\{\bar{m}_q\}, t) = \Pi_q \delta(\bar{m}_q - m_q(t)) \tag{11.37}$$

where $\{\bar{m}_q\}$ is a set of numbers and $\{m_q(t)\}$ are the corresponding values which the long wavelength components of the magnetization take at a time $t$. One obtains a probability distribution from $P(\{\bar{m}_q\}, t)$ by averaging on the time (equivalent to an equilibrium average):

$$\langle P(\{\bar{m}\}, t)\rangle = \langle \Pi_q \delta(\bar{m}_q - m_q(t))\rangle = \mathrm{e}^{-\beta\mathcal{F}(\{\bar{m}\})} \tag{11.38}$$

$-\beta\mathcal{F}(\{\bar{m}\})$ can be identified with the free energy functional above, as suggested by the notation. (Here and in the sequel we sometimes abbreviate $\{\bar{m}_{q<}\}$, which is the set of all the small wave number variables $\bar{m}_{\vec{q}}$, by $\bar{m}$. The product $\Pi_q$ will always extend only over the long wavelength degrees of freedom in the set $\{\bar{m}_{q<}\}$.) Now Mazenko basically postulates that the equation of motion for $P(\{\bar{m}_q\}, t)$ is of a form similar to the equation of motion for $\psi_i$ as described in the first section of this chapter. It is

$$\frac{\partial P(\bar{m}, t)}{\partial t} = -\sum_{q,q'} \frac{\partial}{\partial \bar{m}_q}\left\{\left[V_q \delta_{q,q'} - \Gamma(q, q')\left(\frac{\partial}{\partial \bar{m}_{q'}} + \beta\frac{\partial\mathcal{F}}{\partial \bar{m}_{q'}}\right)\right] P(\bar{m}, t)\right\} + R(\bar{m}, t) \tag{11.39}$$

where

$$V_q = -\sum_{q'}\left[\frac{\partial}{\partial \bar{m}_{q'}} Q_{q,q'} - Q_{q,q'}\beta\frac{\partial\mathcal{F}}{\partial \bar{m}_{q'}}\right] \tag{11.40}$$

and

$$Q_{q,q'} = Zk_{\mathrm{B}}T\langle P(\bar{m}, t)\{m_q, m_{q'}\}\rangle \mathrm{e}^{\beta\mathcal{F}(\bar{m})} \tag{11.41}$$

($Z$ is the partition function, $\{m_q, m_{q'}\}$ is a Poisson bracket.) Equation (11.39) has a form closely related to a Fokker–Planck equation.[6] Notice that the function $P(\bar{m}, t)$

does not involve any averaging and is intrinsically very singular as a function of its arguments. The equation of motion (11.39) is only useful for producing equations for moments of it with respect to powers of the $\bar{m}_q$, and we will only discuss the equation resulting from taking the first moment. The equation for $P$ is linear in $P$ but the corresponding equations for the slow variables $m_q$ will be nonlinear.

Here we reproduce elements of a derivation of (11.39) given by Mazenko (see also reference 7). Some of the assumptions made can only be justified a posteriori (if at all) by reference to the reasonableness of the resulting equations obtained by taking moments of (11.39). The basic steps are similar to those used in deriving the Langevin equation at the beginning of this chapter. Write the time derivative of $P(\{\bar{m}_q\}, t)$ in terms of a term involving a memory function plus a noise term:

$$\frac{\partial P(\{\bar{m}\}, t)}{\partial t} = \int_0^t \mathrm{d}t' \int \mathcal{D}(\{\bar{m}'\}) K(\bar{m}, \bar{m}', t - t') P(\bar{m}', t') = R(t)$$

Separate a mean field like term from the memory function. This mean field term is assumed to be of the form

$$K^{(\mathrm{s})}(\bar{m}, \bar{m}', t - t') = \delta(t - t') \int \mathcal{D}(\bar{m}'') \left\langle P(\bar{m}'') \frac{\partial P(\bar{m})}{\partial t} \right\rangle \langle P(\bar{m}'') P(\bar{m}) \rangle^{-1} \tag{11.42}$$

Assuming that $\langle P(\bar{m}, t = 0) R(\bar{m}, t) \rangle = 0$ as before gives

$$\langle R(\bar{m}, t) R(\bar{m}', t') \rangle = \int \mathcal{D}(\bar{m}'') K^{(\mathrm{d})}(\bar{m}, \bar{m}'', t - t') \langle P(\bar{m}'') P(\bar{m}') \rangle \tag{11.43}$$

where $K^{(\mathrm{d})}(\bar{m}, \bar{m}'', t - t') = K(\bar{m}, \bar{m}', t - t') - K^{(\mathrm{s})}(\bar{m}, \bar{m}'', t - t')$. In (11.42), $\langle P(\bar{m}'') P(\bar{m}) \rangle^{-1}$ is the inverse of the matrix $\langle P(\bar{m}'') P(\bar{m}) \rangle$ with dimension which is the square of the number of low wave vector variables. The matrix

$$\begin{aligned} \langle P(\bar{m}'') P(\bar{m}) \rangle &= \int \mathcal{D}(m_>) \int \mathcal{D}(m_<) \Pi_q \delta(\bar{m}''_q - m_q) \Pi_q \delta(\bar{m}_q - m_q) \mathrm{e}^{-\beta H} \\ &= \mathrm{e}^{-\beta \mathcal{F}(\bar{m})} \Pi_q \delta(\bar{m}''_q - \bar{m}_q)/Z \end{aligned}$$

where $Z$ is the partition function. (Notice that, to take thermal averages, we are integrating over all the variables $m_q$, slow and fast, but that they must be treated differently.) Thus the inverse is

$$\langle P(\bar{m}'') P(\bar{m}) \rangle^{-1} = Z\, \mathrm{e}^{+\beta \mathcal{F}(\bar{m})} \Pi_q \delta(\bar{m}''_q - \bar{m}_q) \tag{11.44}$$

The other factor in $K^{(\mathrm{s})}(\bar{m}, \bar{m}', t - t')$ is $\langle P(\bar{m}'') \partial P(\bar{m}) / \partial t \rangle$. It is rewritten as

$$\frac{\partial P(\bar{m})}{\partial t} = -\sum_{q<} \frac{\partial m_q}{\partial t} \frac{\partial P(\bar{m})}{\partial \bar{m}_q}$$

Therefore

$$\left\langle P(\bar{m}'')\frac{\partial P(\bar{m})}{\partial t}\right\rangle = -\sum_{q<}\left\langle P(\bar{m}'')\frac{\partial m_q}{\partial t}\frac{\partial P(\bar{m})}{\partial \bar{m}_q}\right\rangle$$
$$= -\sum_{q<}\frac{\partial}{\partial \bar{m}_q}\left[\left\langle \Pi_{q''}\delta(\bar{m}''_{q''}-m_{q''})\frac{\partial m_q}{\partial t}\Pi_{q'}\delta(\bar{m}_{q'}-m_{q'})\right\rangle\right]$$
$$= -\sum_{q<}\frac{\partial}{\partial \bar{m}_q}\left[(\Pi_{q''}\delta(\bar{m''}_{q''}-\bar{m}_{q''})\left\langle\frac{\partial m_q}{\partial t}P(\bar{m})\right\rangle\right] \tag{11.45}$$

Now one evaluates the time derivative by use of the Poisson bracket

$$\frac{\partial m_q}{\partial t} = \{H, m_q\} = \sum_{\lambda}\left[\frac{\partial H}{\partial p_\lambda}\frac{\partial m_q}{\partial q_\lambda} - \frac{\partial H}{\partial q_\lambda}\frac{\partial m_q}{\partial p_\lambda}\right]$$

where $q_\lambda$, $p_\lambda$ are canonical coordinates and momenta for the system. We assume that the Hamiltonian depends only on a complete set of the $m_{\vec{q}}$ (not just some low wave vector subset). Then

$$\sum_{\lambda}\left[\frac{\partial H}{\partial p_\lambda}\frac{\partial m_q}{\partial q_\lambda} - \frac{\partial H}{\partial q_\lambda}\frac{\partial m_q}{\partial p_\lambda}\right] = \sum_{q''}\frac{\partial H}{\partial m_{q''}}\sum_{\lambda}\left[\frac{\partial m_{q''}}{\partial p_\lambda}\frac{\partial m_q}{\partial q_\lambda} - \frac{\partial m_{q''}}{\partial q_\lambda}\frac{\partial m_q}{\partial p_\lambda}\right]$$
$$= \sum_{q''}\frac{\partial H}{\partial m_{q''}}\{m_{q''}, m_q\}$$

Therefore the factor $\langle\frac{\partial m_q}{\partial t}P(\bar{m})\rangle$ in the last expression in (11.45) can be written

$$(1/Z)\int \mathcal{D}(\{m\})\Pi_{q'}\delta(\bar{m}_{q'}-m_{q'})\,\mathrm{e}^{-\beta H}\sum_{q''}\frac{\partial H}{\partial m_{q''}}\{m_{q''}, m_q\}$$
$$= (-1/(\beta Z))\int \mathcal{D}(\{m\})\sum_{q''<}\frac{\partial}{\partial \bar{m}_{q''}}\Pi_{q'}\delta(\bar{m}_{q'}-m_{q'})\{m_{q''}, m_q\}\,\mathrm{e}^{-\beta H}$$
$$= -k_{\mathrm{B}}T\sum_{q''<}\frac{\partial}{\partial \bar{m}_{q''}}\langle P(\bar{m})\{m_{q''}, m_q\}\rangle$$

(The integration by parts involved three changes of sign. These integrals $\int\mathcal{D}(\{m\})$ are on *all* the microscopic (unbarred) variables $m_q$.) Thus finally

$$\left\langle P(\bar{m}'')\frac{\partial P(\bar{m})}{\partial t}\right\rangle = k_{\mathrm{B}}T\sum_{q<}\frac{\partial}{\partial \bar{m}_q}\left[\Pi_{q'}\delta(\bar{m}''_{q'}-\bar{m}_{q'})\sum_{q''<}\frac{\partial}{\partial \bar{m}_{q''}}\langle P(\bar{m})\{m_{q''}, m_q\}\rangle\right] \tag{11.46}$$

Now combine (11.42), (11.44) and (11.46):

$$K^{(s)}(\bar{m}, \bar{m}', t-t') = \delta(t-t') \int \mathcal{D}(\bar{m}'') k_B T \sum_{q<} \frac{\partial}{\partial \bar{m}_q} \times \left[ \Pi_{q'} \delta(\bar{m}''_{q'} - \bar{m}_{q'}) \sum_{q''<} \frac{\partial}{\partial \bar{m}_{q''}} \langle P(\bar{m}) \{m_{q''}, m_q\} \rangle \right] \times Z\, \mathrm{e}^{+\beta \mathcal{F}(m_{\bar{q}<})} \Pi_q \delta(\bar{m}''_q - \bar{m}'_q) \tag{11.47}$$

Insert the definition (11.41):

$$\begin{aligned} K^{(s)}(\bar{m}, \bar{m}', t-t') &= \delta(t-t') \int \mathcal{D}(\bar{m}'') k_B T \sum_{q<} \frac{\partial}{\partial \bar{m}_q} \left[ \Pi_{q'} \delta(\bar{m}''_{q'} - \bar{m}_{q''}) \sum_{q''<} \frac{\partial}{\partial \bar{m}_{q''}} (\beta\, \mathrm{e}^{-\beta \mathcal{F}(\bar{m})} Q_{q'',q}/Z) \right] \\ &\quad \times Z\, \mathrm{e}^{+\beta \mathcal{F}(\bar{m}_q)} \Pi q \delta(\bar{m}''_q - \bar{m}'_q) \\ &= \delta(t-t') \int \mathcal{D}(\bar{m}'') \sum_{q<} \frac{\partial}{\partial \bar{m}_q} \\ &\quad \times \left[ \Pi_{q''} \delta(\bar{m}''_{q''} - \bar{m}_{q''}) \sum_{q'<} \frac{\partial}{\partial \bar{m}_{q'}} (Q_{q',q} \mathrm{e}^{-\beta \mathcal{F}(\bar{m})}) \right] \mathrm{e}^{+\beta \mathcal{F}(\bar{m})} \Pi_{q'''} \delta(\bar{m}''_{q'''} - \bar{m}'_{q'''}) \\ &= \delta(t-t') \int \mathcal{D}(\bar{m}'') \sum_{q<} \frac{\partial}{\partial \bar{m}_q} \left[ \Pi_{q''} \delta(\bar{m}''_{q''} - \bar{m}_{q''}) \sum_{q'} \left( \frac{\partial Q_{q',q}}{\partial \bar{m}_{q'}} - \beta \frac{\partial \mathcal{F}(\bar{m})}{\partial \bar{m}_{q'}} Q_{q',q} \right) \right. \\ &\quad \left. \times \Pi_{q'''} \delta(\bar{m}''_{q'''} - \bar{m}'_{q'''}) \right] \\ &= -\delta(t-t') \sum_{q<} \frac{\partial}{\partial \bar{m}_q} [\Pi_{q'} \delta(\bar{m}'_{q'} - \bar{m}_{q'}) V_q] \\ &= -\delta(t-t') \sum_{q,q'} \frac{\partial}{\partial \bar{m}_q} [V_q \delta_{q,q'} P(\bar{m}_q, t)] \end{aligned} \tag{11.48}$$

which is the form given in the first term of the right hand side of (11.39). Notice that this term has the form of the divergence of a current in the multidimensional space defined by the variables $\bar{m}$ with $P$ interpreted as a density and $V$ as a velocity. The expression for $V$ can be evaluated for any choice of Hamiltonian and variables $\bar{m}$. These terms are sometimes called "streaming terms." An equilibrium distribution is unchanged in time by this term, as one can show by proving that

$$\sum_{q,q'} \frac{\partial}{\partial \bar{m}_q} \left[ V_q \delta_{q,q'} \, \mathrm{e}^{-\beta \mathcal{F}} \right] = \sum_q \frac{\partial}{\partial \bar{m}_q} \left[ \sum_{q'} \left( \frac{\partial Q_{q,q'}}{\partial \bar{m}'_q} \mathrm{e}^{-\beta \mathcal{F}} - \beta \frac{\partial \mathcal{F}}{\partial \bar{m}'_q} Q_{q,q'} \, \mathrm{e}^{-\beta \mathcal{F}} \right) \right]$$

$$= \sum_{q,q'} \frac{\partial}{\partial \bar{m}_q} \left[ \frac{\partial Q_{q,q'}}{\partial \bar{m}'_q} \mathrm{e}^{-\beta \mathcal{F}} + \frac{\partial (\mathrm{e}^{-\beta \mathcal{F}})}{\partial \bar{m}'_q} Q_{q,q'} \right]$$

$$= \sum_{q,q'} \frac{\partial^2 (Q_{q,q'} \, \mathrm{e}^{-\beta \mathcal{F}})}{\partial m_q \partial m_{q'}}$$

But the latter is zero because

$$Q_{q,q'} = -Q_{q',q}$$

To choose models taking account of the remaining terms $K^{(\mathrm{d})}$ in the memory function, workers impose this same requirement that an equilibrium distribution be stable. To keep things simple, one usually also assumes that $K^{(\mathrm{d})}$ is local in time. Then writing

$$K^{(\mathrm{d})}(\bar{m}, \bar{m}', t - t') = \delta(t - t') \mathcal{K}^{(\mathrm{d})}(\bar{m}, \bar{m}')$$

we have the requirement

$$\int \mathcal{D}(\bar{m}') \mathcal{K}^{(\mathrm{d})}(\bar{m}, \bar{m}') \, \mathrm{e}^{-\beta \mathcal{F}(\bar{m}')} = 0$$

Choosing the following expression for $\mathcal{K}^{(\mathrm{d})}(\bar{m}, \bar{m}')$ satisfies this condition (it is expressed as a differential operator with respect to the $\bar{m}$ but this can be seen to be a limit of a function of $\bar{m}, \bar{m}'$)

$$\mathcal{K}^{(\mathrm{d})}(\bar{m}, \bar{m}') = -\sum_{q,q'} \Gamma(q, q') \frac{\partial}{\partial \bar{m}_q} \left[ \frac{\partial}{\partial \bar{m}_{q'}} + \frac{\partial \beta \mathcal{F}}{\partial \bar{m}_{q'}} \right] \Pi_{q''} \delta(\bar{m}_{q''} - \bar{m}'_{q''})$$

This form leads to the equation (11.39).

We have emphasized that in (11.39) every term is extremely singular because $P(\bar{m})$ is a product of delta functions. To obtain information about the coarse grained variables we can simply note that if $q$ is within the set of labels for "slow" variables

$$\int \mathcal{D}(\bar{m}) \bar{m}_q P(\bar{m}, t) = \int \mathcal{D}(\bar{m}) \bar{m}_q \delta(\bar{m}_q - m_q(t)) \Pi_{q' \neq q} \delta(\bar{m}_{q'} - m'_q(t)) = m_q(t)$$

That is by averaging over $P$ we get the time dependence of the slow variable. Thus multiplying (11.39) by $\bar{m}_q$ and integrating on all the variables $\bar{m}$ we obtain equations for the slow variables themselves

$$\frac{\partial m_q}{\partial t} = V_q(m) - \sum_{q'} \Gamma(q, q') \frac{\partial \beta \mathcal{F}}{\partial m_{q'}} + f_q(t) \tag{11.49}$$

in which the noise term is

$$f_q(t) = \int \mathcal{D}(\bar{m}) \bar{m}_q R(\bar{m}, t)$$

It is not hard to show using (11.43) that

$$\langle f_q(t) f'_q(t') \rangle = 2\Gamma(q, q')\delta(t - t') \tag{11.50}$$

The equations (11.49) and (11.50) are very similar to the linear Langevin equations (11.19) and (11.5). The important difference is that the terms on the right are nonlinear in the variables $m_q$. Further we have a prescription for calculating their form, though the function $\Gamma(q, q')$ is phenomenological and unknown. (In principle, one could calculate $\Gamma(q, q')$ and check the assumptions about the locality of the memory function in time and in the variables $\bar{m}$ which were made to reach these equations. As far as I know this has not been seriously attempted.)

Finally, we can see how to get the model (11.35) from these considerations. In the case of an Ising model, one can see that $Q_{q,q'}$ must be exactly zero (see (11.41)) because the Poisson brackets between angular momenta are only nonzero for different components of the angular momentum of a given site, and the Hamiltonian only depends on one component. Thus there are no streaming terms for an Ising model. Taking $\Gamma(q, q') = \Gamma_0 \delta_{q,q'}$ we obtain (11.35).

Now let us consider, as an example, how the dynamical critical phenomena can be studied with the model (11.35). First we take the Gaussian model as discussed in Chapter 9. That is, we take $b = 0$ in (11.36). We write the Fourier transform (as defined in (11.23)) of (11.35) in this case:

$$(-i\omega + 2\Gamma_0(q^2 + r))m_q(\omega) = f_q(\omega) \tag{11.51}$$

We multiply this by the corresponding equation for $m_q^*(\omega)$ and take the equilibrium average of the result:

$$\langle m_q^*(\omega) m_q(\omega) \rangle = \langle f_q(\omega)^* f_q(\omega) \rangle / \left(\omega^2 + 4\Gamma_0^2(q^2 + r)^2\right) \tag{11.52}$$

The numerator of the right hand side is just the Fourier transform in time of (11.50) so with our choice of $\Gamma(q, q')$ we obtain

$$\langle m_q^*(\omega) m_q(\omega) \rangle = 2\Gamma_0 / \left(\omega^2 + 4\Gamma_0^2(q^2 + r)^2\right) \tag{11.53}$$

In the language of the dynamical scaling theory, $\langle m_q^*(\omega) m_q(\omega) \rangle$ is to be identified with $\langle\psi\psi\rangle(\vec{q},\omega)$ ((11.23)). Integrating

$$\int_{-\infty}^{\infty}\langle m_q^*(\omega)m_q(\omega)\rangle \mathrm{d}\omega = 1/(q^2+r) = \langle m_q^* m_q\rangle \tag{11.54}$$

or $S(q)$ in the notation of (11.28). Thus the scaling function $f$ in this case obeys the relation ((11.21))

$$2\Gamma_0(q^2+r)/\left(\omega^2+4\Gamma_0^2(q^2+r)^2\right) = \left(2\pi/\omega_{m_q}\right)f\left(\omega/\omega_{m_q}\right) \tag{11.55}$$

where $\omega_{m_q}$ is the scaling frequency. It is plausible to suppose that $\omega_{m_q} = a\Gamma_0(q^2+r)$ where $a$ is a constant. Putting that ansatz into (11.55) one easily finds

$$f\left(\omega/\omega_{m_q}\right) = (1/\pi)\frac{1/a}{\left(\omega/\omega_{m_q}\right)^2+(4/a^2)}$$

and

$$(1/2\pi)\int_1^1 (2/a)\,\mathrm{d}x/(x^2+4/a^2) = 1/2$$

The integral is $(1/\pi)\tan^{-1}(a/2)$ and this is $1/2$ when $a=2$ so for this Gaussian model

$$\omega_{m_q} = 2\Gamma_0(q^2+r)$$

We put this in the form (11.30) by recalling that $\nu = 1/2$ for the Gaussian model (equation (9.108)) and that in the dimensionless units in use here (see after (9.99)) it is therefore the case that $\xi = 1/r^{1/2}$. (We are taking $T > T_c$.) Thus

$$\omega_{m_q} = q^2 2\Gamma_0(1+1/(q\xi)^2)$$

which is written in the form (11.30) by defining $\Omega(q\xi) = 2\Gamma_0(1+1/(q\xi)^2)$. Thus

$$\omega_{m_q} = q^2\Omega(q\xi)$$

and $z=2$ for this model.

Now we consider the case in which $b \neq 0$. The approach is similar to that used for static critical phenomena as described in Chapter 9. We seek a perturbation expansion in $b$. However, the perturbation expansion will not converge below some upper critical dimension $d_c$ above which the renormalized $b$ will be irrelevant. We will consider the artificial case of $d$ infinitesimally less than $d_c$ and make an expansion in $\epsilon = d_c - d$. It is an interesting feature that $d_c$ for the dynamic model is often different from $d_c$ for the static critical properties. However, in the relatively simple model we are considering (Model A with $n=1$ in reference 5) the upper critical dimension turns out to be 4 for dynamics as well as statics.

To produce a perturbation expansion in $b$, it is convenient to introduce a small field time dependent term of the form $-\sum_q h_q(t)m_q$ to the free energy $\beta\mathcal{F}$ and consider the time Fourier transform of the resulting equation of motion (the field here includes a factor $k_{\mathrm{B}}T$ which cancels the $\beta$ in $\beta\mathcal{F}$ in the field term)

$$(-\mathrm{i}\omega + 2\Gamma_0(q^2 + r))m_q(\omega) + 4b\sum_{q_2,q_3}\int(\mathrm{d}\omega_2/(2\pi))\int(\mathrm{d}\omega_3/(2\pi))m_{q_2}(\omega_2)$$
$$\times\, m_{q_3}(\omega_3)m_{q-q_2-q_3}(\omega - \omega_2 - \omega_3))$$
$$= f_q(\omega) + \Gamma_0 h_q(\omega) \qquad (11.56)$$

which is to be compared with (11.51). The field provides a convenient way to organize the calculation. When we take a thermal average of (11.56) the only nonzero terms in $\langle m_q(\omega)\rangle$ will arise because of the presence of the field. We write $\chi_0(\vec{q}\omega) = (\Gamma_0/(-\mathrm{i}\omega + 2\Gamma_0(q^2 + r)))$ and rewrite (11.56) as

$$m_q(\omega) = \frac{\chi_0(\vec{q}\omega)}{\Gamma_0}(f_q(\omega) + \Gamma_0 h_q(\omega)) - (1/\Gamma_0)\chi_0(\vec{q}\omega)(4b)$$
$$\times\sum_{q_2,q_3}\int(\mathrm{d}\omega_2/(2\pi))\int(d\omega_3/(2\pi))m_{q_2}(\omega_2)m_{q_3}(\omega_3)m_{q-q_2-q_3}(\omega-\omega_2-\omega_3)$$
$$(11.57)$$

We can generate a series in $b$ for $m_q(\omega)$ by iterating this equation, inserting the entire expression for $m_q(\omega)$ into each of the three factors in the last term repeatedly. This yields ($\tilde{f} \equiv f/\Gamma_0$)

$$m_q(\omega) = \chi_0(\vec{q}\omega)(\tilde{f}_q(\omega) + h_q(\omega)) + (-4b/\Gamma_0)\chi_0(\vec{q}\omega)\sum_{q_2,q_3}\int((\mathrm{d}\omega_2)/(2\pi))$$
$$\times\int((\mathrm{d}\omega_3)/(2\pi))\Big[\Big\{\chi_0(\vec{q}_2\omega_2)(\tilde{f}_{q_2}(\omega_2) + h_{q_2}(\omega_2)) + (-4b/\Gamma_0)\chi_0(\vec{q}_2\omega_2)$$
$$\times\sum_{q_4,q_5}\int((\mathrm{d}\omega_4)/(2\pi))\int((\mathrm{d}\omega_5)/(2\pi))m_{q_4}(\omega_4)m_{q_5}(\omega_5)m_{q_2-q_4-q_5}$$
$$\times(\omega_2 - \omega_4 - \omega_5))\Big\}\Big\{\chi_0(\vec{q}_3\omega_3)(\tilde{f}_{q_3}(\omega_3) + h_{q_3}(\omega_3)) + (-4b/\Gamma_0)\chi_0(\vec{q}_3\omega_3)$$
$$\times\sum_{q_4,q_5}\int((\mathrm{d}\omega_4)/(2\pi))\int((\mathrm{d}\omega_5)/(2\pi))m_{q_4}(\omega_4)m_{q_5}(\omega_5)m_{q_3-q_4-q_5}$$
$$\times(\omega_2 - \omega_4 - \omega_5))\Big\}\Big\{\chi_0(q - q_2 - q_3, \omega - \omega_2 - \omega_3)(\tilde{f}_{q-q_2-q_3}$$
$$\times(\omega - \omega_2 - \omega_3) + h_{q-q_2-q_3}(\omega - \omega_2 - \omega_3))$$
$$+(-4b/\Gamma_0)\chi_0(q - q_2 - q_3, \omega - \omega_2 - \omega_3)$$
$$\times\sum_{q_4,q_5}\int((\mathrm{d}\omega_4)/(2\pi))\int((\mathrm{d}\omega_5)/(2\pi))m_{q_4}(\omega_4)m_{q_5}(\omega_5)$$
$$\times\, m_{q-q_2-q_3-q_4-q_5}(\omega - \omega_2 - \omega_3 - \omega_4 - \omega_5)\Big\}\Big] \qquad (11.58)$$

where we have just iterated once. The resulting series can be thermally averaged term by term, using the properties $\langle f_q(\omega)\rangle = 0$, $\langle f_{q_1}(\omega_1) f_{q_2}(\omega_2)\rangle = (2\pi)(2\Gamma_0)\delta_{q_1,-q_2}\delta(\omega_1+\omega_2)$. With the assumption that the noise is Gaussian, higher order correlation functions of $f$ can be expressed as products of pairs, using these rules. For example the zeroth order term (first line of (11.58) ) gives

$$\langle m_q(\omega)\rangle = h_q(\omega)\chi_0(\vec{q}\omega) + \cdots$$

which says that $\chi_0(\vec{q}\omega)$ is the susceptibility in zeroth order as the notation suggests. Note that the imaginary part of $\chi_0(\vec{q}\omega)$ is

$$\mathrm{Im}\chi_0(\vec{q}\omega) = \Gamma_0\left(\omega/(\omega^2 + 4\Gamma_0^2(q^2+r)^2)\right) = (\omega/2)\langle m_q(\omega) m_{-q}(-\omega)\rangle_0$$

when $b = 0$ (equation (11.53)). This is the fluctuation–dissipation theorem for the $b = 0$ model. (A factor $k_B T$ appears on the left hand side if one makes the field term in $\beta\mathcal{F}$ equal to $-\beta\sum_q h_q(t)m_q$.) One can show (reference 7 equations 2.32 to 2.46) that, within the model we are using, this fluctuation theorem is also true for the $b \neq 0$ model

$$\mathrm{Im}\chi(\vec{q}\omega) = (\omega/2)\langle m_q(\omega) m_{-q}(-\omega)\rangle \tag{11.59}$$

where $\chi(\vec{q}\omega) = \partial m_q(\omega)/\partial h_q(\omega)$.

At finite order in $-4b$ the terms which contribute to the series all have even numbers of factors $f_q(\omega)$ and one power of $h_q(\omega)$. After iteration to a given order, only the factors $f$ are time dependent and need to be thermally averaged and since the variables $f$ are distributed (by assumption) in a Gaussian way, these averages all factor into products of pairs of $f$. We illustrate by writing down the first order term in $b$ from (11.58):

$$\begin{aligned}
&3(-4b/\Gamma_0)\chi_0(\vec{q}\omega)\sum_{q_2,q_3}\int((\mathrm{d}\omega_2)/(2\pi))\\
&\quad\times\int((\mathrm{d}\omega_3)/(2\pi))\chi_0(\vec{q}_2,\omega_2)\chi_0(\vec{q}_3,\omega_3)\chi_0(q-q_2-q_3,\omega-\omega_2-\omega_3)\\
&\quad\times\langle\tilde{f}_{q_2}(\omega_2)\tilde{f}_{q_3}(\omega_3)\rangle h_{q-q_2-q_3}(\omega-\omega_2-\omega_3)\\
&= 3(-4b/\Gamma_0)\chi_0(\vec{q}\omega)\sum_{q_2,q_3}\int((\mathrm{d}\omega_2)/(2\pi))\\
&\quad\times\int((\mathrm{d}\omega_3)/(2\pi))\chi_0(\vec{q}_2,\omega_2)\chi_0(\vec{q}_3,\omega_3)\chi_0(q-q_2-q_3,\omega-\omega_2-\omega_3)\\
&\quad\times(4\pi/\Gamma_0)\delta_{q_2,-q_3}\delta(\omega_1+\omega_2)h_{q-q_2-q_3}(\omega-\omega_2-\omega_3)\\
&= 3(-4b/\Gamma_0)\chi_0(\vec{q}\omega)\\
&\quad\times\sum_{q_2}\int((\mathrm{d}\omega_2)/(2\pi))(2/\Gamma_0)\chi_0(\vec{q}_2\omega_2)\chi_0(-\vec{q}_2,-\omega_2)\chi_0(\vec{q},\omega)h_q(\omega)
\end{aligned} \tag{11.60}$$

The product

$$(2/\Gamma_0)\chi_0(\vec{q}_2\omega_2)\chi_0(-\vec{q}_2, -\omega_2) = (2/\Gamma_0)\Gamma_0^2/\left(\omega_2^2 + \Gamma_0^2(q^2 + r)^2\right)$$
$$= \langle m_{q_2}(\omega_2)m_{-q_2}(-\omega_2)\rangle_0$$

where $\langle m_{q_2}(\omega_2)m_{-q_2}(-\omega_2)\rangle_0$ is the $b = 0$ value of the correlation function given in (11.53). Thus the first order term is also written

$$3(-4b/\Gamma_0)\chi_0(\vec{q}\omega)\sum_{q_2}\int(\mathrm{d}\omega_2/(2\pi))\langle m_{q_2}(\omega_2)m_{-q_2}(-\omega_2)\rangle_0\chi_0(\vec{q}, \omega)h_q(\omega)$$

The reduction of the terms associated with the averages on the $f$ together with the corresponding factors $\chi_0$ to factors of the form $\langle m_q(\omega)m_{-q}(-\omega)\rangle_0$ occurs repeatedly in the series so that the evaluation of the terms can be reduced to "rules" corresponding to the replacement of pairs of factors of $f$ and the accompanying noninteracting susceptibilities by noninteracting equilibrium correlation functions. These rules differ somewhat for various dynamical models and are formulated differently by various authors. One relatively clear source is the book by Ma.[8] To clarify the calculations further, various authors have introduced diagrammatic notations for the terms in the resulting series. It is convenient to denote

$$\chi(\vec{q}, \omega) = \frac{\partial m_q(\omega)}{\partial h_q(\omega)}$$

and rewrite the series as a series for $\chi(\vec{q}, \omega)$ (which amounts to keeping only linear terms in $h_q(\omega)$ and dropping the factors $h_q(\omega)$). Then the first term in the corresponding series for $\chi(\vec{q}, \omega)$ can be represented by the diagram

$= \chi_0(q,\omega)$

The whole series can then be represented diagrammatically as

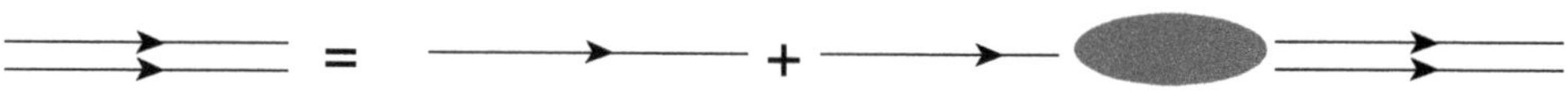

$$\chi_0(q,\omega) \quad = \quad \chi_0(q,\omega) \quad + \quad \chi_0(q,\omega)(\Sigma(q,\omega)/\Gamma_0)\chi(q,\omega)$$

in which the factor labelled $\Sigma(q, \omega)$ contains all the terms in the series with one factor $\chi_0(\vec{q}, \omega)$ factored out on the right and the left. (Some authors define it with the opposite sign.) (Factors $\chi_0(\vec{q}'\omega')$ with arguments not equal to those in the $\chi(\vec{q}, \omega)$ being calculated will occur inside sums on frequency and wavevector in $\Sigma(q, \omega)$.)

The last equation can obviously be formally solved for $\chi(\vec{q},\omega)$

$$\chi(\vec{q},\omega) = \chi_0(\vec{q},\omega)/(1 - \chi_0(\vec{q},\omega)\Sigma(q,\omega)/\Gamma_0)$$

and reintroducing the explicit expression for $\chi_0(\vec{q},\omega)$

$$\chi(\vec{q},\omega) = \Gamma_0/(-\mathrm{i}\omega + 2\Gamma_0(q^2+t) - \Sigma(q,\omega))$$

The "self energy" $\Sigma(q,\omega)$ described in this way thus gives the information about the characteristic frequency used in the scaling theory of dynamical critical phenomena. Since near critical points we are only interested in low frequencies, we can explore how to relate $\Sigma(q,\omega)$ to the characteristic frequency $\omega_{m_q}$ of the scaling theory by expanding $\Sigma(q,\omega)$ about its argument for small $\omega$ giving

$$\begin{aligned}\chi(\vec{q},\omega) &= \Gamma_0/(-\mathrm{i}\omega + 2\Gamma_0(q^2+r) - \Sigma(q,0) - \omega(\mathrm{d}\Sigma(q,\omega)/\mathrm{d}\omega)_{\omega=0} + \cdots) \\ &\approx \frac{\Gamma_0/(1+\mathrm{i}(\mathrm{d}\Sigma(q,\omega)/\mathrm{d}\omega)_{\omega=0})}{-\mathrm{i}\omega + (\omega_q - \Sigma(q,0))/(1+\mathrm{i}(\mathrm{d}\Sigma(q,\omega)/\mathrm{d}\omega)_{\omega=0})}\end{aligned}$$

where $\omega_q = 2\Gamma_0(q^2+t)$. Assuming the fluctuation–dissipation theorem (11.59) this leads to a Lorentzian correlation function $\langle m_q(\omega)m_{-q}(-\omega)\rangle$ at this order, and hence, by the same arguments used for the $b=0$ model, to a characteristic frequency $\omega_{m_q}$ of :

$$\omega_{m_q} = (\omega_q - \Sigma(q,0))/(1+\mathrm{i}(\mathrm{d}\Sigma(q,\omega)/\mathrm{d}\omega)_{\omega=0}) \tag{11.61}$$

The term we calculated explicitly makes a contribution which can be denoted diagrammatically by

where the wavy line stands for a factor $\langle m_{q_2}(\omega_2)m_{-q_2}(-\omega_2)\rangle_0$ and the black dot denotes a factor $(-4b)$. One sums over internal wave vectors and integrates over frequencies so that this contribution is

$$\Sigma_1(\vec{q},\omega) = 3(-4b)\sum_{q_2}\int(\mathrm{d}\omega_2/(2\pi))\langle m_{q_2}(\omega_2)m_{-q_2}(-\omega_2)\rangle_0$$

to the self energy. This term is independent of $q$ and $\omega$. It can be evaluated, but only results in an effective shift in the critical temperature. In particular, one finds

$\Sigma_1(\vec{q},\omega) = Kt^{(d-1)/2}$ which can be added to the term $\Gamma t$ in the denominator of $\chi(\vec{q},\omega)$. By defining a shifted critical temperature $\tilde{T}_c$ by the requirement that $\Gamma t - Kr^{(d-1)/2} = 0$ at $T = \tilde{T}_c$ one can take account of the essential effects of this term inasfar as they affect the critical properties of the model. The first term in the self energy which does more than affect the effective critical temperature is the one denoted by

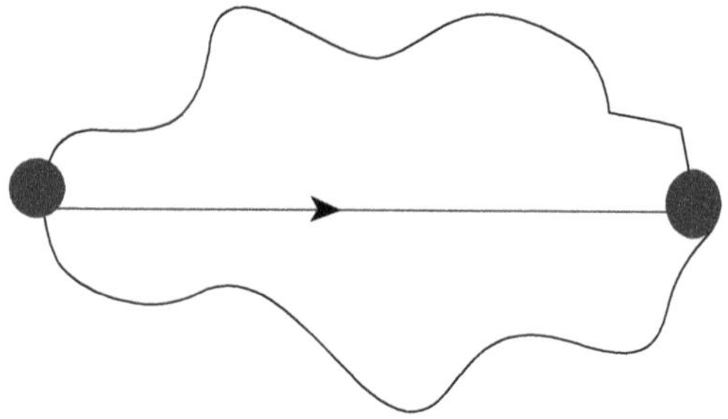

This term may be written as

$$\Sigma_3(\vec{q},\omega) = 18(-4b)^2 \int (\mathrm{d}\omega_2/(2\pi)) \int (\mathrm{d}\omega_3/(2\pi))$$
$$\times \sum_{\vec{q}_2,\vec{q}_3} \left( \frac{1}{-\mathrm{i}(\omega - \omega_2 - \omega_3) + -\omega_{q-q_1-q_2}} \right) \left( \frac{2\Gamma_0}{\omega_2^2 + \omega_{q_2}^2} \right) \left( \frac{2\Gamma_0}{\omega_3^2 + \omega_{q_3}^2} \right) \tag{11.62}$$

where $\omega_{\vec{p}} = 2\Gamma_0(p^2 + r)$. The prefactor 18 can be understood by iterating (11.58) to second order and then noting that the second factor proportional to $(-4b)$ can be selected in three ways, in this second factor the factor proportional to $h$ can be selected in three ways, and the remaining four factors of the fluctating field may be factored in two ways. (The reader is encouraged to do this explicitly.)

One can carry out the integrals on the frequencies in (11.62) giving

$$\Sigma_3(\vec{q},\omega) = -18(-4b)^2(\Gamma_0^2/3) \sum_{\vec{p},\vec{k}} \frac{\omega_{\vec{p}} + \omega_{\vec{k}} + \omega_{\vec{q}-\vec{p}-\vec{k}}}{\omega_{\vec{p}}\omega_{\vec{k}}\omega_{\vec{q}-\vec{p}-\vec{k}}[-\mathrm{i}\omega + (\omega_{\vec{p}} + \omega_{\vec{k}} + \omega_{\vec{q}-\vec{p}-\vec{k}})]} \tag{11.63}$$

To evaluate the effects of this term on the dynamical critical behavior, we use (11.61). We define scaled variables $\vec{q}' = \vec{q}/r^{1/2}$, $\vec{k}' = \vec{k}/r^{1/2}$. Setting $\omega = 0$ we have

$$\Sigma_3(\vec{q},0) = -18(-4b)^2(1/3\Gamma_0)r^{d-3}\left(\Omega^{(\mathrm{d})}/(2\pi)^{\mathrm{d}})\right)^2 \int \mathrm{d}^d\vec{k}'$$
$$\times \int \mathrm{d}^d\vec{p}'(1/(p'^2+1))(1/(k'^2+1))\left(1/\left((\vec{q}/r^{1/2} - \vec{p}' - \vec{q}'\right)^2 + 1\right)\right)$$

which is finite at the critical point if $d \leq 3$. However

$$\begin{aligned}
i(d\Sigma_3(q,\omega)/d\omega)_{\omega=0} &= 18(-4b)^2(1/\Gamma_0^2)r^{d-4}(\Omega^{(d)}/(2\pi)^d))^2 \\
&\times \int d^d\vec{k}' \int d^d\vec{p}'(1/(p'^2+1))(1/(k'^2+1)) \\
&\times (1/((\vec{q}/r^{1/2}-\vec{p}'-\vec{k}')^2+1))(1/[3+p'^2+k'^2 \\
&+ (\vec{q}/r^{1/2}-\vec{p}'-\vec{k}')^2])
\end{aligned}$$

and this diverges as $r \to 0$ for $d < 4$. The contribution to the characteristic frequency is

$$i(d\Sigma_3(q,\omega)/d\omega)_{\omega=0}\omega_q \propto q^2 r^{d-4}$$

(plus a $q$ independent term) so this means that the effective diffusion constant associated with $m_q$ is diverging. However, as in the case of static critical phenomena we cannot use this perturbation theory result to deduce the nature of the divergence because the perturbation theory obviously breaks down when it gives diverging terms.

One copes with this divergence in a way closely analogous to the way the infrared divergences which occured in static critical phenomena were handled. Note that the expression for the characteristic frequency $\omega_{m_q}$ in (11.61) can be written

$$\omega_{m_q} = 2\Gamma_0(q^2 + r - \Sigma(\vec{q},0))/(1 + i(d\Sigma(q,\omega)/d\omega)_{\omega=0})$$

We write this as

$$\omega_{m_q} = \Gamma(q)/\chi(q)$$

in a form suggestive of the discussion of the Van Hove theory. Then we identify

$$\chi(q)^{-1} = q^2 + r - \Sigma(\vec{q},0)$$

a kind of inverse static susceptibility corrected for interactions and

$$1/\Gamma(q) = (1 + i(d\Sigma(q,\omega)/d\omega)_{\omega=0})/2\Gamma_0 \qquad (11.64)$$

which is an inverse "kinetic coefficient" (here a diffusion constant) corrected for interactions. In this model, it is not the effects of interactions on $\chi(q)$ which change the value of the exponent $z$, but the $q$ dependence which enters due to terms in $1/\Gamma(q)$.

We just showed that the first term in a perturbation series of the correction to the kinetic coefficient diverges. The divergence is an infrared one (occurring at the lower, long wavelength limit) in the integrals on intermediate wave vectors, quite analogous to the problem that arose with a perturbation calculation for the effective free energy in the static critical phenomenon problem. One approaches the problem similarly, except that instead of renormalizing the free energy, we have to renormalize the equation of motion. The renormalization occurs in two steps, as it did in the static critical phenomenon problem. One can do these steps in either order but we will begin by describing the "dynamical" step. We will first only take account of those terms in the perturbation series for $1/\Gamma(q)$ for which the intermediate fluctuations have wave vectors in the range $\Lambda/\tilde{b} < k, p, \ldots, < \Lambda$. Here $\Lambda$ is the short wavelength cutoff on the wave vector sums, just as in the static critical phenomenon problem ($\tilde{b}$ was 2 in the discussion in Chapter 9). This is obviously only a partial summation of the perturbation series. The advantage is that we can do this part of the problem without encountering an infinity, because the lower limit of the integrations is not zero. The second step is to rescale the result so that the upper limit of the remaining wave vectors, expressed in scaled form, is the same as it was before. Then if all goes well, we will have a new equation of motion with a rescaled kinetic coefficient and we can do the same perturbation theory calculation again on the next shell of wave numbers, working in from larger to smaller ones, giving a recursion relation. In such a renormalization we must rescale the parameters of the free energy functional at each step in a way which is completely consistent with the result which we got for the analogous renormalization of the free energy in the discussion of static critical phenomena. In the problem at hand, this will turn out to work near the upper critical dimension of 4 and we will evaluate the correction to lowest order in $\epsilon = 4 - d$.

There are two things to note about this. One might think that to cope with dynamical critical phenomena, one would have to do the integrals on $\omega$ (as well as on the wave vector) in the same shell by shell manner, starting with the larger frequencies, but this turned out not to be necessary. Further, one does not actually need to reduce the problem of finding $z$ to an eigenvalue problem of the linearized renormalization group, because it turns out to be sufficient to require that the value of $z$ allow a fixed point of the recursion relation to exist.

To carry out these steps it is convenient to reverse the order and do the rescaling first. To get the scaling right, it is convenient to go back to equation (11.35). We rescale to new length variables $x'$ related to the earlier ones by $x = \tilde{b}x'$. We suppose that $m' = \tilde{b}^a m$. Then the free energy $\mathcal{F}$ in (11.36) becomes

$$\beta\mathcal{F} = \int \mathrm{d}\vec{r}\,' \left(\tilde{b}^{d-2(1+a)-\eta} \mid \nabla_{\vec{r}\,'} m'(\vec{r}\,') \mid^2 + \tilde{b}^{d-2a} r m'(\vec{r})^2 + \tilde{b}^{d-4a} b m'^4\right) \quad (11.65)$$

The exponent $\eta$ will enter after a few rescalings because the critical form of the static correlation function contains the factor $1/r^{d-2+\eta}$. We define

$$r' = \tilde{b}^{d-2a} r$$
$$b' = \tilde{b}^{d-4a} b$$

and require

$$\tilde{b}^{d-2(1+a)-\eta} = 1$$

The last equation arises so that the coefficient of the term involving the gradient will remain 1. An equivalent requirement was imposed in the discussion in Chapter 9. It gives

$$a = (1/2)(d - 2 + \eta) \tag{11.66}$$

and thus

$$r' = \tilde{b}^{2-\eta} r$$
$$b' = \tilde{b}^{4-d-2\eta} b$$

after the rescaling step. These rescalings are consistent with those found in the rescaling step for the renormalization of the free energy described in Chapter 9. For the dynamical part of the equation (11.35) we suppose that frequencies rescale as $\omega = \tilde{b}^{-z}\omega'$ so that the time in (11.35) rescales as $t = \tilde{b}^{z}t'$. We may then write (11.35) as

$$\tilde{b}^{-a}\frac{\partial \bar{m}'}{\tilde{b}^{z}\partial t'} = -2\Gamma_0 \tilde{b}^{a-d}[r'\bar{m} + \nabla'^2\bar{m}' + b'\bar{m}'^3] + \text{field and noise terms} \tag{11.67}$$

Thus if one requires the new equation to have the same form as the old one, the rescaling of the kinetic coefficient $2\Gamma_0$ should be

$$2\Gamma' = \tilde{b}^{z+2a-d} 2\Gamma$$

where we have written $\Gamma$ instead of $\Gamma_0$ in anticipation of the idea that this relation will be iterated. With this rescaling, (11.67) is

$$\frac{\partial \bar{m}'}{\partial t'} = -2\Gamma'[r'\bar{m}' + \nabla'^2\bar{m}' + b'\bar{m}'^3] + \text{field and noise terms}$$

Now to take the dynamical step in the renormalization we can use the results of Chapter 9 for the recursion relations for $r$ (formerly $t$) and $b$ since the renormalization procedure is essentially the same. (Several authors have shown this explicitly, see for example reference 5). For the renormalization of $\Gamma$ we use (11.64) and keep the first relevant term in the perturbation series, evaluated, as described above, only for wave vectors $\Lambda/\tilde{b} < p, k < \Lambda$

$$\begin{aligned} 1/\Gamma' = \tilde{b}^{-(z+2a-d)} \Big( & 1/\Gamma + 18(-4b)^2(1/\Gamma) \\ & \times \int_{\Lambda/b}^{\Lambda} d^d\vec{k} \int_{\Lambda/b}^{\Lambda} d^d\vec{p}(1/(p^2))(1/(k^2))(1/(-\vec{p}-\vec{k})^2) \\ & \times (1/[p^2 + k^2 + (-\vec{p}-\vec{k})^2]) \Big) \end{aligned} \tag{11.68}$$

We have supposed that $\Lambda/\tilde{b} \gg r$ and dropped the terms involving $r$. Following the procedure in the case of static critical phenomena we want the lowest order terms arising from this in a series in $\epsilon = 4 - d$. One can show that the integral in (11.68) is proportional to $\ln b$. Writing the result as $(1/\Gamma)(-4b)^2 c_1 \ln b$ one obtains

$$1/\Gamma' = \tilde{b}^{-(z+2a-d)}(1/\Gamma)(1 + (-4b)^2 c_1 \ln b)$$

At the fixed point $4b$ is of order $\epsilon$ so one can write

$$(1 + (4b)^2 c_1 \ln b) \approx b^{(-4b^*)^2 c_1}$$

and the equation for $1/\Gamma$ has a fixed point if

$$z = 2 - \eta + (-4b^*)^2 c_1$$

$\eta$ is of order $\epsilon^2$ in this model so the corrections to $z = 2$ in this model are of order $\epsilon^2$. (See the references for numerical values.)

## References

1. G. F. Mazenko, in *Correlation Functions and Quasiparticles in Condensed Matter*, ed. J. W. Halley, New York: Plenum, 1978, pp. 151–161.
2. H. Mori, *Progress of Theoretical Physics* **33** (1965) 423.
3. G. F. Mazenko, *Physical Review A* **9** (1974) 360.
4. B. I. Halperin and P. C. Hohenberg, *Physical Review* **177** (1969) 952.
5. P. Hohenberg and B. Halperin, *Reviews of Modern Physics* **49** (1977) 436.
6. For a good elementary account see F. Reif, *Fundamentals of Statistical and Thermal Physics*, New York: McGraw-Hill, 1965.
7. S. Ma and G. F. Mazenko, *Physical Review B* **11** (1975) 4077.
8. S.-K. Ma, *Modern Theory of Critical Phenomena*, Reading, MA: Benjamin-Cummings, 1976.

## Problems

11.1 Show that (11.15) is zero.

11.2 Find the streaming terms $V_q$ for a Heisenberg ferromagnet. Assume the slow degrees of freedom to be low wave vector spatial Fourier transforms $S_q^x$, $S_q^y$, $S_q^z$ of the $x$, $y$ and $z$ components of the spins $S_i^x$, $S_i^y$, $S_i^z$ at sites $i$ and assume the spins at each site to be of the (classical) form $\vec{S}_i = \sum_{\lambda_i} \vec{r}_{\lambda_i} \times \vec{p}_{\lambda_i}$ where $\vec{r}_{\lambda_i}$, $p_{\lambda_i}$ are canonical coordinates and momenta.

11.3 Show that the factor 18 in (11.62) becomes $6(n+2)$ if there are $n$ "components" to the variables $\vec{m}(\vec{q})$ as in the $n$–$d$ model.

# Appendix
# Solutions to selected problems

## Chapter 1

1.1 The three dimensional constant energy surfaces of the two dimensional harmonic oscillator are three dimensional ellipsoids in the $px, py, x, y$ phase space and are described by

$$p_x^2/2m + p_y^2/2m + K_x x^2/2 + K_y y^2/2 = E$$

If the trajectory filled this surface, then all the values of $x, y$ in the range

$$K_x x^2/2 + K_y y^2/2 < E$$

would be passed through by the trajectory, for example. But, in general, this cannot occur. Suppose one starts at $x_0$, $y_0$, with zero momenta. The energies

$$E_x = p_x^2/2m + (K_x/2)x^2$$

and

$$E_y = p_y^2/2m + (K_y/2)y^2$$

are separately conserved by the equations of motion so that the maximum value that $x$ can take during the trajectory is $x_0$ and not $\sqrt{(x_0^2 + (K_y/K_x)y_0^2)}$ which is included in the ellipse in the first equation. In fact, in the $xy$ plane, the trajectories are confined to a rectangle bounded by $\pm x_0$ and $\pm y_0$ which fits inside the ellipse. Inside this rectangle, though, one can say something about a kind of limited ergodicity. If the frequencies of motion in the two directions are in rational ratio so that $\sqrt{(K_x/K_y)}$ is a rational number then the trajectories close and do not even fill the rectangle, whereas if this ratio is irrational, then the trajectories do fill the rectangle but not the ellipse.

We illustrate numerically with a couple of cases in Figure A.1. The Fortran program used to generate the figures is also attached.

1.3 A particle confined to a box of length $a$ by elastic walls:

$$q = q_0 + (p_0/m)t$$

until reflection at walls $p \to -p_0$.

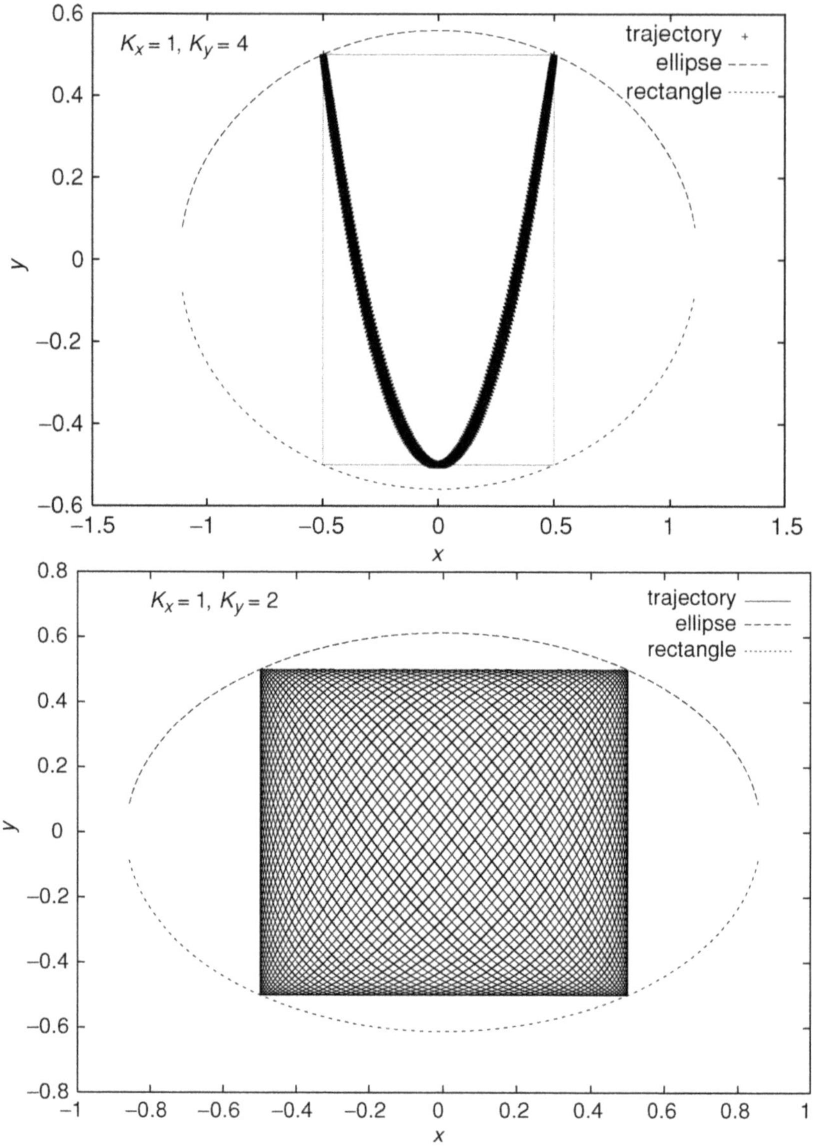


Figure A.1

The time $\Delta q$ is $m\Delta q/\mid p_0 \mid$ while the period is $2ma/\mid p_0 \mid$. Thus

$$\begin{aligned}\frac{\Delta}{t} &= \frac{m\Delta q/\mid p_0\mid}{2ma/\mid p_0\mid}\left[\delta(p-\mid p_0\mid)+\delta(p+\mid p_0\mid)\right]\Delta p\Theta(q)\Theta(a-q)\\ &= \frac{\Delta q}{2a}\left[\delta(p-\mid p_0\mid)+\delta(p+\mid p_0\mid)\right]\Delta p\Theta(q)\Theta(a-q)\end{aligned}$$

```
c       this  does the ho for problem 1.1,2003
        open(1,file='xyhob3.dat')
        open(2,file='ellipse3.dat')
        open(3,file='rectangle3.dat')
        dt2=.001
        ntimes=10000
        fkx=1.
        fky=2.
        x00=0.5
        y00=0.5
        x0=x00
        y0=y00
        x1=x0
        y1=y0
        e=fkx*(x0**2)/2.+fky*(y0**2)/2.
        write(1,*) x0,y0
        write(1,*) x1,y1
        do i=1,ntimes
         xnew=2.0*x1-x0-fkx*x1*dt2
         ynew=2.0*y1-y0-fky*y1*dt2
         write(1,*) xnew,ynew
         x0=x1
         x1=xnew
         y0=y1
         y1=ynew
        enddo
        xmax=sqrt((2.*e)/fkx)
        write(*,*) xmax, e
        dx=xmax/100.
        do ix=-99,99
         x=ix*dx
         yp=sqrt((2.*e)/fky-(fkx/fky)*(x**2))
         ym=-sqrt((2.*e)/fky-(fkx/fky)*(x**2))
         write(2,*) x,yp,ym
        enddo
        dx=x00/100.
        do ix=-99,99
         x=ix*dx
         write(3,*) x,y00
        enddo
        do ix=-99,99
         x=ix*dx
         write(3,*) x,-y00
        enddo
        dy=y00/100.
        do iy=-99,99
         y=iy*dy
         write(3,*) x00,y
        enddo
        do iy=-99,99
         y=iy*dy
         write(3,*) -x00,y
        enddo
        stop
        end
```

and

$$\rho(q, p; q_0, p_0) = \frac{1}{2a}\left[\delta(p- \mid p_0 \mid) + \delta(p+ \mid p_0 \mid)\right]\Theta(q)\Theta(a-q)$$

A one dimensional harmonic oscillator. Here the system is ergodic and periodic for all initial conditions. The energy is

$$E = \frac{{p_0}^2}{2m} + \frac{k{q_0}^2}{2} = \frac{p^2}{2m} + \frac{kq^2}{2}$$

so

$$p(q) = \pm\sqrt{2m(E - kq^2/2)} = m\frac{\mathrm{d}q}{\mathrm{d}t}$$

so that over one period in $(\Delta q, \Delta p)$ at $(q, p)$ we spend the time

$$\Delta t = \frac{m\Delta q}{\sqrt{2m(E - kq^2/2)}}\left[\delta(p - \sqrt{2m(E - kq^2/2)}) + \delta(p + \sqrt{2m(E - kq^2/2)})\right]\Delta p$$

whereas one whole period takes $T = 2\pi\sqrt{m/k}$. Thus, ignoring the finite time spent in the fraction of a period immediately after the initial conditions (which is negligible in the large time limit),

$$\begin{aligned}\rho(q, p; q_0, p_0) &= \frac{\Delta t}{T\Delta q\Delta p}\\ &= \frac{m}{2\pi}\sqrt{\frac{k}{m}}\frac{\delta(p - \sqrt{2m(E - kq^2/2)}) + \delta(p + \sqrt{2m(E - kq^2/2)})}{\sqrt{2m(E - kq^2/2)}}\end{aligned}$$

A ball in the Earth's gravitational field bouncing elastically from a floor. Here the system is ergodic and periodic for all initial conditions. The energy is

$$E = \frac{{p_0}^2}{2m} + mgq_0 = \frac{p^2}{2m} + mgq$$

thus

$$p(q) = \pm\sqrt{2m(E - mgq)} = m\frac{\mathrm{d}q}{\mathrm{d}t}$$

so that over one period, in $(\Delta q, \Delta p)$ at $(q, p)$, we spend the time

$$\Delta t = \frac{m\Delta q}{\sqrt{2m(E - mgq)}}\left[\delta(p - \sqrt{2m(E - mgq)}) + \delta(p + \sqrt{2m(E - mgq)})\right]\Delta p$$

whereas one whole period takes $T = 2t_{\mathrm{drop}}$, the time it takes for the ball to fall from its maximum height, $q_{\max}$. This is found by solving

$$mgq_{\max} = E$$

$$0 = q_f = q_{\max} - gt_{\mathrm{drop}}^2/2$$

which give

$$T = 2\sqrt{\frac{2E}{mg^2}}$$

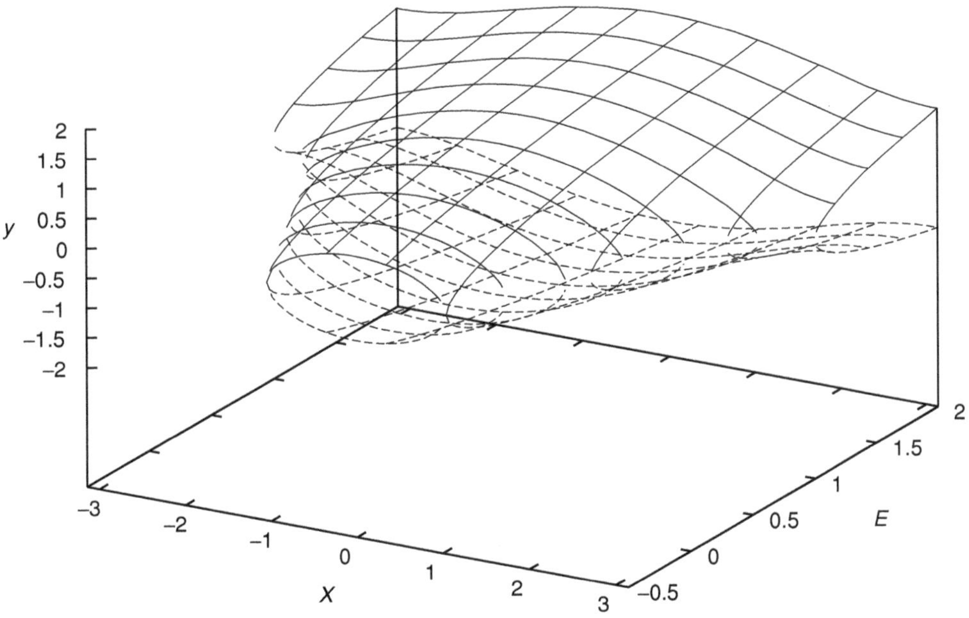


Figure A.2

Thus, ignoring the finite time spent in the fraction of a period immediately after the initial conditions (which is negligible in the large time limit),

$$\rho(q, p; q_0, p_0) = \frac{\Delta t}{T \Delta q \Delta p} = \frac{mg}{4} \frac{\left[\delta(p - \sqrt{2m(E - mgq)}) + \delta(p + \sqrt{2m(E - mgq)})\right]}{\sqrt{E(E - mgq)}}$$

A pendulum with arbitrary amplitude. Here the system is periodic, but only ergodic for some initial conditions: for small energies, the pendulum's angular position is bounded and it traces a nearly elliptical path in phase space similar to that of the harmonic oscillator (Figure A.2). For large enough energies, however, the pendulum rotates continuously around the pivot, maintaining the sign of its momentum.

In the coordinate system $\theta$, $p_\theta = ml^2 \mathrm{d}\theta/\mathrm{d}t$, the energy takes the form

$$E = \frac{p_{\theta,0}^2}{2m} - mgl\ \cos(\theta_0) = \frac{p_\theta^2}{2m} - mgl\ \cos(\theta)$$

(note $E \geq -mgl$). Thus the energetically allowed momenta are

$$p_\theta(\theta) = \pm\sqrt{2m(E + mgl\ \cos(\theta))} = ml^2 \frac{\mathrm{d}\theta}{\mathrm{d}t}$$

Now we can find the bounds on the coordinate $\theta$ by looking for the condition in which the momentum is zero, i.e. $E = -mgl\ \cos(\theta_{\max})$, or

$$\theta_{\max} = \cos^{-1}\left(\frac{-E}{mgl}\right)$$

which becomes undefined when its argument passes $\pm 1$. For $E > mgl$, the motion is unbounded in $\theta$.

For $|E| < mgl$, over one period in $(\Delta q, \Delta p)$ at $(q, p)$ we spend the time

$$\Delta t = \frac{ml^2 \Delta\theta}{\sqrt{2m(E + mgl\,\cos(\theta))}} \Big[\delta(p - \sqrt{2m(E + mgl\,\cos(\theta))}) + \delta(p + \sqrt{2m(E + mgl\,\cos(\theta))})\Big] \Delta p_\theta$$

as the system samples both directions of momentum. On the other hand, for $E > mgl$,

$$\Delta t = \frac{ml^2 \Delta\theta}{\sqrt{2m(E + mgl\,\cos(\theta))}} \Big[\delta(p \pm \sqrt{2m(E + mgl\,\cos(\theta))})\Big] \Delta p_\theta$$

where the $\pm$ depends on the sign of the initial condition, $p_{\theta,0}$.

The whole period for the motion can be expressed as

$$T = \int_{\text{oneperiod}} \mathrm{d}t = \oint \frac{\mathrm{d}\theta}{\sqrt{2m(E + mgl\,\cos(\theta))}}$$

which can be written for each case (using the symmetry of the integrand)

$$T_{|E|<mgl} = 4 \int_0^{\cos^{-1}(-E/mgl)} \frac{\mathrm{d}\theta}{\sqrt{2m(E + mgl\,\cos(\theta))}}$$

and

$$T_{E>mgl} = 2 \int_0^{\pi} \frac{\mathrm{d}\theta}{\sqrt{2m(E + mgl\,\cos(\theta))}}$$

This integral can be evaluated numerically and one can then combine the above pieces using (for each energy case separately)

$$\rho = \frac{\Delta t}{T \Delta\theta \Delta p_\theta}$$

(What happens in the special case when $E = mgl$?)

## Chapter 2

2.3 The eigenfunctions are

$$\psi_n^+(x) = \sqrt{\frac{2}{a}} \cos k_n x \qquad k_n = \left(\frac{\pi}{a}\right)(2n - 1)$$

$$\psi_n^-(x) = \sqrt{\frac{2}{a}} \sin k_n x \qquad k_n = \left(\frac{\pi}{a}\right)(2n)$$

$n = 0, 1, 2, \ldots$. Energy eigenvalues are $E_n = \hbar^2 k_n^2/2m$. The general form of the density matrix is $(s = \pm)$

$$\rho_{n,s;n',s'} = |a_n(0)|^2 \delta_{n,n'} \delta_{s,s'}$$

but if you start with an energy eigenstate $n_0, s_0$ it is

$$\rho_{n,s;n',s'} = \delta_{n,s;n',s'} \delta_{n,s;n_0,s_0}$$

For an operator $O(x)$ the expectation value, for $s_0 = +$, is

$$\langle O\rangle = (2/a)\int_{-a/2}^{a/2} \cos^2 k_{n_0}x\, O(x)\,\mathrm{d}x$$

$$= (2/a)\int \frac{\mathrm{d}p}{\hbar}\int_{-a/2}^{a/2} \mathrm{d}x\,(1/2)(\delta(p-\hbar k_0)+\delta(p+\hbar k_0))\cos^2 px/\hbar\; O(x)$$

with $\cos^2 px/\hbar = (1/2)(1+\cos 2px/\hbar)$ this is

$$\langle O\rangle = \int \mathrm{d}p \int \mathrm{d}x(1/2a)(\delta(p-\hbar k_0)+\delta(p+\hbar k_0))[1+\cos(2px/\hbar)]O(x)$$

The first term in the square bracket gives the classical result and the second term gives the correction. For $s_0 = -$ it is the same except that one uses $\sin^2 px/\hbar = (1/2)(1-\cos 2px/\hbar)$ and the correction term has the opposite sign. For functions of momentum $O(\frac{\hbar}{\mathrm{i}}\frac{\mathrm{d}}{\mathrm{d}x})$ we have, with $s_0 = +$

$$\langle O\rangle = (2/a)\int_{-a/2}^{a/2} \cos(k_{n_0}x)O\left(\frac{\hbar}{\mathrm{i}}\frac{d}{dx}\right)\cos k_{n_0}x\,\mathrm{d}x$$

$$= (2/a)\int_{-a/2}^{a/2} \cos(k_{n_0}x)(1/2)[O(\hbar k_0)\,\mathrm{e}^{\mathrm{i}k_0x} + O(-\hbar k_0)\,\mathrm{e}^{-\mathrm{i}k_0x}]$$

$$= (2/a)\int \frac{\mathrm{d}p}{\hbar}\int_{-a/2}^{a/2} \mathrm{d}x\,\cos(px/\hbar)(1/2)(\delta(p-\hbar k_0)+\delta(p+\hbar k_0))\,\mathrm{e}^{\mathrm{i}px/\hbar}\,O(p)$$

But $\cos(px/\hbar)\,\mathrm{e}^{\mathrm{i}px/\hbar} = (1/2)[1+\mathrm{e}^{2\mathrm{i}px/\hbar}]$ so

$$\langle O\rangle = \int \mathrm{d}p \int \mathrm{d}x(1/2a)(\delta(p-\hbar k_0)+\delta(p+\hbar k_0))\big[1+\mathrm{e}^{2\mathrm{i}px/\hbar}\big]O(p)$$

and the correction term is again the second one in the square bracket. (However, in this case, with one eigenfunction and a function of only the momentum, the correction term actually integrates to zero.) Finally for $s_0 = -$ and functions of the momentum one gets the last result with a minus sign in front of the correction term. The correction terms are small if (i) the wavelength $\hbar/p_0$ is small compared to the size $a$ of the box and (ii) the operator $O(x)$ varies slowly over the wavelength. In terms of the uncertainty principle language, the uncertainty in $x$ here is of order $a$ and the uncertainty in $p$ is $2p_0$ so the criterion (i) is $a \gg \hbar/p_0$ or $\Delta x \gg 2\hbar/\Delta p$ consistent with the expected requirement for classical considerations to work; (ii) just means that you can focus down more on the particle and determine its position (with $O(x)$) down to some $\Delta x < a$ as long as $\Delta x \gg \hbar/\Delta p$. The more general case $O(p,x)$ and a general starting wave function, corresponds classically to averaging over a variety of classical trajectories with a probability distribution of starting energies. Let the momentum width of the starting wave function be $\Delta p$ and the position width be $\Delta x$. Then as long as $O(p,x)$ varies slowly over both of these, and the average momentum $\bar{p}$ satisfies $\hbar/\bar{p} \gg a$, the classical result can be used with a momentum distribution

$$\rho(p, x) = (1/4a)|a_{|p|a/h}|^2$$

(Here one should use the nearest integer to $|p|a/h$.)

## Chapter 3

3.5 The needed derivatives are

$$\frac{\partial E_{\{\vec{p}_i\}}}{\partial V} = -(2/3)E_{\{\vec{p}_i\}}/V$$
$$\frac{\partial^2 E_{\{\vec{p}_i\}}}{\partial V^2} = (10/9)E_{\{\vec{p}_i\}}/V^2$$

Thus

$$\left\langle \left(\frac{\partial H}{\partial V}\right)^2 \right\rangle - \left\langle \frac{\partial H}{\partial V} \right\rangle \left\langle \frac{\partial H}{\partial V} \right\rangle = (4/9V^2)(\bar{E}^2 - \bar{E}^2) = (4/9V^2)k_B T^3 \left(\frac{\partial S}{\partial T}\right)_V$$
$$= (4/9V^2)k_B T^2 C_V = (4/9V^2)k_B T^2 (3/2)N k_B$$
$$= (2/3)k_B^2 T^2 N/V^2$$

so the left hand side of (3.78) is $(2/3)k_B T N/V^2$. Similarly the second quantity on the right of (3.78) is

$$\left\langle \frac{\partial^2 H}{\partial V^2} \right\rangle = (10/9V^2)\bar{E} = (10/9V^2)(3/2)N k_B T = (5/3)N k_B T/V^2$$

and the first term on the right of (3.78) is $-N k_B T/V$ so the equality is verified for an ideal gas and the fluctuations behave as claimed.

3.6 Use

$$\left(\frac{\partial w}{\partial x}\right)_y = \left(\frac{\partial w}{\partial x}\right)_z + \left(\frac{\partial w}{\partial z}\right)_x \left(\frac{\partial z}{\partial x}\right)_y$$

with $w = S, x = T, y = P, z = V$ giving

$$C_P - C_V = T\left(\frac{\partial S}{\partial V}\right)_T \left(\frac{\partial V}{\partial T}\right)_P$$

From $\mathrm{d}F = -S\,\mathrm{d}T - P\,\mathrm{d}V + \mu\,\mathrm{d}N$ one gets the Maxwell relation

$$\left(\frac{\partial S}{\partial V}\right)_T = \left(\frac{\partial P}{\partial T}\right)_V$$

giving

$$C_P - C_V = T\left(\frac{\partial P}{\partial T}\right)_V \left(\frac{\partial V}{\partial T}\right)_P$$

All the quantities on the right hand side get small at low temperatures in a solid. But using

$$\left(\frac{\partial z}{\partial y}\right)_x \left(\frac{\partial y}{\partial x}\right)_z \left(\frac{\partial x}{\partial z}\right)_y = -1$$

with $x = V, y = T, z = P$ gives

$$\left(\frac{\partial V}{\partial T}\right)_P = -\left(\frac{\partial P}{\partial T}\right)_V \left(\frac{\partial V}{\partial P}\right)_T$$

and thus

$$C_P - C_V = T\left(\frac{\partial P}{\partial T}\right)_V^2 \Big/ -\left(\frac{\partial P}{\partial V}\right)_T$$

The denominator (inversely proportional to the compressibility) goes to zero at a gas–liquid phase critical point, while the numerator remains finite, so this quantity diverges.

## Chapter 4

4.2

$$Z_{\mathrm{gc}} = \sum_N (1/h^{3N}N!)\left[\int \mathrm{e}^{-p^2\beta/2m}\mathrm{d}p\right]^{3N}\left[\int \mathrm{e}^{-K\beta x^2/2}\mathrm{d}x\right]^{3N} \mathrm{e}^{\beta\mu N}$$

If $\sqrt{2\pi/\beta K} \ll V^{1/3}$ then the limits on the integrals on $x$ can be extended to $\pm\infty$ and

$$Z_{\mathrm{gc}} = \exp\left[\left(\lambda_K^3/\lambda_T^3\right)\mathrm{e}^{\beta\mu}\right]$$

where $\lambda_T = \sqrt{h^2/2m\pi k_{\mathrm{B}}T}$ and $\lambda_K = \sqrt{2\pi/\beta K}$. Then

$$\Omega = -k_{\mathrm{B}}T\left(\lambda_T^3/\lambda_K^3\right)\mathrm{e}^{\beta\mu}$$

so that

$$\langle N\rangle = -(\partial\Omega/\partial\mu)_{T,K} = \left(\lambda_T^3/\lambda_K^3\right)\mathrm{e}^{\beta\mu}$$

and

$$\mu = k_{\mathrm{B}}T\ln\left(\langle N\rangle\left(\lambda_T^3/\lambda_K^3\right)\right)$$

which depends on $\langle N\rangle$. We have

$$\Omega = -k_{\mathrm{B}}T\langle N\rangle$$
$$\mathrm{d}\Omega = -S\,\mathrm{d}T + (\partial\Omega/\partial K)_{T,\mu}\,\mathrm{d}K - \langle N\rangle\mathrm{d}\mu$$

and

$$S = -(\partial\Omega/\partial T)_{K,\mu} = \langle N\rangle\left(4k_{\mathrm{B}} + k_{\mathrm{B}}\ln\left(\lambda_K^3/\lambda_T^3\langle N\rangle\right)\right)$$
$$(\partial\Omega/\partial K)_{T,\mu} = (3/2)(k_{\mathrm{B}}T/K)\langle N\rangle$$

is an *area*. The specific heat is

$$T(\partial S/\partial T)_{K,N} = 3k_{\mathrm{B}}\langle N\rangle$$

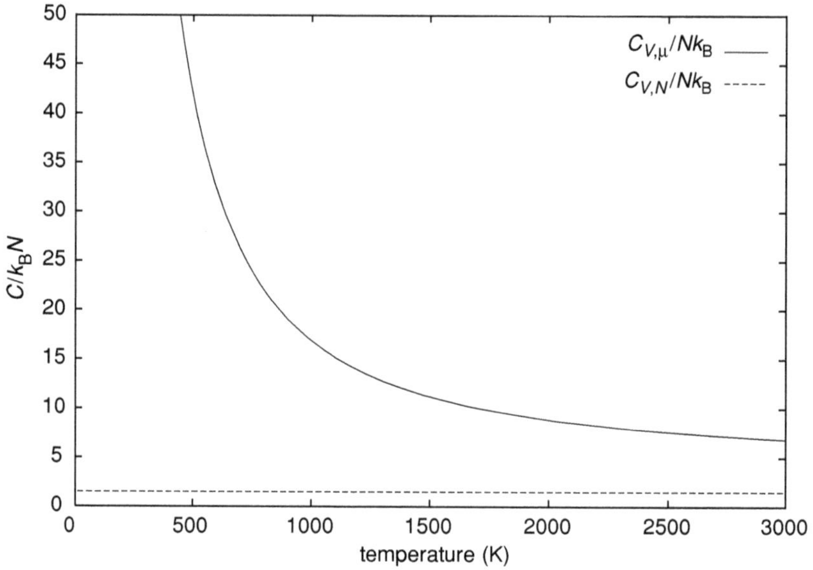


Figure A.3

## Chapter 5

5.2

$$
\begin{aligned}
S &= Vk_B[e^{\beta\mu/}\lambda^3](5/2 - \mu/k_BT) \\
(\partial S/\partial T)_{\mu,V} &= (Vk_B[e^{\beta\mu/}\lambda^3])(\mu/k_BT^2 + (3/2T - \mu/k_BT^2))(5/2 - (\mu/k_BT)) \\
&= Nk_B\big[(15/4T - 3(\mu/k_BT^2) + (\mu^2/k_B^2T^3)\big]
\end{aligned}
$$

So the specific heat

$$
C_{V,\mu}/Nk_B = 15/4 - 3\mu/k_BT + \mu^2/k_B^2T^2
$$

with $\mu/k_BT = \ln(N\lambda^3/V)$ and $\lambda = \sqrt{2\pi\hbar^2/mk_BT}$ one can calculate $C_{V,\mu}/Nk_B$ and compare it with $C_{V,N}/Nk_B = 3/2$

We show an example in Figure A.3 of the specific heat at constant $\mu$ and volume as a function of temperature for helium gas at a chemical potential corresponding to 1 atmosphere at 300 K.

Now we show algebraically the relation between $C_{V,\mu}$ and $C_{V,N}$:

$$
\begin{aligned}
(\partial S/\partial T)_{N,V} &= (\partial S/\partial T)_{\mu,V} + (\partial S/\partial\mu)_{T,V}(\partial\mu/\partial T)_{N,V} \\
(\partial S/\partial\mu)_{T,V} &= (Vk_B[e^{\beta\mu}/\lambda^3]((1/k_BT)(5/2 - \mu/k_BT) - (1/k_BT)) \\
&= (N/T)(3/2 - \mu/k_BT) \\
(\partial\mu/\partial T)_{N,V} &= (\mu/T - (3/2)k_B) \\
(\partial S/\partial T)_{N,V} &= (Nk_B/T)(15/4 - 3\mu/k_BT + \mu^2/(k_BT)^2 + (3/2 - \mu/k_BT)(\mu/k_BT - 3/2)) \\
&= (Nk_B/T)(3/2) \\
C_{V,N} &= 3Nk_B/2
\end{aligned}
$$

as expected.

5.3 $n = 2$, $M = 2$ and there are $4!/(2!^2)(2!) = 3$ assignments.

| Permutation | Exchange center of mass | Sign |
|---|---|---|
| For assignment of 12 to one molecule and 34 to the other: | | |
| 12 34 | 34 12 | + |
| 21 34 | 34 21 | − |
| 21 43 | 43 21 | + |
| 12 43 | 43 12 | − |
| For assignment of 13 to one molecule and 24 to the other: | | |
| 13 24 | 24 13 | − |
| 31 24 | 24 31 | + |
| 31 42 | 42 31 | − |
| 13 42 | 42 13 | + |
| For assignment of 23 to one molecule and 14 to the other: | | |
| 32 14 | 14 32 | − |
| 23 41 | 41 23 | − |
| 23 14 | 14 23 | + |
| 32 41 | 41 32 | + |

The approximation associated with (5.22) would correspond to just using the first eight of these. In the semiclassical approximation for the centers of mass one is effectively using just four of these eight (e.g. the first column).

5.4 The vibrational and translational contributions to the specific heat are the same in the equilibrium and mixed states. Here I evaluate only the rotational part.

In each case, total $I$ of the molecule is 0, 1, 2 with nuclear wave functions respectively even, odd and even and multiplicities 1, 3, 5. Since deuterons are bosons the total wave function is even so the corresponding rotational states must have $L$ even (with weight $1+5=6$) for the even nuclear states and $L$ odd (with weight 3) for the odd nuclear states.

Let $Z_{\text{even, odd}} = \sum_{L \text{ even, odd}}(2L+1)\exp(-\hbar^2 L(L+1)/k_B T)$

Equilibrium:

$$F^e = -k_B NT \ln[6Z_{\text{even}} + 3Z_{\text{odd}}]$$

$$S^e = -(\partial F^e/\partial T)_{V,N} = k_B N\left[\ln[6Z_{\text{even}} + 3Z_{\text{odd}}] + T\left[\frac{6(\partial Z_{\text{even}}/\partial T) + 3(\partial Z_{\text{odd}}/\partial T)}{6Z_{\text{even}} + 3Z_{\text{odd}}}\right]\right]$$

$$C_v^e = T(\partial S^e/\partial T)_{V,N}$$

$$= k_B N\left[2\left(\frac{6(\partial Z_{\text{even}}/\partial T) + 3(\partial Z_{\text{odd}}/\partial T)}{6Z_{\text{even}} + 3Z_{\text{odd}}}\right) + T\left(\frac{6(\partial^2 Z_{\text{even}}/\partial T^2) + 3(\partial^2 Z_{\text{odd}}/\partial T^2)}{6Z_{\text{even}} + 3Z_{\text{odd}}}\right)\right.$$

$$\left. - T\left(\frac{6(\partial Z_{\text{even}}/\partial T) + 3(\partial Z_{\text{odd}}/\partial T)}{6Z_{\text{even}} + 3Z_{\text{odd}}}\right)^2\right]$$

Mixture:

$$F^{\rm m} = -k_{\rm B}NT\,((2/3)\ln Z_{\rm even} + (1/3)\ln Z_{\rm odd})$$

$$S^{\rm m} = -(\partial F^{\rm m}/\partial T)_{V,N} = k_{\rm B}N\Bigg[(2/3)\ln Z_{\rm even} + (1/3)\ln Z_{\rm odd} + T\left[(2/3)\left(\frac{(\partial Z_{\rm even}/\partial T)}{Z_{\rm even}}\right) + (1/3)\left(\frac{(\partial Z_{\rm odd}/\partial T)}{Z_{\rm odd}}\right)\right]\Bigg]$$

$$\begin{aligned} C_v^{\rm m} &= T(\partial S^{\rm m}/\partial T)_{V,N} \\ &= k_{\rm B}TN\Bigg[2\left((2/3)\left(\frac{(\partial Z_{\rm even}/\partial T)}{Z_{\rm even}}\right) + (1/3)\left(\frac{(\partial Z_{\rm odd}/\partial T)}{Z_{\rm odd}}\right)\right) \\ &\quad + \left((2/3)T\left(\frac{(\partial^2 Z_{\rm even}/\partial T^2)}{Z_{\rm even}}\right) + (1/3)T\left(\frac{(\partial^2 Z_{\rm odd}/\partial T^2)}{Z_{\rm odd}}\right)\right) \\ &\quad - \left((2/3)T\left(\frac{(\partial Z_{\rm even}/\partial T)}{Z_{\rm even}}\right)^2 + (1/3)T\left(\frac{(\partial Z_{\rm odd}/\partial T)}{Z_{\rm odd}}\right)^2\right)\Bigg] \end{aligned}$$

For numerical purposes I wrote these in terms of the sums

$$\Sigma_m^{\rm e,o}(x) = \sum_{L\ \text{even, odd}} (2L+1)(L(L+1))^m \exp(-L(L+1)x)$$

$m = 0, 1, 2$ giving ($x = \hbar^2/2Ik_{\rm B}T$)

$$C_v^{\rm e} = Nk_{\rm B}x^2\left[\left(\frac{2\Sigma_2^{\rm e} + \Sigma_2^{\rm o}}{2\Sigma_0^{\rm e} + \Sigma_0^{\rm o}}\right) - \left(\frac{2\Sigma_1^{\rm e} + \Sigma_1^{\rm o}}{2\Sigma_0^{\rm e} + \Sigma_0^{\rm o}}\right)^2\right]$$

$$C_v^{\rm m} = Nk_{\rm B}x^2\left[(2/3)\left(\frac{\Sigma_2^{\rm e}}{\Sigma_0^{\rm e}} - \left(\frac{\Sigma_1^{\rm e}}{\Sigma_0^{\rm e}}\right)^2\right) + (1/3)\left(\frac{\Sigma_2^{\rm o}}{\Sigma_0^{\rm o}} - \left(\frac{\Sigma_1^{\rm o}}{\Sigma_0^{\rm o}}\right)^2\right)\right]$$

The calculated rotational specific heat as a function of absolute temperature in kelvins is shown in Figure A.4.

5.7 In each case choose a partition of the energy axis such that there are $G_\alpha$ levels in the partition $\alpha$. We require $G_\alpha \gg 1$ and that the energy width of the partition be much less than $k_{\rm B}T$. (This is possible for a macroscopic system where the level spacing is small as long as there are no localized states.) Now consider a set $\{N_\alpha\}$ where $N_\alpha = \sum_{\nu\ \text{in}\ \alpha} n_\nu$ and ask: how many ways can the $n_\nu$ be assigned, consistent with this set $\{N_\alpha\}$? $k_{\rm B}$ times the ln of this number is the entropy. For convenience we say that $n_\nu$ "particles" have been assigned to $\nu$ for each set $\{n_\nu\}$ though the particles are all indistinguishable. We calculate the number of ways separately for fermions and for bosons.

**Fermions** For each $\alpha$, assign the first particle in $G_\alpha$ ways. For each of these, there are $G_\alpha - 1$ ways to assign the second particle, etc. until $N_\alpha$ particles have been assigned for a total of

$$G_\alpha!/(G_\alpha - N_\alpha)!$$

However, permutations of the labels on the $N_\alpha$ particles in these partitions give identical states, so the total number of distinguishable states is

$$G_\alpha!/((G_\alpha - N_\alpha)!N_\alpha!)$$

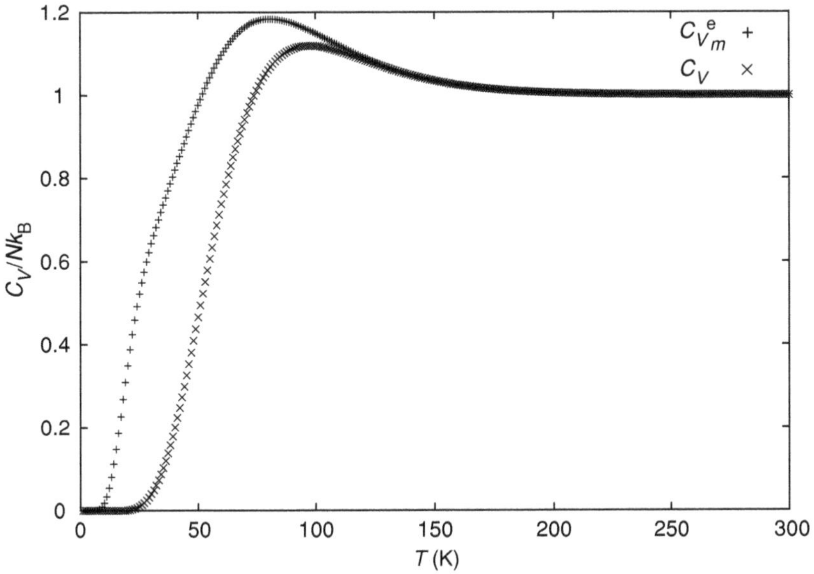


Figure A.4

(To be sure you understand this, working out an example, say with $G_\alpha = 5$ and $N_\alpha = 2$ is advised.) Now we suppose that both $G_\alpha$ and $N_\alpha$ are $\gg 1$ and apply Stirling's approximation to the ln:

$$\begin{aligned} S/k_B &= \sum_\alpha \ln(G_\alpha!/((G_\alpha - N_\alpha)!N_\alpha!)) \\ &= \sum_\alpha [G_\alpha \ln G_\alpha - (G_\alpha - N_\alpha)\ln(G_\alpha - N_\alpha) - N_\alpha \ln N_\alpha] \\ &= \sum_\alpha G_\alpha [\ln G_\alpha - (1 - n_\alpha)[\ln G_\alpha + \ln(1 - n_\alpha)] - n_\alpha[\ln n_\alpha + \ln G_\alpha] \\ &= \sum_\alpha G_\alpha[\ln G_\alpha(1 - (1 - n_\alpha) - n_\alpha) - (1 - n_\alpha)\ln(1 - n_\alpha) - n_\alpha \ln n_\alpha)] \\ &= \sum_\alpha G_\alpha((n_\alpha - 1)\ln(1 - n_\alpha) - n_\alpha \ln n_\alpha) \end{aligned}$$

Here $n_\alpha = N_\alpha/G_\alpha$. Finally, if the energy width of the partition is much less than $k_B T$, one can replace $\sum_\alpha G_\alpha \to \sum_\nu$ and $n_\alpha \to n_\nu$ in the summand, yielding (5.40) for fermions.

**Bosons** In this case, one can think of the distribution of states within $\alpha$ as follows. Lay out all the $N_\alpha$ particles, like marbles, in a row. Insert $G_\alpha - 1$ partitions between them in all possible ways. Each partitioning corresponds to a possible assignment of the particles to states $\nu$ except that the particles and the partitions are indistinguishable. Thus there are $(N_\alpha + G_\alpha - 1)!$ ways to lay out particles and partitions if particles and partitions are distinguishable and a total of

$$(N_\alpha + G_\alpha - 1)!/N_\alpha!(G_\alpha - 1)!$$

possible states taking indistinguishability into account. Applying Stirling's approximation to the ln, a calculation very similar to the fermion one gives (5.40) with the boson signs.

5.8 The energy levels in a box with periodic boundary conditions are $\hbar^2k^2/2m$ and the smallest nonzero value of $k$ is $2\pi/V^{1/3}$ (taking a cubic box for convenience; the argument works for any shaped box as long as it goes to $\infty$ in all directions in the thermodynamic limit). Thus we require $\mu \ll \hbar^2 4\pi^2/2mV^{2/3}$ in order to ignore $\mu$ in the second term in (5.64). On the other hand, combining (5.65) and (5.61) and assuming that $-\mu \ll k_\mathrm{B}T$ gives

$$-\mu = k_\mathrm{B}T/\left(\bar{N}\left(1-(T/T_0)^{3/2}\right)\right) = \left(k_\mathrm{B}T/\left(\rho\left(1-(T/T_0)^{3/2}\right)\right)\right)(1/V)$$

which varies as $1/V$ at any temperature a finite amount below $T_0$ in the thermodynamic limit. Thus

$$-\mu/(\text{smallest } \epsilon_k) = \left[\frac{\left(k_\mathrm{B}T/\left(\rho\left(1-(T/T_0)^{3/2}\right)\right)\right)(1/V)}{\hbar^2 4\pi^2/2mV^{2/3}}\right] \propto 1/V^{1/3} \to 0$$

in the thermodynamic limit.

5.10 (a)

$$Z_{\mathrm{gc}} = \sum_{\{n_\nu\}} \mathrm{e}^{\left(N\mu - \sum_\nu n_\nu \epsilon_\nu\right)\beta} = \sum_{\{n_\nu\}} \Pi_\nu \mathrm{e}^{n_\nu(\mu-\epsilon_\nu)\beta} = \Pi_\nu\left(1+\mathrm{e}^{(\mu-\epsilon_\nu)\beta} + 2\mathrm{e}^{2(\mu-\epsilon_\nu)\beta}\right)$$

with $z = \mathrm{e}^{\mu\beta}$:

$$\Omega = -k_\mathrm{B}T\sum_\nu \ln(1+z\,\mathrm{e}^{-\epsilon_\nu\beta} + 2z^2\,\mathrm{e}^{-2\epsilon_\nu\beta})$$

(b)

$$N = \sum_\nu \frac{1+2z\,\mathrm{e}^{-\epsilon_\nu\beta}}{z^{-1}\,\mathrm{e}^{\epsilon_\nu\beta}+1+z\,\mathrm{e}^{-\epsilon_\nu\beta}}$$

$$E = \sum_\nu \epsilon_\nu\left(\frac{1+2z\,\mathrm{e}^{-\epsilon_\nu\beta}}{z^{-1}\,\mathrm{e}^{\epsilon_\nu\beta}+1+z\,\mathrm{e}^{-\epsilon_\nu\beta}}\right)$$

The function $n_\nu = (1+2z\mathrm{e}^{-\epsilon_\nu\beta})/(z^{-1}\mathrm{e}^{\epsilon_\nu\beta}+1+z\mathrm{e}^{-\epsilon_\nu\beta})$ is plotted in Figure A.5, where it is compared with the corresponding function $2/(1+z\mathrm{e}^{-\epsilon_\nu\beta})$ which describes the occupancy of electrons with spin 1/2. We have used the value $\mu(0) = (\hbar^2/2m)(3\pi^2\rho)^{2/3}$ which can quite easily be shown to be the same for the two cases. We also show the function $1/(1+z\mathrm{e}^{-\epsilon_\nu\beta})$ with $\mu(0) = 2^{2/3}(\hbar^2/2m)(3\pi^2\rho)^{2/3}$ which is the occupancy at the same particle density for spinless fermions. The cases of spin 1/2 fermions and three state anyons are different near the Fermi level as shown in Figure A.6. (Figures are for free particles of density $0.5\times10^{23}$ cm$^{-3}$ and temperature 300 K.)

(c) As in the text, consider a function $f(\epsilon)$ (we will use $f=\mathcal{N}(\epsilon)$ and $f=\epsilon\mathcal{N}(\epsilon)$ as before). We need to evaluate

$$I = \int \mathrm{d}\epsilon f(\epsilon)\frac{1+2z\,\mathrm{e}^{-\epsilon\beta}}{z^{-1}\,\mathrm{e}^{\epsilon\beta}+1+z\,\mathrm{e}^{-\epsilon\beta}}$$

and introducing $x=(\mu-\epsilon)\beta$ we note that

$$(1+2\mathrm{e}^x)/(\mathrm{e}^{-x}+1+\mathrm{e}^x) - 2 = -(1+2\mathrm{e}^{-x})/(\mathrm{e}^{-x}+1+\mathrm{e}^x)$$

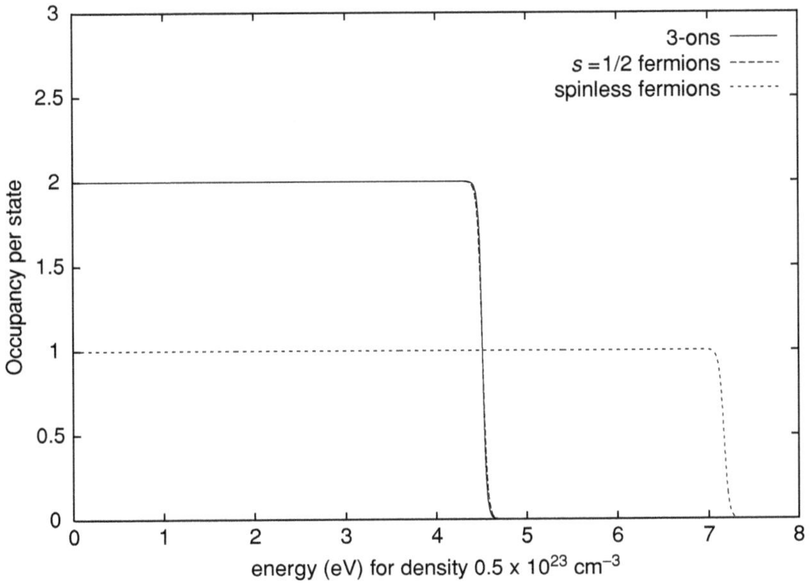


Figure A.5

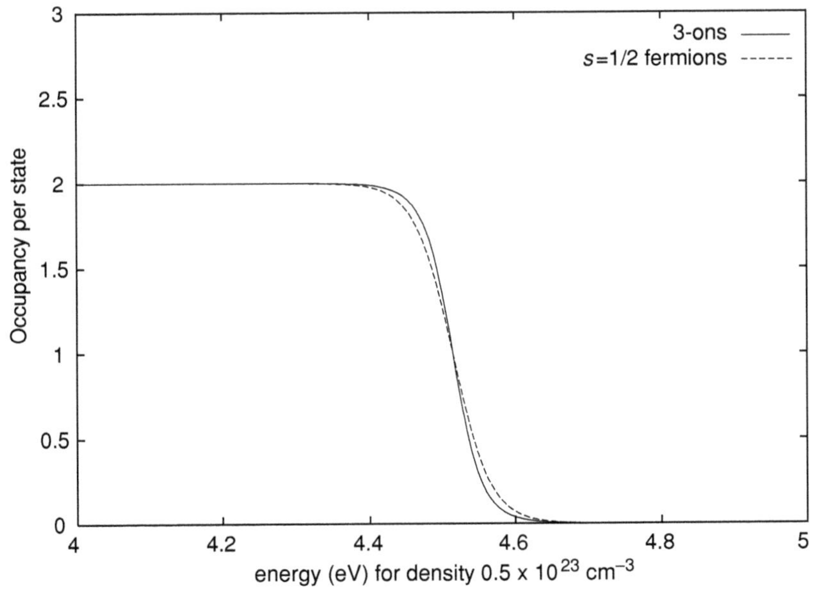


Figure A.6

Using this we rewrite $I$ as

$$I = 2k_{\mathrm{B}}T \int_0^{\mu\beta} f(\mu - k_{\mathrm{B}}Tx)\,\mathrm{d}x + k_{\mathrm{B}}T \int_0^{\infty} \frac{[f(\mu + k_{\mathrm{B}}Tx) - f(\mu - k_{\mathrm{B}}Tx)](1 + 2\mathrm{e}^{x})}{(\mathrm{e}^{-x} + 1 + \mathrm{e}^{x})}\,\mathrm{d}x$$

Using $f = \mathcal{N}(\epsilon)$ we have

$$N = 2\int_0^{\mu(0)} \mathcal{N}(\epsilon)\,\mathrm{d}\epsilon + 2(\mu(T) - \mu(0))\mathcal{N}(\mu(0)) + 2(k_{\mathrm{B}}T)^2\mathcal{N}'(\mu(0))\int_0^{\infty} \frac{x(1 + 2\mathrm{e}^{-x})}{\mathrm{e}^{-x} + 1 + \mathrm{e}^{x}}\,\mathrm{d}x$$

Denote the integral

$$\int_0^{\infty} \frac{x(1 + 2\mathrm{e}^{-x})}{\mathrm{e}^{-x} + 1 + \mathrm{e}^{x}}\,\mathrm{d}x \equiv I_1$$

We have

$$\mu(T) - \mu(0) = -(k_{\mathrm{B}}T)^2 \frac{\mathcal{N}'(\mu(0))}{\mathcal{N}(\mu(0))} I_1$$

Now with $f = \epsilon\mathcal{N}(\epsilon)$ we get the energy

$$E = 2\int_0^{\mu(0)} \epsilon\mathcal{N}(\epsilon)\,\mathrm{d}\epsilon + 2(\mu(T) - \mu(0))\mu(0)\mathcal{N}(\mu(0)) + 2(k_{\mathrm{B}}T)^2\mathrm{d}(\epsilon\mathcal{N}(\epsilon))/\mathrm{d}\epsilon|_{\mu(0)} \int_0^{\infty} \frac{x(1 + 2\mathrm{e}^{-x})}{\mathrm{e}^{-x} + 1 + \mathrm{e}^{x}}\,\mathrm{d}x$$

and using $\mathcal{N} \propto \epsilon^{1/2}$

$$E = E(0) + 2(k_{\mathrm{B}}T)^2\mathcal{N}(\mu(0))I_1$$

This is exactly the same as the low temperature energy of a gas of free spin 1/2 fermions except that the factor $I_1$ has replaced the integral $\int_0^{\infty} \frac{x}{\mathrm{e}^{x}+1}\mathrm{d}x = \pi^2/12$. Numerically these are close but not the same: $\pi^2/12 = 0.822467\ldots$, $I_1 = 1.09629\ldots$

(d) We show the function $\langle n_\nu\rangle(\epsilon)$ for the cases in which the maximum value of $n_\nu$ is 1, 2, 3 and 4 in Figure A.7 (same parameter choices as for the preceding figures).

Analytically,

$$\langle n_\nu\rangle = \sum_{k=0}^{k=n} k\,\mathrm{e}^{-kx} \Bigg/ \sum_{k=0}^{k=n} \mathrm{e}^{-kx}$$

with $x = (\epsilon - \mu)\beta$ and $\mu = \mu_2(2/n)^{2/3}$ where $\mu_2$ is the chemical potential when $n = 2$.

5.12 First evaluate the next term in $\mu(T) - \mu(0)$.

Let $\mu(T) - \mu(0) = aT^2 + bT^4$. (It is not hard to see that only even powers come in.) Then from (5.84) with $I = N$ and $f = \mathcal{N} = K\epsilon^{1/2}$ density of states and the $\mathcal{O}((\mu(T) - \mu(0))^2)$ term written explicitly we have:

$$N = \int_0^{\mu(0)} f(\epsilon)\,\mathrm{d}\epsilon + (\mu(T) - \mu(0))f(\mu(0)) + (1/2)(\mu(T) - \mu(0))^2 f'(\mu(0)) + 2f'(\mu(T))(k_{\mathrm{B}}T)^2 \int_0^{\infty} z\,\mathrm{d}z/(\mathrm{e}^{z} + 1) + (2/3!)f'''(\mu(T))(k_{\mathrm{B}}T)^4 \int_0^{\infty} z^3\mathrm{d}z/(\mathrm{e}^{z} + 1)$$

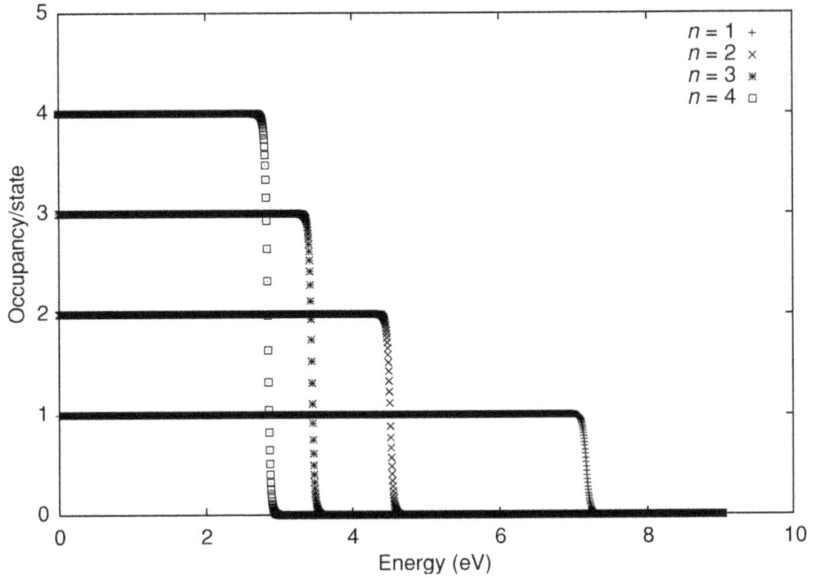


Figure A.7

Now expand $f'(\mu(T)) = f'(\mu(0)) + (\mu(T) - \mu(0))f''(\mu(0))$ in the fourth term. (No such expansion is needed in the fifth term if we are only working to order $T^4$.) Inserting $\mu(T) - \mu(0) = aT^2 + bT^4$ and writing the resulting equations order by order:

$T^0$

$$N = \int_0^{\mu(0)} f(\epsilon)\,\mathrm{d}\epsilon$$

$T^2$

$$aT^2 f(\mu(0)) + 2f'(\mu(0))(k_\mathrm{B}T)^2 \int_0^\infty z\,\mathrm{d}z/(\mathrm{e}^z + 1) = 0$$

$T^4$

$$bT^4 f(\mu(0)) + (1/2)a^2T^4 f'(\mu(0)) + 2f''(\mu(0))aT^2(k_\mathrm{B}T)^2 \times \int_0^\infty z\,\mathrm{d}z/(\mathrm{e}^z + 1) + (2/3!)f'''(\mu(0))(k_\mathrm{B}T)^4 \int_0^\infty z^3\,\mathrm{d}z/(\mathrm{e}^z + 1) = 0$$

The order $T^0$ equation gives (5.85). The order $T^2$ equation gives (5.88) or

$$a = -2(f'(\mu(0))/f(\mu(0)))k_\mathrm{B}^2 \int_0^\infty z\,\mathrm{d}z/(\mathrm{e}^z + 1)$$

The order $T^4$ equation gives

$$b = (1/f(\mu(0))\Big[-(1/2)a^2 f'(\mu(0)) - 2af''(\mu(0))k_\mathrm{B}^2 \times \int_0^\infty z\,\mathrm{d}z/(\mathrm{e}^z + 1) - (1/3)f'''(\mu(0))k_\mathrm{B}^4 \int_0^\infty z^3\,\mathrm{d}z/(\mathrm{e}^z + 1)\Big]$$

Now with density of states $\mathcal{N}(\epsilon) = f(\epsilon) = K\epsilon^{1/2}$ one has $f'(\epsilon) = (1/2\epsilon)\mathcal{N}$, $f''(\epsilon) = (-1/4\epsilon^2)\mathcal{N}, f'''(\epsilon) = (3/8\epsilon^3)\mathcal{N}$. Let $I_1 = \int_0^\infty z\,\mathrm{d}z/(\mathrm{e}^z + 1)$ and $I_3 = \int_0^\infty z^3\,\mathrm{d}z/(\mathrm{e}^z + 1)$. Then the equation for $a$ becomes

$$a = -k_\mathrm{B}^2 I_1/\mu(0)$$

Inserting this in the equation for $b$ and simplifying one finds

$$b = -\left(k_\mathrm{B}^4/\mu(0)^3\right)\left[(3/4)I_1^2 + (1/8)I_3\right]$$

Now consider the energy. We use (5.84) again with the $\mathcal{O}(\mu(T) - \mu(0))^2)$ term made explicit and with $f = \epsilon\mathcal{N} \equiv F$:

$$\begin{aligned} E = \int_0^{\mu(0)} & F(\epsilon)\,\mathrm{d}\epsilon + (\mu(T) - \mu(0))F(\mu(0)) + (1/2)(\mu(T) - \mu(0))^2 F'(\mu(0)) \\ & + 2F'(\mu(0)) + (\mu(T) - \mu(0))F''(\mu(0))(k_\mathrm{B}T)^2 I_1 + (2/3!)F'''(\mu(T))(k_\mathrm{B}T)^4 I_3 \end{aligned}$$

Insert $\mu(T) - \mu(0) = aT^2 + bT^4$ and keep only terms up to $\mathcal{O}(T^4)$:

$$\begin{aligned} E(T) - E(0) = {} & aT^2 F(\mu(0)) + 2F'(\mu(0))(k_\mathrm{B}T)^2 I_1 + bT^4 F(\mu(0)) + (1/2)a^2T^4 F'(\mu(0)) \\ & + 2aT^2 F''(\mu(0))(k_\mathrm{B}T)^2 I_1 + (1/3)F'''(\mu(0))(k_\mathrm{B}T)^4 I_3 \end{aligned}$$

With $\mathcal{N} = K\epsilon^{1/2}$ one easily finds $F'(\mu(0)) = (3/2)\mathcal{N}(\mu(0))$, $F''(\mu(0)) = (3/4)\mathcal{N}(\mu(0))/\mu(0)$, $F'''(\mu(0)) = (-3/8)\mathcal{N}(\mu(0))/\mu(0)^2$. Then from the equation for $E(T) - E(0)$ one finds the value of the $T^3$ term in $C_V$ as

$$\begin{aligned} C_V = \text{linear term} + T^3\big[ & 4bF(\mu(0)) + 2a^2F'(\mu(0)) \\ & + 8ak_\mathrm{B}^2 F''(\mu(0))I_1 + (4/3)F'''(\mu(0))k_\mathrm{B}^4 I_3\big] \end{aligned}$$

Inserting the expressions of the derivatives of $F$ and for $a$ and $b$ and simplifying one obtains

$$C_V = \text{linear term} - T^3\left(k_\mathrm{B}^4\mathcal{N}(\mu(0))/\mu(0)^2\right)\left[6I_1^2 + I_3\right]$$

The integrals are

$$\begin{aligned} I_1 &= (1/2)\Gamma(2)\zeta(2) = \pi^2/12 \\ I_3 &= (7/8)\Gamma(4)\zeta(4) \end{aligned}$$

The value of $\zeta(4)$ is not given in the table in the text. It can be deduced from a formula in Gradshteyn and Ryzhik to be $\pi^4/90$ giving

$$I_3 = (7/120)\pi^4$$

Then

$$C_V = \text{linear term} - T^3\left(k_\mathrm{B}^4\mathcal{N}(\mu(0))/\mu(0)^2\right)(1/10)\pi^4$$

5.13 For $T \leq T_0$,

$$\bar{N} = N_0 + \sum_{n_x,n_y,n_z \neq 0} \frac{1}{\mathrm{e}^{\beta\epsilon_{n_x,n_y,n_z}} - 1}$$

where $\epsilon_{n_x,n_y,n_z} = \hbar\sqrt{K/m}(n_x + n_y + n_z)$ and $n_x, n_y, n_z$ run from 0 to $\infty$. When the number of particles is large, the spacing of energy levels is small compared to the temperature near $T_0$ and the sums can be approximated by integrals giving ($\omega_0 = \sqrt{K/m}$)

$$\bar{N} = N_0 + (k_B T/\hbar\omega_0)^3 \int_0^\infty \int_0^\infty \int_0^\infty \mathrm{d}x\,\mathrm{d}y\,\mathrm{d}z(1/(\mathrm{e}^{x+y+z} - 1))$$

Denote

$$I = \int_0^\infty \int_0^\infty \int_0^\infty \mathrm{d}x\,\mathrm{d}y\,\mathrm{d}z(1/(\mathrm{e}^{x+y+z} - 1))$$

Then $T_0$ is found by setting $N_0 = 0$ and $T = T_0$ in this:

$$T_0 = (\hbar\omega_0/k_B)(N/I)^{1/3}$$

and for $T < T_0$

$$N_0/\bar{N} = 1 - (T/T_0)^3$$

## Chapter 6

6.1

$$b_2 = \left(\Omega^{(d)}/2\lambda^d\right) \int_0^\infty r^{(d-1)} \left(\exp\left(-\epsilon \left(\frac{b}{r}\right)^n \beta\right) - 1\right) \mathrm{d}r$$

Introduce $x = (b/r)(1/(\epsilon\beta)^{1/n})$ so that

$$b_2 = \left(\Omega^{(d)}/2\lambda^d\right)\left(b^d/(\epsilon\beta)^{d/2}\right) \int_0^\infty \frac{\mathrm{d}x}{x^{d+1}} \left(\mathrm{e}^{-x^n} - 1\right)$$

The integral is always convergent for $x \to \infty$ corresponding to short distances, but for $x \to 0$ the integrand becomes $-x^{n-d-1}$ and gives a finite result only if $n - d - 1 > -1$ or $n > d$.

The equation of state is

$$P/k_B T = \rho - b_2\lambda^d\rho^2 = \rho - \rho^2 b^d I_d/(\epsilon\beta)^{d/2}$$

in which

$$I_d = \Omega^{(d)} \int_0^\infty \frac{\mathrm{d}x}{x^{d+1}} \left(\mathrm{e}^{-x^n} - 1\right)$$

6.2

$$b_2 = (4\pi/2\lambda^3) \int_0^\sigma r^2\left(\mathrm{e}^{V_0/k_B T} - 1\right) \mathrm{d}r = (2\pi/3)(\sigma/\lambda)^3\left(7\mathrm{e}^{V_0/k_B T} - 8\right)$$

$$P/k_B T = \rho + (2\pi/3)(\sigma^3\rho)\rho\left(8 - 7\mathrm{e}^{V_0/k_B T}\right)$$

The second term of $P$ versus $1/\rho$ changes sign when

$$8 - 7\mathrm{e}^{V_0/k_B T} = 0$$

that is when

$$T = T_0 \equiv V_0/k_B \ln(8/7)$$

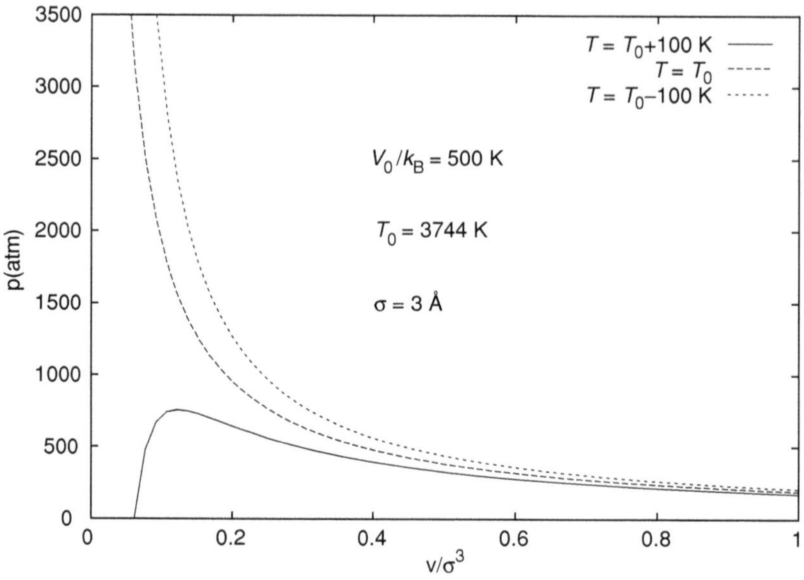


Figure A.8

we show the equation of state for some indicated values in Figure A.8.

$$\Omega = -(k_B T V/\lambda^3)\sum_l b_l z^l$$

$$N = -\left(\frac{\partial \Omega}{\partial \mu}\right)_{T,V} = (k_B T V/\lambda^3)\sum_l \beta b_l z^l$$

so to second order

$$\rho\lambda^3 = z + 2z^2 b_2$$

solving for $z$:

$$z = \rho\lambda^3 - 2b_2(\rho\lambda^2)^2 + \cdots$$

then

$$F = \Omega + \mu N = N k_B T(\ln(\rho\lambda^3) - 1 - b_2(\lambda^3)\rho)$$

and using $S = -(\partial F/\partial T)_{T,V}$ gives

$$S = N k_B\left(-\ln(\rho\lambda^3) + \left(\frac{5}{2}\right) + \rho T\frac{\partial}{\partial T}(b_2\lambda^3)\right)$$

and we find the specific heat

$$C_{V,N} = N k_B\left[\frac{3}{2} + \frac{14\pi\sigma^3\rho}{3}\left(\frac{V_0}{k_B T}\right)^2 e^{V_0/k_B T}\right]$$

6.3 The relevant integrals are

$$b_2 = (1/2\lambda^3) \int d\vec{r}_{12} \left(e^{-v(r_{12})/k_B T} - 1\right)$$

and

$$b_3 = (1/6\lambda^6) \int d\vec{r}_{21}\, d\vec{r}_{31} \left(e^{-v(r_{21})/k_B T} - 1\right)\left(e^{-v(r_{31})/k_B T} - 1\right)\left(e^{-v(r_{32})/k_B T} - 1\right) + 2b_2^2$$

The integrals can be done using the Monte Carlo method (or otherwise). Results are shown in Figure A.9.

The equation of state from the first 2 cluster integrals is

$$P_{\text{vir}} = k_B T(\rho + \lambda^3 a_2 \rho^2 + \lambda^6 a_3 \rho^3)$$

in which $a_2 = -b_2$, $a_3 = 4b_2^2 - 2b_3$ and $\lambda$ is the thermal wavelength. In Figures A.10 and A.11 the van der Waals equation of state is compared with the ideal gas law and $P_{\text{vir}}$. At higher temperatures and/or higher volumes, the virial expansion is an improvement on the ideal gas equation of state (Figure A.10).

Near the critical point, neither the ideal gas law nor the virial equation of state does very well (Figure A.11).

6.5 (a) $U_1(\vec{r}) = \sum_\alpha e^{-\beta\epsilon_\alpha} |\psi_\alpha(\vec{r})|^2$

With periodic boundary conditions $\psi_\alpha(\vec{r}) = (1/\sqrt(V))e^{i\vec{k}_\alpha \cdot \vec{r}}$ so

$$U_1 = (1/V) \sum_{\vec{k}} e^{-\beta\epsilon_{\vec{k}}}$$

with $\epsilon_{\vec{k}} = \hbar^2 k^2/2m$. As long as there is no Bose condensation one changes the sum to an integral on the energy and obtains

$$U_1 = 1/\lambda^3$$

where $\lambda$ is the thermal wavelength.

(b)

$$W_2(1,2) = \sum_\alpha \psi_\alpha^*(1,2)\, e^{-(T_1+T_2+V(1,2))\beta} \psi_\alpha(1,2) \to \sum_\alpha \psi_\alpha^*(1,2)\, e^{-(T_1+T_2)\beta} \psi_\alpha(1,2)$$

because $V(1,2)$ goes to zero at large separation. $T_1$ and $T_2$ are the single particle kinetic energies. In the limit of large separation the two particle Schrödinger equation is

$$(T_1 + T_2)\psi_\alpha(1,2) = \epsilon_\alpha \psi_\alpha(1,2)$$

which is separable with solutions

$$\psi_\alpha(1,2) = \phi_{\nu_1}(1)\phi_{\nu_2}(2)$$

in which

$$T_1 \phi_{\nu_i}(1) = \epsilon_{\nu_i} \phi_{\nu_i}(1)$$

Imposing the Pauli principle we have, again at large separations,

$$\psi_\alpha(1,2) = (1/\sqrt{2})(\phi_{\nu_1}(1)\phi_{\nu_2}(2) \pm \phi_{\nu_1}(2)\phi_{\nu_2}(1))$$

which is still an eigenstate at large separations. Insert this back in the formula for $W_2(1,2)$ at large separations and use periodic boundary conditions so that the $\phi_{\nu_i}$ are plane waves.

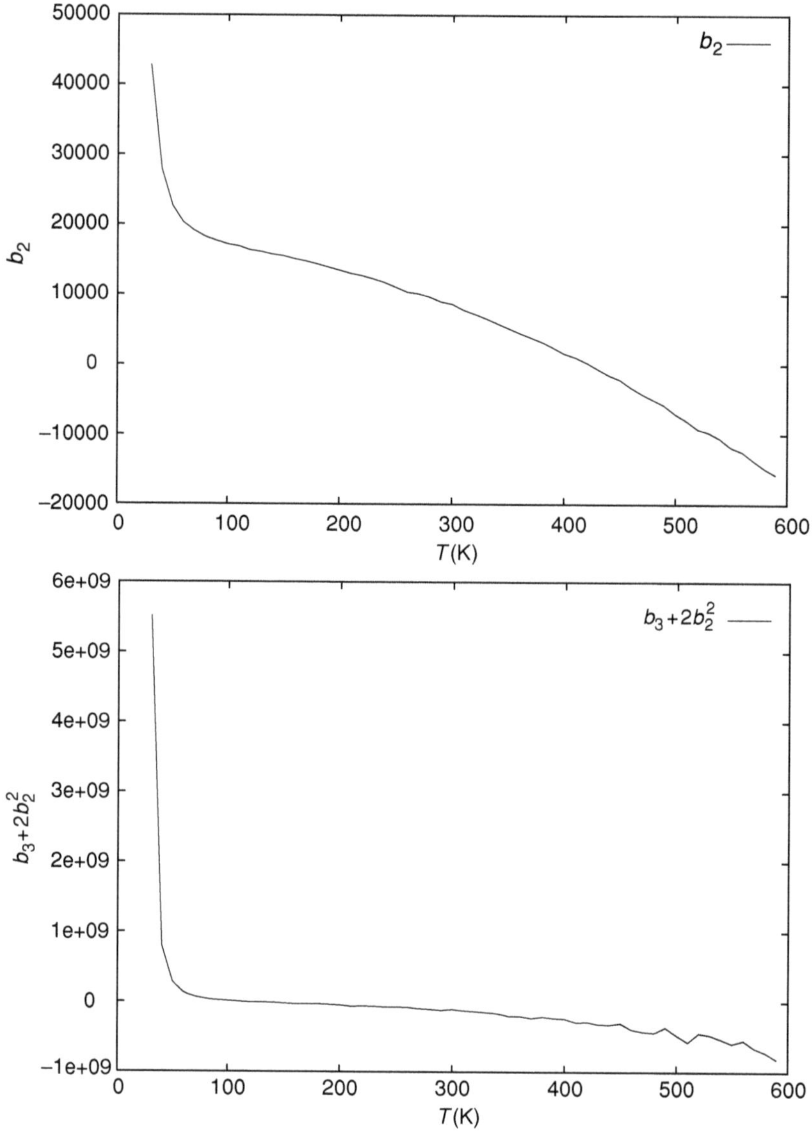


Figure A.9

(There is a problem of normalization here which goes away as the normalization volume becomes large.)

$$W_2(1,2) \to \sum_{\nu_1} \mathrm{e}^{-\beta\epsilon_{\nu_1}} \sum_{\nu_2} \mathrm{e}^{-\beta\epsilon_{\nu_2}} \pm \left(\sum_{\vec{k}_1} \mathrm{e}^{\mathrm{i}\vec{k}_1\cdot(\vec{r}_1-\vec{r}_2)} \mathrm{e}^{-\beta\epsilon_{\vec{k}_1}} / V\right)\left(\sum_{\vec{k}_2} \mathrm{e}^{-\mathrm{i}\vec{k}_2\cdot(\vec{r}_1-\vec{r}_2)} \mathrm{e}^{-\beta\epsilon_{\vec{k}_2}} / V\right)$$

The first term on the right is the one given in (6.97). In the second term, I have explicitly inserted the plane wave single particle solutions and one sees the same integrals that

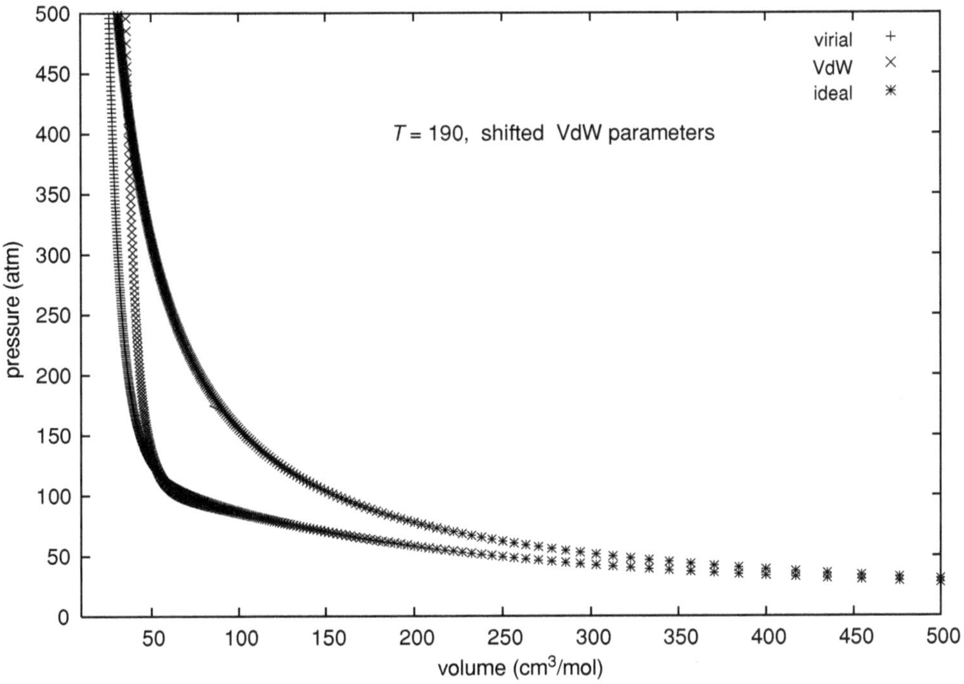


Figure A.10

appeared, for example, in Chapter 4 (equation (4.15)) which give, for the second term

$$(1/\lambda^6)\exp(-2\pi|\vec{r}_1 - \vec{r}_2|^2/\lambda^2)$$

which goes to zero at large separations.

## Chapter 7

7.1 The Yvon–Born–Green equation is

$$\ln g(r_{12}) = \frac{-v(r_{12})}{k_B T} + \frac{\rho}{k_B T}\int_{r_{12}}^{\infty} dr'_{12} \int d\vec{r}'_3\, g(r'_{13})g(r'_{23})\frac{\partial v(r'_{13})}{\partial r'_{13}}(\hat{r}'_{12}\cdot\hat{r}'_{13})$$

The attached code solves it numerically by iteration. The code starts at a higher temperature of 125 K and "works down" to 85 K to improve the convergence of the iterative procedure. The second term has been written

$$\frac{\rho}{k_B T}\int_{r_{12}}^{\infty} dr'_{12} 2\pi \int dr'_{13}\, r'^2_{13} g(r'_{13})\frac{\partial v(r'_{13})}{\partial r'_{13}} \int d\mu'_{23}\, \mu'_{23}\, g\left(\sqrt{r'^2_{12} + r'^2_{13} - 2\mu'_{23} r'_{12} r'_{13}}\right)$$

We show a self consistent solution with $T = 85$ K in Figure A.12.

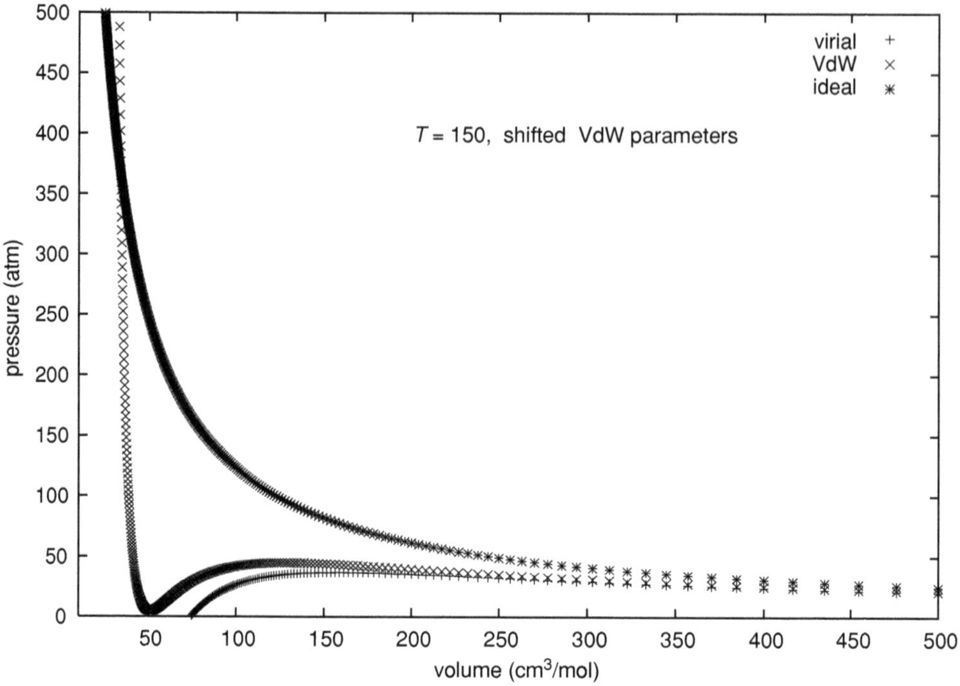


Figure A.11

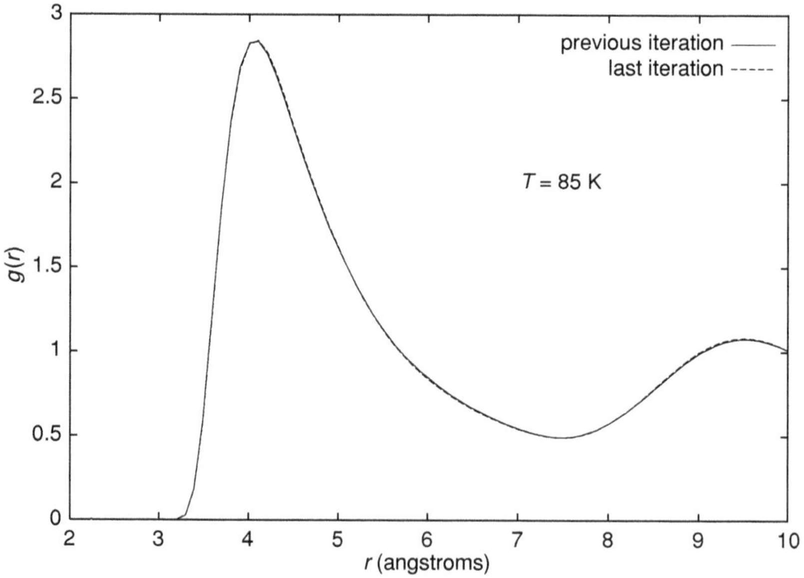


Figure A.12

```
c       this will do ybg equation for  hw1
        double precision gnew(1000),g(1000)
        dx=0.1
        pi=4.0*atan(1.0)
c       B=0.912e12
        sigma=3.45
        epsilon=125.
        rho=1.407/(40*1.66)
        t=150.
        open(1,file='goutps1.dat')
        do 1 i=1,300
         r=i*dx
         sigmap=1.2*sigma
         if(r.le.sigma)then
             g(i)=0.0
         else
c        g(i)=1.0
         x=sigma/r
         v=4*epsilon*(x**12-x**6)
             g(i)=exp(-v/t)
         endif
1        continue
        do 200 it=1,3
        t=145-it*20
        write(*,*) 't=',t
        do 100 iterate=1,10
        test2=0.0
        do 2 i=25,100
         r12=i*dx
c        write(*,*) 'r12=',r12
         fintegral12p=0.0
        do 3 j=i,100
         r12p=j*dx
c        write(*,*)'r12p=',r12p
         fint13old=0.0
         fintegral13=0
         do 4 k=1,100
            r13p=k*dx
c           write(*,*)'r13p=',r13p
            finside=0.0
            fmuold=-1.0
            argold=sqrt(r12p**2+r13p**2-2.0*fmuold*r12p*r13p)
            iold=argold/dx
            fintold=fmuold*(g(iold)-1.0)
           do 5 l=1,50
             fmu=-1.0+float(l)*dx/(2.5)
             arg=sqrt(r12p**2+r13p**2-2.0*fmu*r12p*r13p)
             if(arg.ge.0.5*sigma)then
             int1=arg/dx
             fint=fmu*(g(int1)-1.0)
             else
               fint=-fmu
             endif
             finside=finside +(dx/5.0)*(fintold+fint)
             fintold=fint
5           continue
c           write(*,*) 'finside=',finside,'r12p=',r12p,'r13p',r13p
            x=sigma/r13p
            if(x.le.2.0)then
            f13=(epsilon/sigma)*(-48.*(x**13)+24.0*(x**7))
            fint13=(r13p**2)*g(k)*f13*finside
            else
```

```
              fint13=0.0
           endif
c          write(*,*) 'r13p,g(k),f13,finside',r13p,g(k),f13,finside
           fintegral13=fintegral13+(dx/2.0)*(fint13+fint13old)
           fint13old=fint13
4        continue
         if(j.eq.i)then
           f12pold=fintegral13
           go to 3
         else
           f12p=fintegral13
           fintegral12p=fintegral12p+(dx/2)*(f12pold+f12p)
           f12pold=f12p
         endif
c       write(*,*)'r12p,f12p, fintegral12p',r12p,f12p,fintegral12p
3       continue
c       write(*,*) 'finished 3'
        x=sigma/r12
c       write(*,*) 'computed x'
        v=4*epsilon*(x**12-x**6)
c       write(*,*) 'computed v=',v
c       write(*,*) 'fintegral12p=',fintegral12p
c       write(*,*) 'rho=',rho,'pi=','t=',t
        argexp=(v+rho*2.0*pi*fintegral12p)/t
        correction=rho*2.0*pi*fintegral12p
c       write(*,*)'rho,pi,fintegral12p',rho,pi,fintegral12p
c       write(*,*) v,correction
c       write(*,*) 'computed argexp'
        test = abs(argexp)
        if(test.ge.100)then
             gnew(i)=0.0
        else
             gnew(i)=exp(-argexp)
        endif
        test2=test2+(g(i)-gnew(i))**2
        gold=g(i)
        g(i)=0.5*gnew(i)+(1-0.5)*gold
c       write(1,*)r12,g(i),gnew(i)
2       continue
        write(*,*) 'iterate=',iterate, 'test2=', test2
        if(test2.le.1.0E-4) go to 200
100     continue
200     continue
        do 300 i=25,100
         r12=i*dx
        write(1,*)r12,g(i),gnew(i),g(i)
300     continue
        stop
        end
```

7.3

$$\langle K^2\rangle = \int \mathrm{d}^n\vec{p}\left(\sum_i \vec{p}_i^2/2m\right)^2 \mathrm{e}^{-\beta\sum_i \vec{p}_i^2/2m} \Big/ \int \mathrm{d}^n\vec{p}\,\mathrm{e}^{-\beta\sum_i \vec{p}_i^2/2m}$$
$$= \int \mathrm{d}^n\vec{p}\,\mathrm{e}^{-\beta\sum_i \vec{p}_i^2/2m}\left(\left(\vec{p}_1^2/2m\right)^2 + \left(\vec{p}_2^2/2m\right)^2 + \cdots + 2p_1^2p_2^2/4m^2 + \cdots\right)\Big/$$
$$\int \mathrm{d}^n\vec{p}\,\mathrm{e}^{-\beta\sum_i \vec{p}_i^2/2m}$$
$$= N\int \mathrm{d}\vec{p}\,\mathrm{e}^{-\beta\vec{p}^2/2m}(\vec{p}^2/2m)^2 \Big/ \int \mathrm{d}\vec{p}\,\mathrm{e}^{-\beta\vec{p}^2/2m} + (N(N-1)/2)$$
$$\times\left(\int d\vec{p}\,\mathrm{e}^{-\beta\vec{p}^2/2m}\vec{p}^2/2m \Big/ \int d\vec{p}\,\mathrm{e}^{-\beta\vec{p}^2/2m}\right)^2$$
$$\equiv N\langle p^4/4m^2\rangle + N(N-1)\langle p^2/2m\rangle^2$$

Similarly

$$\langle K\rangle^2 = N^2\langle p^2/2m\rangle^2$$

so

$$\langle K^2\rangle - \langle K\rangle^2 = N(\langle p^4/4m^2\rangle - \langle p^2/2m\rangle^2)$$

Evaluating the integrals,

$$\langle p^4/4m^2\rangle = (15/4)N(k_\mathrm{B}T)^2$$
$$\langle p^2/2m\rangle = (3/2)Nk_\mathrm{B}T$$

(which is equipartition) and

$$\langle K^2\rangle - \langle K\rangle^2 = N(3/2)(k_\mathrm{B}T)^2$$
$$\sqrt{(\langle K^2\rangle - \langle K\rangle^2)}/\langle K\rangle = \sqrt{(2/3N)}$$

## Chapter 8

8.1 Expand (8.10) by writing $\tilde{\epsilon}_{\vec{p},\sigma} - \mu = \epsilon^{(0)}_{\vec{p},\sigma} - \mu_0 + \delta\epsilon_{\vec{p},\sigma} + \sum_{\vec{p}',\sigma'} f_{\vec{p},\sigma;\vec{p}\,',\sigma'}\langle\delta n_{\vec{p}\,',\sigma'}\rangle - \delta\mu$ where $\epsilon^{(0)}_{\vec{p},\sigma} = \hbar^2p^2/2m$ is the single particle energy for noninteracting electrons $\delta\epsilon_{\vec{p},\sigma} = (\hbar^2p^2/2)(1/m^* - 1/m)$ and $\delta\mu$ is the shift in the chemical potential caused by the interactions. (We keep the temperature dependence of the chemical potential in the noninteracting model in the first term. We drop the $\langle\rangle$ on $\langle\delta n_{\vec{p}',\sigma'}\rangle$ in the sequel because we will always be referrring to the thermal average.) We denote $n^{(0)}_{\vec{p},\sigma} \equiv 1/\left(\mathrm{e}^{(\epsilon^{(0)}_{\vec{p},\sigma}-\mu^{(0)})\beta} + 1\right)$ and $\delta n_{\vec{p},\sigma} = n_{\vec{p},\sigma} - n^{(0)}_{\vec{p},\sigma}$.

We have

$$\delta n_{\vec{p},\sigma} = (\delta\epsilon_{\vec{p},\sigma} - \delta\mu)\left(\mathrm{d}n^{(0)}_{\vec{p},\sigma}/\mathrm{d}\epsilon^{(0)}_{\vec{p},\sigma}\right) + \sum_{\vec{p}',\sigma'} f_{\vec{p},\sigma;\vec{p}',\sigma'}\delta n_{\vec{p}',\sigma'}\left(\mathrm{d}n^{(0)}_{\vec{p},\sigma}/\mathrm{d}\epsilon^{(0)}_{\vec{p},\sigma}\right)$$

Rearranging,

$$\sum_{\vec{p}',\sigma'}\left(\delta_{\vec{p},\sigma;\vec{p}'\sigma'} - f_{\vec{p},\sigma;\vec{p}',\sigma'}\left(\mathrm{d}n^{(0)}_{\vec{p},\sigma}/\mathrm{d}\epsilon^{(0)}_{\vec{p},\sigma}\right)\right)\delta n_{\vec{p}'} = (\delta\epsilon_{\vec{p},\sigma} - \delta\mu)\left(\mathrm{d}n^{(0)}_{\vec{p},\sigma}/\mathrm{d}\epsilon^{(0)}_{\vec{p},\sigma}\right)$$

We invert the matrix $\delta_{\vec{p},\sigma;\vec{p}'\sigma'} - f_{\vec{p},\sigma;\vec{p}',\sigma'}\left(\mathrm{d}n^{(0)}_{\vec{p},\sigma}/\mathrm{d}\epsilon^{(0)}_{\vec{p},\sigma}\right)$ to lowest order in the interaction term giving

$$\delta n_{\vec{p},\sigma} = (\delta\epsilon_{\vec{p},\sigma} - \delta\mu)\left(\mathrm{d}n^{(0)}_{\vec{p},\sigma}/\mathrm{d}\epsilon^{(0)}_{\vec{q},\sigma}\right) + \sum_{\vec{p}',\sigma'} f_{\vec{p},\sigma;\vec{p}',\sigma'}(\delta\epsilon_{\vec{p}',\sigma} - \delta\mu)\left(\mathrm{d}n^{(0)}_{\vec{p},\sigma}/\mathrm{d}\epsilon^{(0)}_{\vec{p},\sigma}\right)\left(\mathrm{d}n^{(0)}_{\vec{p}',\sigma}/\mathrm{d}\epsilon^{(0)}_{\vec{p}',\sigma'}\right)$$

The chemical potential shift is evaluated using the requirement

$$\sum_{\vec{p},\sigma} \delta n_{\vec{p},\sigma} = 0$$

$$\delta\mu = \left(\frac{\int \mathrm{d}\epsilon(\mathrm{d}n^{(0)}(\epsilon)/\mathrm{d}\epsilon)\mathcal{N}(\epsilon)\epsilon(1/m^* - 1/m) + \int \mathrm{d}\epsilon\left(\mathrm{d}n^{(0)}(\epsilon)/\mathrm{d}\epsilon\right)\int \mathrm{d}\epsilon'\left(\mathrm{d}n^{(0)}(\epsilon')/\mathrm{d}\epsilon'\right)\mathcal{N}(\epsilon)\mathcal{N}(\epsilon')\epsilon(1/m^* - 1/m)F_0}{\int \mathrm{d}\epsilon\left(\mathrm{d}n^{(0)}(\epsilon)/\mathrm{d}\epsilon\right)\mathcal{N}(\epsilon) + \int \mathrm{d}\epsilon\left(\mathrm{d}n^{(0)}(\epsilon)/\mathrm{d}\epsilon\right)\int \mathrm{d}\epsilon'\left(\mathrm{d}n^{(0)}(\epsilon')/\mathrm{d}\epsilon'\right)\mathcal{N}(\epsilon)\mathcal{N}(\epsilon')F_0}\right)$$

Here $\mathcal{N}$ is the noninteracting density of states and we have summed over spin indices assuming that the $\epsilon_{\vec{p},\sigma}$ are spin independent (no magnetic field). The integrals in the last equation are evaluated in lowest nonvanishing order in the temperature by use of

$$\int \mathrm{d}\epsilon\, F(\epsilon)\left(\mathrm{d}n^{(0)}(\epsilon)/\mathrm{d}\epsilon\right) = -F(\mu) - (k_{\mathrm{B}}T)^2 I\left(\mathrm{d}^2F/\mathrm{d}\epsilon^2)|_\mu - (1/\mu)(\mathrm{d}F/\mathrm{d}\epsilon)|_\mu\right)$$

which is easily derived from results in Chapter 5. ($I = \int \mathrm{d}z\, z/(\mathrm{e}^z + 1)$ and $\mu$ is the zero temperature noninteracting chemical potential (Fermi energy).) Then expanding the result to lowest nonvanishing order $(k_{\mathrm{B}}T)^2$ and $\mathcal{N}(\epsilon')F_0$ we find

$$\delta\mu = \mu(1/m^* - 1/m)(1 + (k_{\mathrm{B}}T/\mu)^2 I)$$

(The terms involving interactions drop out to this order.) Then putting this back in the expression for $\delta n$

$$\delta n(\epsilon) = (1/m^* - 1/m)\left((\epsilon - \mu)\left(\mathrm{d}n^{(0)}(\epsilon)/\mathrm{d}\epsilon\right) - (k_{\mathrm{B}}T/\mu)^2 I\mu\left(\mathrm{d}n^{(0)}(\epsilon)/\mathrm{d}\epsilon\right)\right)$$

Again, the quasiparticle interaction terms have dropped out to this order. It is not hard to check that the integral $\int \mathrm{d}\epsilon\mathcal{N}(\epsilon)\delta n(\epsilon) = 0$ as required.

8.4

```
c       a  Monte Carlo code for problem 8.4
        dimension fmag(1000)
        integer is(1000,1000)
        open(1,file='problem84.dat')

        latticesize=100
        imcpersite=10000
        call srand(1257)
        imc=(latticesize**2)*imcpersite
        ianneal=imc/2.
        write(*,*) 'imc,ianneal',imc,ianneal
c  initialize
c       fmagtot=0.0
        do ix=1,latticesize
        do iy=1,latticesize
           test=rand(0)
           if(test.gt.0.5)then
            is(ix,iy)=1
           else
            is(ix,iy)=-1
           endif
           fmagtot=fmagtot+is(ix,iy)
        enddo
        enddo
        write(*,*) 'starting mag=',fmagtot/(latticesize**2)
c  loop on temperature
        do  1 it=1,400
          t=float(it)/100.
          fmag(it)=0.0
          itotaccept=0
         do  2 itry=1,imc
           ix=latticesize*rand(0)+1
           iy=latticesize*rand(0)+1
           istemp=-is(ix,iy)
           ixnp=mod(ix,latticesize)+1
           ixnm=mod(ix-2,latticesize)+1
              if(ixnm.eq.0)ixnm=latticesize
           iynp=mod(iy,latticesize)+1
           iynm=mod(iy-2,latticesize)+1
              if(iynm.eq.0)iynm=latticesize
c          write(*,*) 'ix,iy,ixnp,ixnm,iynp,iynm'
c          write(*,*) ix,iy,ixnp,ixnm,iynp,iynm
           field=is(ixnp,iy)+is(ixnm,iy)+is(ix,iynp)+is(ix,iynm)
           de=-(istemp-is(ix,iy))*field
c          write(*,*)'ix,iy,is(ix,iy)',ix,iy,is(ix,iy)
c          write(*,*) 'field,de',field,de
           iaccept=0
           if(de.lt.0)then
              is(ix,iy)=istemp
              iaccept=1
           else
            fact=exp(-de/t)
            test=rand(0)
            if(test.lt.fact)then
              is(ix,iy)=istemp
              iaccept=1
            endif
           endif
c          write(*,*) 'iaccept,is(ix,iy),istemp'
```

```
c            write(*,*) iaccept,is(ix,iy),istemp
             if(itry.eq.ianneal+1)then
                 fmag(it)=0.0
                 write(*,*) 'finished anneal for it',it
                 do ix=1,latticesize
                 do iy=1,latticesize
                   fmag(it)=fmag(it)+float(is(ix,iy))
                 enddo
                 enddo
                 write(*,*)'mag right after anneal=',fmag(it)
                 netchange=0
              elseif(itry.gt.ianneal+1)then
                 netchange=netchange+2.*istemp*iaccept
              endif
              itotaccept=itotaccept+iaccept
2         continue
c end mc steps for this temp
             write(*,*) itotaccept,'acceptances'
             write(*,*) 'netchange=',netchange
             fmag(it)=fmag(it)+float(netchange)/float(imc-ianneal)
              fmag(it)=fmag(it)/float(latticesize**2)
             write(*,*) 'mag for it',it,'is',fmag(it)
             write(1,*)t, fmag(it),abs(fmag(it))
1          continue
c end of temperature loop
          stop
          end
```

The magnetization calculated using this code is shown in Figure A.13. Results are good away from the critical region, but fluctuations around $T_c$ are still too large to characterize accurately the average magnetization near the critical point with this number of MC moves per site. Even if this problem is resolved by running longer, finite size effects will round the calculated transition near $T_c$. This run took about 12 hours on a middle aged work station (2004). (The absolute value of the magnetization is plotted. In fact the average magnetization changed sign once near the critical point in this run as temperature was increased.)

## Chapter 9

9.3

$$Z = \sum_{\{\sigma\}} \exp\left(\sum_i (K\sigma_i\sigma_{i+1} - h\sigma_i)\right) = \sum_{\{\sigma\}} \exp\left(\sum_i (K\sigma_i\sigma_{i+1} - (h/2)(\sigma_i + \sigma_{i+1})\right)$$

Let $\mathcal{S}_i = \sigma_{2i}$ and $S_i = \sigma_{2i+1}$. Then

$$Z = \sum_{\{\mathcal{S}\}}\sum_{\{S\}} \exp\left(\sum_i K(\mathcal{S}_i S_i + S_i\mathcal{S}_{i+1}) - (h/2)(\mathcal{S}_i + 2S_i + \mathcal{S}_{i+1})\right)$$

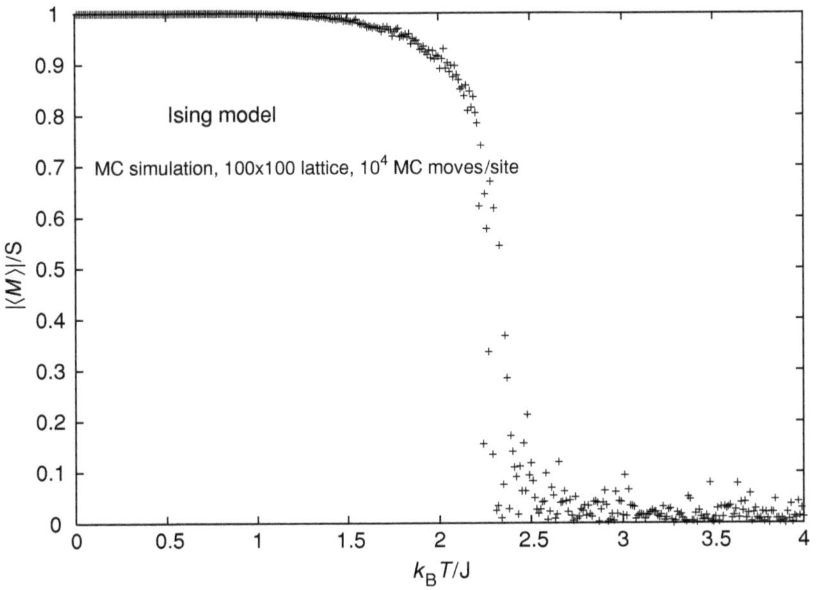


Figure A.13

Do the sum on the $S_i$:

$$Z = \sum_{\{\mathcal{S}\}} \Pi_i \big(\exp(K(\mathcal{S}_i + \mathcal{S}_{i+1}) - (h/2)(\mathcal{S}_i + \mathcal{S}_{i+1} + 2))$$
$$+ \exp(-K(\mathcal{S}_i + \mathcal{S}_{i+1}) - (h/2)(\mathcal{S}_i + \mathcal{S}_{i+1} - 2))\big)$$
$$= \sum_{\{\mathcal{S}\}} \Pi_i \big(\exp(K(\mathcal{S}_i + \mathcal{S}_{i+1}) - h) + \exp(-K(\mathcal{S}_i + \mathcal{S}_{i+1}) + h)\big)\exp(-(h/2)(\mathcal{S}_i + \mathcal{S}_{i+1}))$$

The can be rewritten as the partition function of an Ising model with renormalized coupling $K'$ and field $h'$ plus a constant $C$

$$Z = \sum_{\{\mathcal{S}\}} \Pi_i \exp(K'\mathcal{S}_i\mathcal{S}_{i+1} - (h'/2)(\mathcal{S}_i + \mathcal{S}_{i+1}) + C)$$

To find the relation between the primed parameters and the original ones, write the values of one term in the product for $\mathcal{S}_i\mathcal{S}_{i+1} = ++, +-, -+, --$ in the form of a matrix:

$$\begin{pmatrix} (e^{2K-h} + e^{-2K+h})e^{-h} & e^{-h} + e^{h} \\ e^{-h} + e^{h} & (e^{-2K-h} + e^{2K+h})e^{h} \end{pmatrix} = \begin{pmatrix} e^{K'-h'+C} & e^{-K'+C} \\ e^{-K'+C} & e^{K'h'+C} \end{pmatrix}$$

These are three independent equations which may be solved for the variables $K'$, $h'$ and $C$ with the result

$$K' = (1/4)\ln\left[\frac{\cosh(2K - h)\cosh(2K + h)}{\cosh^2 h}\right]$$
$$h' = h + (1/2)\ln\left[\frac{\cosh(2K + h)}{\cosh(2K - h)}\right]$$
$$C = (1/4)\ln\left[16\cosh(2K - h)\cosh(2K + h)\cosh^2 h\right]$$

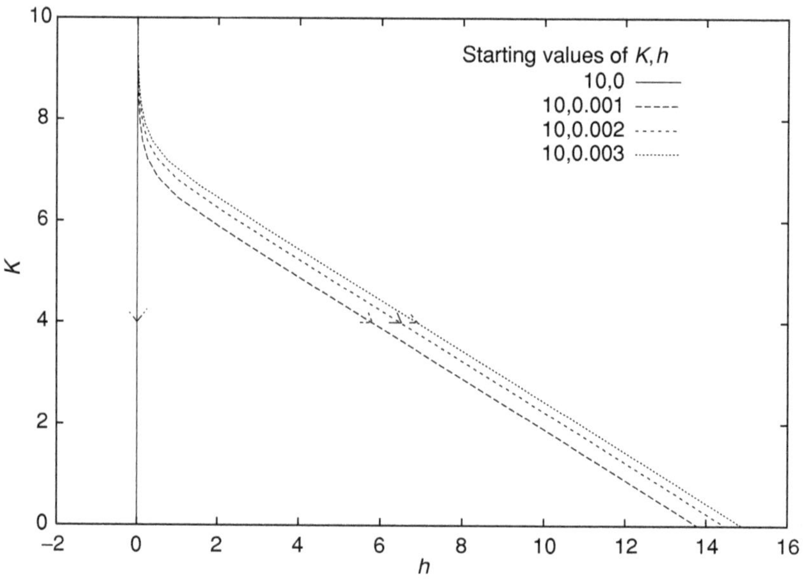


Figure A.14

The fixed point equations are

$$K^* = (1/4)\ln\left[\frac{\cosh(2K^* - h^*)\cosh(2K^* + h^*)}{\cosh^2 h^*}\right]$$

$$h^* = h^* + (1/2)\ln\left[\frac{\cosh(2K^* + h^*)}{\cosh(2K^* - h^*)}\right]$$

The second equation is satisfied if $K^* = 0$ or $h^* = 0$. The first equation is satisfied for $K^* = 0$ and any $h^*$ so, in the $K$–$h$ plane there is a line of fixed points along the $K = 0$ axis. For $h^* = 0$ the first equation is also satisfied for $K^* \to \infty$ so there is another fixed point at $h^* = 0$, $K^* \to \infty$. It is quite easy to see that, near the $K = 0$ axis, the renormalization equations drive the solution toward the $K = 0$ axis so those fixed points are stable, whereas the renormalization equations drive the solution away from the fixed point at $h^* = 0$, $K^* \to \infty$ so it is unstable. For $K$ large and $h = 0$ the renormalization equation for $K$ becomes $K' = K - (1/2)\ln 2 \approx K$ so the eigenvalue associated with the temperature is $\Lambda_1 = 1$ which would give $\nu = \ln 2/\ln \Lambda_1 \to \infty$. This means that the coherence length does not diverge as a power law as one approaches the critical point at $K \to \infty$ ($T = 0$) but goes faster than any power law. (Indeed the coherence length diverges as $e^K$.)

It is easy to write a little code to iterate the equations. Such a code is shown below and some trajectories are displayed in Figure A.14.

```
c       a  code for problem 9.3
        dimension fh(10000,10),fk(10000,10)

        open(1,file='problem94.dat')
        test=log(10.)
        write(*,*) 'test=',test
        iterate=35
        fkstart=10.
        dh=.001
        do ih=1,4
         fh(1,ih)=(ih-1)*dh
          fk(1,ih)=fkstart
        do i=2,iterate
        arg=cosh(2.*fk(i-1,ih)-fh(i-1,ih))*
     xcosh(2.*fk(i-1,ih)+fh(i-1,ih))
          argd=cosh(fh(i-1,ih))**2
        argh=cosh(2.*fk(i-1,ih)+fh(i-1,ih))/
     xcosh(2.*fk(i-1,ih)-fh(i-1,ih))
          write(*,*) 'i,ih,arg,argd,argh',i,ih,arg,argd,argh
          fk(i,ih)=0.25*log(arg/argd)
          fh(i,ih)=fh(i-1,ih)+0.5*log(argh)
        enddo
        enddo
        do i=1,iterate
          write(1,2) i,(fh(i,ih),fk(i,ih),ih=1,4)
        enddo
2       format(i2,8f8.4)
        stop
        end
```

9.4 (a)

$$F = -k_{\mathrm{B}}T \ln \int \mathcal{D}(\{m_{\vec{q}}\}) \exp\left(-\int \frac{\mathrm{d}^d q}{(2\pi)^d}(q^\mu + t)m_{-\vec{q}}m_{\vec{q}}\right)$$

Let $m_{\vec{q}} = \zeta m'_{\vec{q}}, q' = 2q$

$$F = -k_{\mathrm{B}}T \ln \int \mathcal{D}(\{m_{\vec{q}'}\})\exp\left(-\int \frac{\mathrm{d}^d q'}{(2\pi)^d}((q'/2)^\mu + t)\right)\zeta^2 m'_{-\vec{q}}m'_{\vec{q}} \times \text{ constant}$$

$$= -k_{\mathrm{B}}T \ln \int \mathcal{D}(\{m_{\vec{q}'}\}) \exp\big(-\mathcal{F}'(\{m_{\vec{q}'}\})\big)$$

so that the renormalized $\mathcal{F}$ is

$$\beta\mathcal{F}' = \int \frac{\mathrm{d}^d q'}{(2\pi)^d}(q'^2 + 2^\mu t)\frac{\zeta^2}{2^{d+\mu}}m'_{-\vec{q}}m'_{\vec{q}}$$

giving, for the Gaussian model, the transformations

$$t' = 2^\mu t$$

$$\zeta = 2^{d/2+\mu/2}$$

Thus $\Lambda_1 = 2^{\mu}$ and the exponent $\nu$ in the Gaussian model in this case is

$$\nu = \ln 2/\ln 2^{\mu} = 1/\mu$$

The renormalization of the field is $h' = 2^{d/2+\mu/2}h$ so the exponent $x = d/2 + \mu/2$ from which we can get all the exponents of this "Gaussian" model:

$$\alpha = 2 - d\nu = 2 - d/\mu$$
$$\beta = \nu(d - x) = \nu(d - (d/2 + \mu/2)) = (d - \mu)/2\mu$$
$$\gamma = \nu(2x - d) = (2(d/2 + \mu/2) - d)/\mu = 1$$
$$\delta = x/(d - x) = (d/2 + \mu/2)/(d - (d/2 + \mu/2)) = (d + \mu)/(d - \mu)$$
$$\eta = 2 - \gamma/\nu = 2 - \mu$$

The upper critical dimension is reached when the factor

$$\xi^4/2^{3d} = 1$$

(using the analogue to (9.125)). With the value $\xi = 2^{d/2+\mu/2}$, found above, this gives upper critical dimension $d = 2\mu$. These results all coincide with the Gaussian model results when $\mu = 2$.

Proceeding as in the text with finite $b$ we obtain the renormalization group equations

$$t' = 2^{\mu} + 6 \cdot 2^{\mu} \int \frac{\mathrm{d}q_3}{(2\pi)^d} \frac{1}{\left(q_3^{\mu} + t\right)}$$

$$b' = 2^{\epsilon} \left[ b + 36 \int \frac{\mathrm{d}q_3}{(2\pi)^d} \frac{1}{\left(q_3^{\mu} + t\right)^2} \right]$$

where $\epsilon = 2\mu - d$. Linearizing, we obtain, from the eigenvalue for temperature

$$\nu = 1/\mu + \epsilon/3\mu^2 \tag{A.1}$$

This also agrees with equation (9.134) when $\mu = 2$. This model is of some interest for the study of certain kinds of random walks.[2] The RNG behavior is studied in more detail in reference 3. (However, these references give $\epsilon/4\mu^2$ for the second term in (A.1). I have not traced the origin of this discrepancy.)

9.5 (a) For $t > 0$ one expands around $\langle m_{\vec{q}} \rangle = 0$ so, in zero field

$$\langle m_{\vec{q},\nu} m_{-\vec{q},\nu} \rangle = \frac{\int \mathcal{D}(\{m\}) \exp\left( - \sum_{\vec{q}} \vec{m}_{\vec{q}} \cdot \vec{m}_{-\vec{q}} \right) m_{\vec{q},\nu} m_{-\vec{q},\nu}}{\int \mathcal{D}(\{m\}) \exp\left( - \sum_{\vec{q}} \vec{m}_{\vec{q}} \cdot \vec{m}_{-\vec{q}} \right)}$$

Changing the variables to $x_{\vec{q},\nu} = m_{\vec{q},\nu}/\sqrt{(q^2 + t)}$ the integrals are easy to do

$$\langle m_{\vec{q},\nu} m_{-\vec{q},\nu} \rangle = K/(q^2 + t)$$

for each component, where

$$K = \int \mathrm{d}x_{\vec{q},\nu}\, \mathrm{d}x_{-\vec{q},\nu}\, \mathrm{e}^{-(x_{-\vec{q},\nu} x_{\vec{q},\nu})} x_{-\vec{q},\nu} x_{\vec{q},\nu} \Big/ \int \mathrm{d}x_{\vec{q},\nu}\, \mathrm{d}x_{-\vec{q},\nu}\, \mathrm{e}^{-(x_{-\vec{q},\nu} x_{\vec{q},\nu})}$$

is a constant with respect to $q$. But $\langle m_{\vec{q},\nu} m_{-\vec{q},\nu} \rangle$ is the spatial Fourier transform of $g(r)$ which, near the critical point, is of form $g(r) = \text{constant} \times \mathrm{e}^{-r/\xi}/r^{d-2+\eta}$ so

$$\langle m_{\vec{q},\nu} m_{-\vec{q},\nu} \rangle = \text{constant} \times 2/(q^2 + 1/\xi^2) = K/(q^2 + t)$$

requiring $\eta = 0$, $\nu = 1/2$.

(b) By differentiating the free energy with respect to the $q = 0$ field one has

$$\chi = \langle m_{\vec{q}=0,\nu} m_{-\vec{q}=0,\nu} \rangle = K/t$$

where $\nu$ is the field direction, giving $\gamma = 1$

(c) For $t < 0$ the Gaussian approximation requires expanding the free energy functional about its minimum and keeping only the quadratic terms in the fluctuations. For $t > 0$ the minimum was at all $\langle m_{\vec{q},\nu} \rangle = 0$ but for $t < 0$ this is not the case and we must find the minimum and expand around it. We let $h$ be in the $\nu = 1$ direction and differentiate with respect to $m_{\vec{q}=0,\nu=1} \equiv m_1$ giving

$$m_1(2t + 4b(\vec{m} \cdot \vec{m})) = h$$

and with respect to $m_{\vec{q}=0,\nu \neq 1} \equiv m_\nu$

$$(2t + 4b(\vec{m} \cdot \vec{m}))m_\nu = 0$$

The second equation has solutions $m_\nu = 0$ and $2t + 4b(\vec{m} \cdot \vec{m}) = 0$. However, the latter is excluded because inserting it in the first equation leads to a contradiction since we assume $h \neq 0$. Thus the solutions are $m_{\nu \neq 1} = 0$, $m_1$ solutions to the cubic $m_1(2t + 4bm_1^2) - h = 0$. The Gaussian free energy of interest is the quadratic terms obtained by expanding the Landau–Ginzburg free energy about this solution. We write $\vec{m} = \hat{i}_1 m_1 + \delta\vec{m}$ and expand, giving

$$\beta\mathcal{F} = \sum_{\vec{q}} \left[ (q^2 + t + 6bm_1^2)\delta m_{\vec{q},1}\delta m_{-\vec{q},1} + (q^2 + t + 2bm_1^2) \sum_{\nu \neq 1} \delta m_{\vec{q},\nu}\delta m_{-\vec{q},\nu} \right]$$

Using $m_1(2t + 4bm_1^2) - h = 0$ this is rewritten

$$\sum_{\vec{q}} \left[ (q^2 + 4bm_1^2 + h/2m_1)\delta m_{\vec{q},1}\delta m_{-\vec{q},1} + (q^2 + h/2m_1) \sum_{\nu \neq 1} \delta m_{\vec{q},\nu}\delta m_{-\vec{q},\nu} \right]$$

Now we take the limit $h \to 0$ and find

$$\langle \delta m_{\vec{q},1}\delta m_{-\vec{q},1} \rangle = K/(q^2 - 2t)$$

so that $\nu = 1/2$ for the correlation function associated with this component of the magnetization. Also, by taking $q \to 0$ we get $\gamma = 1$ for the susceptibility associated with a field in the magnetization direction. However, for the susceptibility associated with a field in a $\nu \neq 1$ direction perpendicular to the spontaneous magnetization we get

$$\langle \delta m_{\vec{q},\nu}\delta m_{-\vec{q},\nu} \rangle = K/q^2$$

in the $h_1 \to 0$ limit. Therefore the $q = 0$ susceptibility with respect to fields in directions perpendicular to the spontaneous magnetization *diverges* for all $t$ and $\gamma$ is not defined.

## Chapter 10

10.1(a)–(c) The continuity equation (10.4) is derived in the same way as in the text. Since it is linear, averaging is straightforward:

$$\frac{\partial \langle \rho \rangle}{\partial t} = -\nabla \cdot \langle \vec{J} \rangle / m$$

However, because we are assuming that energy is not locally conserved, we do not write the corresponding equation (10.6), but instead relate $\langle J\rangle$ to the gradient of the only density which has appeared at this level, namely $\langle\rho\rangle$:

$$\langle\vec{J}\rangle = -D\nabla\langle\rho\rangle$$

$D$ is the diffusion coefficient. The hydrodynamic equation is

$$\frac{\partial\langle\rho\rangle}{\partial t} = D\nabla^2\langle\rho\rangle$$

which is the diffusion equation. From the relation given in the problem statement (which is the same as equation (10.32))

$$\langle\rho\rho\rangle(\vec{q},\omega) = \frac{2\hbar\omega V}{(1-\mathrm{e}^{-\beta\hbar\omega})}\mathrm{Re}(\delta\rho(\vec{q},z\to\omega+\mathrm{i}\epsilon)\langle\rho\rangle/\delta p(\vec{q}))$$

Laplace transform of the diffusion equation gives

$$\begin{aligned}[-\mathrm{i}z\delta\rho(z,\vec{q})+q^2D\delta\rho(z,\vec{q})] &= \rho(t=0,\vec{q})-\rho(t\to\infty,\vec{q})\\ &= \delta\rho(\vec{q},t=0) = \left(\frac{\partial\rho}{\partial p}\right)_T\delta p(\vec{q})\end{aligned}$$

giving

$$\delta\rho(z,\vec{q}) = \frac{\left(\frac{\partial\rho}{\partial p}\right)_T\delta p(\vec{q})}{(-\mathrm{i}z+Dq^2)}$$

and inserting this in the equation for $\langle\rho\rho\rangle(\vec{q},\omega)$

$$\langle\rho\rho\rangle(\vec{q},\omega) = \frac{2\hbar\omega\rho_0}{(1-\mathrm{e}^{-\hbar\omega\beta})}\left(\frac{\partial\rho}{\partial p}\right)_T\frac{Dq^2}{(\omega^2+(Dq^2)^2)}$$

In the classical limit in which $\hbar\omega \ll k_{\mathrm{B}}T$,

$$\langle\rho\rho\rangle(\vec{q},\omega) = 2k_{\mathrm{B}}T\rho_0\left(\frac{\partial\rho}{\partial p}\right)_T\frac{Dq^2}{(\omega^2+(Dq^2)^2)}$$

Thus one can get the transport coefficient $D$ from two different limits of the time and space dependent correlation function

$$\lim_{\omega\to0}\lim_{q\to0}\frac{\langle\rho\rho\rangle(\vec{q},\omega)}{2k_{\mathrm{B}}T\rho_0\left(\frac{\partial\rho}{\partial p}\right)_T}\frac{\omega^2}{q^2} = D$$

and

$$\lim_{q\to0}\lim_{\omega\to0}\frac{\langle\rho\rho\rangle(\vec{q},\omega)}{2k_{\mathrm{B}}T\rho_0\left(\frac{\partial\rho}{\partial p}\right)_T}q^2 = 1/D$$

10.2 It is convenient to express the linearized equations in terms of the entropy $s^v \equiv S/V$ per unit volume (instead of per particle) and to use this quantity instead of $q$ which was used to characterize the heat flow in the linearized Navier–Stokes equations. Using $G = \mu N = E + PV - ST$ one sees easily that $\mu\rho = e + P - s^vT$. Further,

$d(E/V) = Tds^v + \mu d\rho$ follows from the standard thermodynamic relations in the canonical ensemble. The energy equation becomes

$$T\frac{\partial s^v}{\partial t} + \mu\frac{\partial \rho}{\partial t} = -\mu\nabla\cdot\vec{J}/m - \vec{J}\cdot\nabla\mu - \nabla(Ts^v\vec{v}_n) + \Lambda\nabla^2 T$$

The second term on the left cancels the first on the right by use of the equation of continuity. The second term on the right is nonlinear and in the third term the linear form only needs the divergence of $\vec{v}_n$ giving

$$\frac{\partial s^v}{\partial t} = -s^v\nabla\cdot\vec{v}_{\rm n} + \Lambda\nabla^2 T/T$$

A convenient form for the superfluid equation (3.71) is obtained by use of the Clausius–Clapeyron equation in the form $d\mu = -(s^v/\rho)dT + dP/\rho$ giving

$$\frac{\partial \vec{v}_{\rm s}}{\partial t} = (s^v/\rho)\nabla T - \nabla P/\rho - \nabla h$$

The other equations are already in essentially linear form.

Dropping the dissipative terms they take the form

$$m\frac{\partial \rho}{\partial t} = -\nabla\cdot\vec{J} \tag{A.2}$$

$$\frac{\partial \vec{J}}{\partial t} = -\nabla P$$

$$m\frac{\partial \vec{v}_{\rm s}}{\partial t} = (-1/\rho)\nabla P + (s^v/\rho)\nabla T \tag{A.3}$$

$$m\frac{\partial s^v}{\partial t} = -s^v\nabla\cdot\vec{v}_{\rm n} \tag{A.4}$$

$$\vec{J}/m = \rho_{\rm s}\vec{v}_{\rm s} + \rho_{\rm n}\vec{v}_{\rm n}$$

Combining the first two equations, just as in the case of normal fluids, gives

$$m\frac{\partial^2 \rho}{\partial t^2} = \nabla^2 P \tag{A.5}$$

Take a second time derivative of equation (A.4)

$$\frac{\partial^2 s^v}{\partial t^2} = -s^v\frac{\partial\nabla\cdot\vec{v}_{\rm n}}{\partial t}$$

Solve $\vec{J}/m = \rho_{\rm s}\vec{v}_{\rm s} + \rho_{\rm n}\vec{v}_{\rm n}$ for $\vec{v}_{\rm n}$ and insert it in the right hand side:

$$\frac{\partial^2 s^v}{\partial t^2} = -(s^v/\rho_{\rm n})\frac{\partial\nabla\cdot}{\partial t}(\nabla\cdot\vec{J}/m - \rho_{\rm s}\nabla\cdot\vec{v}_{\rm s})$$

Use the continuity equation (A.2) to express the first term on the right hand side in terms of the density

$$\frac{\partial^2 s^v}{\partial t^2} = (s^v/\rho_{\rm n})\frac{\partial^2 \rho}{\partial t^2} + (\rho_{\rm s}/\rho_{\rm n})\frac{\partial\nabla\cdot\vec{v}_{\rm s}}{\partial t}$$

Take the divergence of A.3 and use the result in the last term

$$\frac{\partial^2 s^v}{\partial t^2} = (s^v/\rho_{\rm n})\frac{\partial^2 \rho}{\partial t^2} + (\rho_{\rm s}s^v/\rho_{\rm n}M)\left(-\nabla^2 P/\rho + (s^v/\rho)\nabla^2 T\right) \tag{A.6}$$

Bearing in mind that $s^v = S/V$ one can obtain the thermodynamic identities

$$dT = ds^v/c_v - ((s^v/c_v\rho) + 1/\alpha\rho)d\rho$$
$$dP = -ds^v/\alpha + ((s^v/\alpha\rho) + 1/\beta)\,d\rho$$

where $\alpha = (1/V)(\partial V/\partial T)_S$ is the adiabatic thermal expansion coefficient, $\beta = -(1/V)(\partial V/\partial P)_S$ is the adiabatic compressibility and $c_v = (T/V)(\partial S/\partial T)_V$ is the specific heat per unit volume at constant density. We use these relations to express the terms involving Laplacians of $T$ and $P$ in terms of Laplacians of $s^v$ and $\rho$ in both (A.6) and (A.4)

$$\frac{\partial^2 s^v}{\partial t^2} - (s^v/\rho_n)(1-\rho_s/\rho)\frac{\partial^2\rho}{\partial t^2} = \left(\rho_s s^{v2}/\rho_n m\rho\right)\left(T\nabla^2 s^v/c_v - ((s^v T/c_v\rho) + 1/\alpha\rho)\nabla^2\rho\right)$$
$$\frac{\partial^2\rho}{\partial t^2} = (1/m)(-\nabla^2 s^v/\alpha + \nabla^2\rho\,((s^v/\alpha\rho) + 1/\beta))$$

These are two simultaneous wave equations in $\rho$ and $s^v$ so the existence of two sound wave modes is established. The equations simplify significantly if one transforms to the same heat variable which was used in the solution the analogous problem for the normal fluid, namely $dq = T dS/V$ in place of $ds^v = d(S/V)$. The two variables are related by $ds^v = dq/T + s^v d\rho/\rho$. With this transformation, the two equations become:

$$\frac{\partial^2 q}{\partial t^2} = \left(T\rho_s s^{v2}/c_v\rho_n m\rho\right)\nabla^2 q - \left(\rho_s s^{v2} T/\rho_n m\alpha\right)\nabla^2\rho$$
$$\frac{\partial^2\rho}{\partial t^2} = -(1/\alpha m T)\nabla^2 q + (1/m\beta)\nabla^2\rho$$

which clearly describe two sound like modes, one closely analogous to the sound in a normal fluid and termed first sound and the other involving the transport of heat and termed second sound. They are coupled by terms involving the thermal expansion coefficient which makes good physical sense. In fact the coupling terms can be shown to be small in liquid helium at low temperatures. If they are ignored, then the first sound velocity is seen directly given by the same expression which determines it in the normal fluid, and the second sound velocity is

$$c_2 = \sqrt{T\rho_s s^{v2}/c_v\rho_n m\rho}$$

It vanishes when the superfluid density is zero and has other intuitively satisfying features. The velocity is always lower than the first sound velocity in liquid helium. What is remarkable is that the hydrodynamics induced by the existence of a condensate has qualitatively changed the mechanism of heat transport in the fluid from diffusive propagation to wave propagation. Second sound is observed experimentally and much studied (see for example reference 4).

## References

1. I. S. Gradshteyn and I. M. Ryzhik, *Table of Integrals, Series and Products*, New York: Academic Press, 1980, Formula 9.541.1 and Table 9.7.1.
2. J. W. Halley and H. Nakanishi, *Physical Review Letters* **55** (1984) 551.
3. M. Fisher, S.-K. Ma and B. Nickel, *Physical Review Letters* **29** (1972) 917.
4. J. Wilks, *The Properties of Liquid and Solid Helium*, Oxford: Clarendon Press, 1967.

# Index

For EU product safety concerns, contact us at Calle de José Abascal, 56–1°,
28003 Madrid, Spain or eugpsr@cambridge.org.

www.ingramcontent.com/pod-product-compliance
Ingram Content Group UK Ltd.
Pitfield, Milton Keynes, MK11 3LW, UK
UKHW052257271225
466426UK00009B/568

* 9 7 8 0 5 2 1 8 2 5 7 5 7 *